T0271268

Real-Time Environmental Monitoring

Written 10 years after the publication of the first edition, this updated edition of *Real-Time Environmental Monitoring: Sensors and Systems* introduces the fundamentals of environmental monitoring based on electronic sensors, instruments, systems, and software that allow continuous and long-term ecological and environmental data collection. It accomplishes two objectives: explains how to use sensors for building more complex instruments, systems, and databases, and introduces a variety of sensors and systems employed to measure environmental variables in air, water, soils, vegetation canopies, and wildlife observation and tracking. This second edition is thoroughly updated in every aspect of technology and data, and each theoretical chapter is taught parallel with a hands-on application lab manual.

- Emphasizes real-time monitoring as an emerging area for environmental assessment and compliance and covers the fundamentals on how to develop sensors and systems.
- Presents several entirely new topics not featured in the first edition, including remote sensing and GIS, machine learning, weather radar and satellites, groundwater monitoring, spatial analysis, and habitat monitoring.
- Includes applications to many environmental and ecological systems.
- Uses a practical, hands-on approach with the addition of laboratory exercises students can do to deepen their understanding, based on the author's 40 years of academic experience.

Intended for upper level undergraduate and graduate students, taking courses in electrical engineering, civil and environmental engineering, mechanical engineering, agricultural and biological engineering, geosciences, and environmental sciences, as well as professionals working in environmental services, and researchers and academics in engineering.

Real-Time Environmental Monitoring

Monitoring
Sensors and Systems – Textbook

Second Edition

Miguel F. Acevedo

CRC Press
Taylor & Francis Group
Boca Raton London New York

CRC Press is an imprint of the
Taylor & Francis Group, an **informa** business

Second edition published 2024
by CRC Press
6000 Broken Sound Parkway NW, Suite 300, Boca Raton, FL 33487-2742

and by CRC Press
4 Park Square, Milton Park, Abingdon, Oxon, OX14 4RN

CRC Press is an imprint of Taylor & Francis Group, LLC

© 2024 Miguel F. Acevedo

First edition published by CRC Press 2015

Library of Congress Cataloging-in-Publication Data
Names: Acevedo, Miguel F., author.
Title: Real-time environmental monitoring : sensors and systems /
Miguel F. Acevedo.
Description: Second edition. | Boca Raton, FL : CRC Press, 2024. | Includes
bibliographical references and index.
Identifiers: LCCN 2023008254 (print) | LCCN 2023008255 (ebook) | ISBN
9781032545714 (hbk) | ISBN 9781032545721 (pbk) |
ISBN 9781003425496 (ebk)
Subjects: LCSH: Environmental monitoring–Data processing. | Environmental
monitoring–Equipment and supplies.
Classification: LCC QH541.15.M64 A24 2024 (print) | LCC QH541.15.M64
(ebook) | DDC 363.7/063–dc23/eng/20230703
LC record available at https://lccn.loc.gov/2023008254
LC ebook record available at https://lccn.loc.gov/2023008255

ISBN: 978-1-032-54571-4 (hbk)
ISBN: 978-1-032-54572-1 (pbk)
ISBN: 978-1-003-42549-6 (ebk)

DOI: 10.1201/9781003425496

Typeset in Times
by codeMantra

Access the Instructor and Student Resources/Support Material: http://www.routledge.com/9781032545714

Contents

Preface to the Second Edition xv
Preface to the First Edition xvii
Author xix

Chapter 1   Principles of Environmental Monitoring 1

Introduction 1
Why is Environmental Monitoring Necessary? 1
Environmental Systems, Ecosystems, and Planet Earth 1
Human Interactions with the Environment 2
Continuous Real-Time Monitoring 3
Data Management and the World Wide Web 3
Example: Global Monitoring 4
Statistics and Data Analysis 5
    Random Variables: Distributions and Moments 5
    Normal or Gaussian Distribution 8
    Covariance and Correlation 10
    Exploratory Data Analysis 12
Statistical Inference 14
    Hypothesis Testing 14
    Confidence Intervals 15
    Parametric Methods 15
    Nonparametric Methods 16
Simple Linear Regression 17
From Measuring to Knowing, Analysis, and Modeling 20
Scales and Resolution 21
Precision and Accuracy 21
Remote Sensing: Airborne and Spaceborne Platforms 22
More on Applications 23
Examples of Programs and Agencies 24
Environmental Monitoring Books 24
Exercises 25
References 26

Chapter 2   Programming and Single-Board Computers 29

Introduction 29
Computer Organization and Architecture 29
Single Board Computers 29
A/D and D/A Channels 30
Serial Communications 31
    Universal Asynchronous Receiver Transmitter 32
    RS-232 33
    USB 34
Networks 36
    Open Systems Interconnection (OSI) Model 36
    TCP/IP 37

Ethernet 37
Wireless Fidelity (Wi-Fi).. 38
Internet and World Wide Web (WWW) 38
Internet of Things (IoT).. 39
SBC, System on a Chip (SoC), and MCU.. 39
SoC 39
ARM Architectures 39
SBC Example: Raspberry Pi 39
Microcontrollers 39
MCU Example 41
In-Circuit Serial Programming (ICSP) 41
MCU-Based SBC Example: Arduino.. 42
Example MCU-Based SBC 43
Concepts of Programming.. 43
Python.. 44
Arduino.. 44
HTML... 45
CSS 46
PHP... 47
JS 49
SQL... 49
Exercises.. 49
References 51

Chapter 3 Sensors and Transducers: Basic Circuits.. 53

Introduction 53
Principles of Electrical Quantities... 53
Circuits: Nodes and Loops 57
Measuring Voltages, Currents, and Resistances 60
Sensors 62
From Sensors to Transducers 62
Sensor Specifications: Static 62
Resistive Sensors 63
Thermistors: Temperature Response 63
Example: From Thermistor to Temperature Transducer 66
Calculating Sensitivity and Linearity Error.. 66
Reading Output Voltage with A Digital Device 68
Inverting the Transducer Output to Obtain Temperature........................... 69
Self-Heating Effect.. 70
Example: Wider Temperature Range... 71
Example: A Temperature Transducer for Air, Soil, and Water.................. 73
Example: Thermocouples.. 73
Examples: Using Thermocouples... 76
Exercises.. 77
References 78

Chapter 4 Bridge Circuits and Signal Conditioning 79

Introduction 79
Linearized Thermistor: Small Variation Analysis 79

Voltage Divider with Linearized Thermistor 80
Balanced Source Voltage Divider 81
One-Sensor Circuit: Quarter-Bridge 84
Two-Sensor Circuit: Half-Bridge 86
Two sensors with opposite effect: Half-Bridge 87
Four-sensor Circuit: Full-bridge 88
Zero Adjust and Range Adjust 89
Sensor Specifications 90
Electrochemical Sensors 91
Example: Dynamic Specifications and a Potentiometer-Based Wind Direction 93
Dielectric Properties 93
Example: Piezoelectric Sensors 94
Example: Soil Tensiometer 94
Signal Conditioning 95
Operational Amplifiers 95
Linearization of the Bridge Circuit Output 97
Common-Mode Rejection 100
Instrumentation Amplifier 100
Spectrum 101
Noise 101
Electric Field and Electrostatic Shielding 102
Isolation 104
Cold-Junction Compensation 104
A/D Converter (ADC) 104
Current Loop: 4–20 mA 105
Pulse Sensors 107
Exercises 107
References 108

Chapter 5   Dataloggers and Sensor Networks 109

Introduction 109
DAS 109
Dataloggers 109
Applications in Environmental Monitoring 110
Analog Channels 110
RTC 111
Pull-Up Resistors 113
Serial Communication, Dataloggers, and Sensor Networks 113
RS-232 114
RS-485 114
SPI 115
I2C 116
SDI-12 117
MCUs as DAS 118
Conditions and Enclosures 118
A Datalogger Example: The CR1000 120
Remote Telemetry Unit or Remote Terminal Unit (RTU) and Supervisory
Control and Data Acquisition (SCADA) 127
Exercises 127
References 128

Chapter 6   Wireless Technologies: Telemetry and Wireless Sensor Networks 129

  Introduction 129
  Radio Wave Concepts... 129
    Electromagnetic Waves 129
    Radio Waves 130
    Propagation... 130
  Propagation Models... 132
    Free-space Propagation Model 132
    Two-Ray Propagation Model... 134
  Fresnel Zones 135
  Antennas and Cables 136
  Fade Margin... 137
  Polarization.. 137
  Radio Links: Communication Channel 138
    Modulation: Digital Signals 138
    Channel Performance 138
    Multiplexing 138
    Spread Spectrum 139
  Wi-Fi.. 139
  Cellular Phone Network and Satellite Links 141
  Wireless Sensor Networks (WSNs)... 141
    WSN Standards and Technologies 141
    IoT.. 143
    WSN Nodes 143
  Examples of Devices for WSN Nodes and IoT.. 144
    Moteino.. 144
    ESP8266 and ESP32... 144
    Xbee... 145
    LoRa 145
  Network Protocols... 146
    Media Access Control (MAC)... 146
    Multi-Hop Wireless Communication 146
    LoRa-Based Protocols.. 146
    Network Protocol for Environmental Monitoring... 146
    MQTT Protocol 147
  WSN and Environmental Monitoring: Practical Considerations and Examples...... 148
    Radio Propagation and WSN... 148
    Radio Propagation Experiments for WSN 149
    Example: WSN for Soil Moisture in a Hardwood Bottomland Forest............... 149
    Example: WSN for Soil Moisture using Moteino.. 151
    Example: Soil Monitoring Using ESP8266 and MQTT................................... 152
  Exercises.. 153
  References 154

Chapter 7   Environmental Monitoring and Electric Power 157

  Introduction 157
  PV Panels 157
    PV Cells... 157
    PV-Cell Model and I-V Characteristics... 159

From Cell to Module 161
Load and Power 164
Charging a Battery from a Solar Panel 164
Using a Voltage Regulator 164
Using a Buck Converter 165
Using a Buck-Boost Converter: MPP Tracking (MPPT) 166
Tilting the Panel 167
Atmospheric Effects 169
Sun Path 170
Impact of Temperature on Solar Panel 171
Power Budget and Power System Sizing 173
Powering WSN Nodes 174
Environmental Monitoring of Renewable Power Systems 175
Solar Radiation 175
Wind Speed 176
Hydroelectric 182
Exercises 183
References 184

Chapter 8   Remote Monitoring of the Environment 185

Introduction 185
Remote Sensing of the Environment 185
Optical Remote Sensing 185
Pixel, Raster, and Image 186
Imagery Specifications: Resolution and Quality 187
Spaceborne Remote Sensing: Types of Orbits 187
Platforms and Imagery 188
Geodetic Datum or System 188
World Reference System (WRS) 189
UTM Coordinate System 189
Bands and Image Display 192
Analysis Using Indices 195
Reclassification 198
Multivariate Analysis and Machine Learning 198
Reducing Dimensionality 200
PCA 200
SVD and Biplots 203
PCA Applied to Remote Sensing Images 207
Unsupervised Classification: Cluster Analysis 207
Hierarchical Cluster Analysis 209
K-Means 210
Unsupervised Classification of Remote Sensing Images Using K-Means 212
Exercises 213
References 214

Chapter 9   Probability, Statistics, and Machine Learning 215

Introduction 215
Probability 215
Algebra of Events 215

Combinations..216
Probability Trees 217
Conditional Probability 218
Binary (2-Class) Classification..219
Confusion Matrix 220
Bayes' Theorem and Classification 220
Generalization of Bayes' Rule to Many Events...222
Biosensing of Water Quality...223
Bayes' Rule and ML 223
Naïve Bayes Classifier 224
Decision Trees 224
Discrete RVs..224
Probability Mass Function (PMF)..225
Cumulative Mass Function (CMF)...226
First Moment or Mean...226
Second Central Moment or Variance 227
Binomial Distribution..228
Bivariate Discrete Random Variables..228
Information Theory 229
Counts and Proportions 230
χ^2 (Chi-Square) Test 230
Contingency Tables and Cross-Tabulation 232
Supervised Classification: Confusion Matrix..233
Multiple Linear Regression 236
Matrix Approach 236
Evaluation and Diagnostics 240
Variable Selection..241
CART..242
Classification Trees..243
Regression Trees...244
Model Complexity..245
Cross-Validation 246
CART Applied to Supervised Classification for Remote Sensing 246
Exercises...248
References 249

Chapter 10 Databases and Geographic Information Systems 251

Introduction 251
Databases..251
Server Client: Datalogging and DB...251
Relational databases 253
Data Models and Entity Relation Diagrams..255
SQL 257
DDL..257
DML 259
XML..260
GIS 263
GIS Software 263
GIS Layers..264
Raster Layers..264

Raster Analysis: Entry-Wise Calculations 266
Raster Analysis: Neighborhood and Zonal Calculations 267
Vector Layers 270
Vector Analysis 270
Backup 274
Web Services 274
Metadata, Standards, Interoperability, Preservation 275
Example: Data Collected from Distributed Sensor Systems 275
Exercises 276
References 277

Chapter 11 Atmospheric Monitoring 279

Introduction 279
Earth's Atmosphere 279
Composition and Vertical Structure 279
Direct and Diffuse Solar Radiation 281
Greenhouse Effect 281
Increasing Atmospheric CO_2 Concentration 282
Doubly Exponential 285
Nonlinear Regression 285
Global Temperature: Increasing Trend 286
Atmosphere – Near-Surface Air Quality 288
Standards 288
Air Monitoring Stations 289
Optical Devices 289
Linear Photodiode Array (PDA) and Charged Coupled Devices (CCD) 289
Dispersive Spectrometers 290
Photomultiplier Tubes 290
Beam Splitter 291
Fourier Transform Interferometer 291
Fiber Optics 292
Measurement Methods Using Samples in Closed Path 292
OAS 293
Chemiluminescent Analyzer 294
Fluorescence Instruments 295
Non-Dispersive Infrared 295
Measurement Methods Using Open Path 296
Total Column Estimation from the Ground 297
Example: Measuring UV and Total Column Ozone Concentration by
OAS and DOAS 300
Atmospheric Gases and Air Quality from Remote Sensing 301
Atmosphere – Weather 301
Air Temperature 301
Precipitation 301
Relative Humidity 302
Solar Radiation 302
Wind Velocity and Direction: Sonic Anemometers 302
Next-Generation Weather Radar (NEXRAD) 304
Weather Satellites 305
Exercises 306
References 307

Chapter 12 Water Monitoring 309

    Introduction 309
    Water 309
    Water Level and Depth 309
    Water Velocity and Flow 311
    Water Quality Parameters 313
        Temperature 314
        Electrical Conductivity 315
        TDS and Salinity 317
        pH and ORP 317
        Dissolved Oxygen 318
        Turbidity 318
        Fluorometer 319
        Multiple Parameter Probes 320
        Importance of Ionic Profile of Water Quality 321
    Light as a Function of Depth 322
    Productivity and Respiration 323
    Automated Real-Time Biomonitoring 324
    Modeling and Monitoring of Surface Water 325
        Hydrodynamic Models 325
        Water Quality Models 326
        Hydrological Models 326
    Remote Sensing of Water Quality 327
    Ocean Monitoring 327
    Groundwater Monitoring 328
    Autoregressive Analysis of Time Series 329
    Exercises 335
    References 335

Chapter 13 Terrestrial Ecosystems Monitoring 339

    Introduction 339
    Soil Moisture 339
        TDR 339
        Capacitance Probes 341
    Soil Tension: Tensiometer 341
    Infiltrometers 342
    Soil EC 343
    Evapotranspiration 344
    Sap Flow 347
    Lysimeters 347
    Productivity 348
        Gas Exchange 348
        Chlorophyll Fluorescence Combined with Gas Exchange 348
        Canopy Gas Exchange 349
        Micrometeorological Flux Measurements 349
        Covariance: A Review of Basic Concepts 349
        Eddy Covariance 350
    Tree Growth, Dendrometers 351
    Leaf Area 351

   Leaf Level 352

   Canopy Analyzer 353

  Solar Radiation and Spectral Measurements 354

  IR Thermometer 355

  Ground Penetrating Radar 357

  Remote Sensing of Terrestrial Ecosystems 358

  Exercises 359

  References 360

Chapter 14 Wildlife Monitoring 363

  Introduction 363

  Radio Monitoring 363

   Terrestrial 363

   Aquatic 365

  Acoustic Monitoring 367

   Terrestrial 367

   Aquatic 368

  Satellite 369

   Global Positioning System 369

   Pop-Up Satellite Archival Tags 370

   Proximity Sensors 371

  Data Storage Tags 371

  Camera and Video 371

  Autonomous Vehicles 373

  Habitat Monitoring 373

   Habitat Suitability Index 373

   Habitat Fragmentation Analysis 374

   Vertical Structure 376

  Spatial Analysis 376

   Testing Spatial Patterns: Cell Count Methods 376

   Nearest Neighbor Analysis 378

   Geostatistics 381

  Population and Community Modeling 389

  Exercises 390

  References 391

Index 393

Preface to the Second Edition

I wrote the first edition of this book ten years ago inspired by our experiences developing real-time environmental monitoring systems, offering mini-courses, and starting a regularly offered course on the subject for undergraduate and graduate students in the Electrical Engineering program at the University of North Texas (UNT). Several factors motivated me to write a second edition, primarily that there have been many advances in technology that helped move forward environmental monitoring programs by leaps and bounds.

I also developed a feeling that the textbook would be enhanced by adding a component in remote sensing which I had deliberately excluded in the first edition to focus on observations from the ground. Along with adding remote sensing, it became clear that the textbook needed a component on geographic information systems (GIS) to accompany the concepts of databases. After teaching this class for several years using the textbook, it became evident that it was important to add the laboratory exercises that I was using for the students to have a hands-on experience on the subject.

These factors determined the concept and structure of this second edition which includes new material on hardware, programming, data analysis and statistics, machine learning, remote sensing, and GIS. Moreover, a few years back, we developed a parts kit that the students could take home to conduct the hands-on experiments on their own and individually spending more quality time than the short session in the campus laboratory. The laboratory manual part of this second edition reflects this approach and is written such that the skills learned in each lab guide corresponds to a chapter in the textbook.

Following the structure of the first edition, the textbook and lab manual are organized in two parts, the first part with a focus on methods (the first ten chapters and lab guides) yet provides some practical examples to illustrate the applications. The second part focuses on application domains, the last four chapters and lab guides, namely atmospheric, water, terrestrial, and wildlife monitoring, yet uses opportunities to introduce more methods.

I would like to express my gratitude to many individuals that have made this second edition possible. Students that have taken the environmental monitoring course following this material, particularly the lab manual, provided feedback that contributed to its continued improvement. This edition as well as the first, benefited from examples inspired by the work of students who developed projects in this subject, including senior capstone projects, master thesis, and doctoral dissertations. UNT's electrical engineering department supported experimenting with the idea of providing a lab kit for the students to develop the laboratory exercises individually. Sanjaya Gurung helped developing some of the hands-on exercises in wireless and wireless sensor networks. Breana Smithers helped organize and maintain the laboratory kits over several semesters as well as preparing figures for this book. At CRC Press, Irma Britton, editor of Environmental Science and Engineering, was supportive of this project through the entire process, and Chelsea Reeves, editorial assistant, provided help preparing the materials for production. Anonymous reviewers provided excellent feedback that helped to improve the approach of this second edition.

Miguel F. Acevedo, Denton, Texas, January 2023

Preface to the First Edition

My aim in writing this book is to introduce the fundamentals of environmental monitoring based on electronic sensors, instruments, systems, and software that allow continuous and long-term ecological and environmental data collection. I have tried to accomplish two objectives, as reflected in the two major parts of this book. In the first part, I develop a story of how starting with sensors, we progressively build more complex instruments, leading to entire systems, and ending on database servers, web servers, and repositories. In the second part, once I lay out this foundation, I cover a variety of sensors and systems employed to measure environmental variables in air, water, soils, vegetation canopies, and wildlife observation and tracking.

I have attempted to present the state-of-the-art technology, while at the same time using a practical approach, and being comprehensive including applications to many environmental and ecological systems. My preference has been to explain the fundamentals behind the many sensors and systems so that the reader can gain an understanding of the basics. As in any other endeavor, specialized references would supplement this basic material according to specific interests.

I have based this material on my experience developing systems for ecological and environmental studies, particularly those leading to ECOPLEX and the Texas Environmental Observatory (TEO). I have tried to provide a wide coverage and offer a broad perspective of environmental monitoring; naturally, I emphasize those topics with which I am more familiar. In the last few years, I have employed successive drafts of this book while developing a course in environmental modeling for undergraduate and graduate students in electrical engineering and environmental science.

Although my target is a textbook, I have also structured the material in such a way that could serve as a reference book for the monitoring practitioner. The material is organized into 14 chapters; therefore, when used as a textbook, it can be covered on a chapter-per-week basis in a typical 14-week semester. Part I includes problems that can be assigned as homework exercises.

I hope to reach out to students and practitioners worldwide interested and engaged in efforts to develop, employ, and maintain environmental monitors. The book includes examples of low-cost and open-access systems that can serve as the basis of learning tools for the concepts and techniques described in the book.

I would like to thank many individuals with whom I have shared experiences in this field, in a variety of projects, such as monitoring and assessment methods in lakes and estuaries, the start-up of Ecoplex, developing a cyber-infrastructure approach to monitoring and TEO (NSF-funded projects CI-TEAM, CRI) and the TEO, and the NSF RET (Research Experiences for Teachers) on sensor networks. These individuals include faculty and students of several units of the UNT, such as the Institute of Applied Science, Electrical Engineering Department, Computer Science Department, School of Library and Information Sciences, University Information Technology (UIT), as well as colleagues of the City of Denton, and the University of the Andes (Venezuela). Among many, I would like to mention Ken Dickson, Tom Waller, Sam Atkinson, Bruce Hunter, Rudy Thompson, David Hunter, Shengli Fu, Xinrong Li, Yan Huang, Bill Moen, Duane Gustavus, Phillip Baczewski, Ermanno Pietrosemoli, Michele Ataroff, Wilfredo Franco, Jue Yang, Carlos Jerez, Gilbert Nebgen, Chengyang Zhang, Mitchel Horton, Jennifer Williams, Andrew Fashingbauer, and Jarred Stumberg.

As an outcome of the CI-TEAM mini-courses, I developed a pilot of the environmental monitoring class that I currently teach using this book. I say thanks to several individuals who contributed guest lectures: Shengli Fu, Xinrong Li, David Hunter, Kuruvilla John, Rudy Thompson, Carlos Jerez, Sanjaya Gurung, and Jason Powell.

I would like to say special thanks to Breana Smithers for providing help in the field to maintain monitoring equipment and preparing the figures in this book.

I am very grateful to Irma Shagla-Britton, editor for Environmental Science and Engineering, at CRC Press for her enthusiasm for this project. Several reviewers provided excellent feedback that shaped the final version and approach of the manuscript.

Miguel F. Acevedo, Denton, Texas, May 2015.

Author

Miguel F. Acevedo has over 40 years of academic experience, the last 27 of these at the University of North Texas (UNT) where he currently serves as a Regents Professor. His career has been interdisciplinary and especially at the interface of science and engineering. He has served UNT as faculty member in the Department of Geography, the Graduate Program in Environmental Sciences of the Biology Department, and the Electrical Engineering Department. He obtained his PhD in Biophysics from the University of California, Berkeley (1980) and master's degrees in Electrical Engineering and Computer Science from Berkeley (M.E., 1978) and the University of Texas at Austin (M.S., 1972). Before joining UNT, he was at the Universidad de Los Andes, Merida, Venezuela, where he served in the School of Systems Engineering, the graduate program in tropical ecology, and the Center for Simulation and Modeling. He has served on the Science Advisory Board of the U.S. Environmental Protection Agency and on many review panels of the U.S. National Science Foundation. He has received numerous research grants, and written several textbooks, numerous journal articles, as well as many book chapters and proceeding articles. In addition to the Regents Professor rank, UNT has recognized him with the Citation for Distinguished Service to International Education, and the Regent's Faculty Lectureship. His research interests focus on environmental systems and sustainability. He has published four textbooks with CRC Press.

1 Principles of Environmental Monitoring

INTRODUCTION

As we humans developed our capacity to modify and exploit our environment for food and shelter, we have become more and more aware of the importance of managing and preserving the quantity and quality of natural resources upon which we depend to sustain our livelihood. It is very difficult to manage something without understanding it, and to be able to comprehend we need information as well as ways of interpreting and integrating that information. *Environmental monitoring* is about measuring aspects of the environment in a repetitive manner so that we can learn about its structure and functioning. Once we understand it better, we can apply that knowledge in multiple ways to manage it wisely. According to Artiola et al. (2004a, p. 2), "Environmental monitoring is the observation and study of the environment. In scientific terms, we wish to collect data from which we can derive knowledge".

This chapter is an overview of many topics that will be covered throughout the book, such as sensors, data management, data analysis, instruments, precision, and accuracy. In addition, this chapter introduces random variables, exploratory data analysis, statistical inference, and linear regression, concepts that will be employed in Lab 1 of the Lab Manual companion of this book (Acevedo, 2024).

WHY IS ENVIRONMENTAL MONITORING NECESSARY?

Our natural environment is complex and changes continuously at varying paces. Sometimes, we can note these changes and build an awareness of the rhythms and patterns involved in those changes. For example, we note subjectively how *weather* changes during the day, from day to day, week to week, month to month, season to season, and year to year. We make comparisons between those changes year to year and even decade to decade. However, weather records over many years and decades allow a much less subjective comparison. Indeed, we can calculate averages, maximum, minimum, and trends, for various timescales such as daily, monthly, annually, and seasonally. These weather statistics now become *climate* and help us build an understanding of the patterns of change over the *long term*.

Note from the example that the key to build this understanding is the accumulation of careful records of weather in a database. Management and problem-solving benefit from the prediction of environmental changes, and this requires continuous and long-term monitoring with archiving in a database and making it readily available for retrieval.

ENVIRONMENTAL SYSTEMS, ECOSYSTEMS, AND PLANET EARTH

When we refer to environmental systems, we not only consider our surrounding air, water, and soil but also living entities sharing these resources with us. Thus, ecological interactions become part of environmental systems and its monitoring. A concept that helps framework ecological interactions is that of *ecosystems*, which emphasize relations of *biotic* components (living) such as animals and plants with *abiotic* factors (nonliving) such as air, light, soil, and water. Key aspects are functional relationships among species focusing on the transfer of material and energy among them, and interactions with the abiotic factors. As a generalization, we are concerned with how materials *cycle* among components of the system and how energy *flows* from one component to another (Figure 1.1).

DOI: 10.1201/9781003425496-1

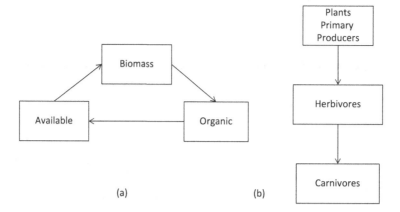

FIGURE 1.1   Nutrient cycle (a) and energy flow (b) in ecosystems.

Many environmental monitoring programs are designed to protect human health. For example, we monitor concentration of ozone in the air we breathe, to help prevent excessive exposure and thus harmful effects on our respiratory systems. However, we are also concerned with monitoring to protect the health of the ecosystem from our actions on the environment; for example, protecting organisms from excessive exposure to contaminants we release into the environment.

Ecosystem concepts are important at planetary scale as well. We typically group biotic components in the *biosphere* and abiotic components in the *geosphere* comprising *atmosphere* (gaseous envelope of the planet), *hydrosphere* (water in oceans, lakes, rivers, and glaciers), and *lithosphere* (rocks and mineral matter). Cycles and flows are then *global* or occurring at planetary scale and include relations of the geosphere and biosphere. For example, the global carbon cycle has received considerable attention because of its relationship with climate control. Carbon is an important part of the biosphere and plays a key role in global climate control because the concentration of carbon dioxide (CO_2) in the atmosphere contributes to the greenhouse effect, which in turn affects global air temperature.

CO_2 is used by primary producers (e.g., terrestrial plants and algae) to make carbohydrates by photosynthesis, utilizing sunlight. Some of the carbon goes back to atmospheric CO_2 by respiration and emission from these compartments; the rest is stored, consumed, decomposed, and part is recycled. At slower rates, carbon transfers to fossil organic matter, sediments, and sedimentary rocks. Sedimentary deposits contain most of the carbon. Human action accelerates the release of CO_2 by burning fossil fuels (coal, oil, and gas) and from the terrestrial biota by deforestation. The timescale is mixed, some processes occur rapidly as the exchanges between atmosphere and biota, and others slowly such as sedimentation.

There are multiple systems in the Earth's biosphere and geosphere and multiple interactions among its components. For example, the hydrological cycle and the subject of atmosphere–vegetation–soil interaction are of great importance. Atmospheric water as vapor is condensed and falls as rain, feeding the soil with water that can be stored for use by the vegetation. Transpiration by plants and evaporation from the soil returns water vapor to the atmosphere. Monitoring these processes (e.g., rainfall, runoff, soil moisture, evapotranspiration, and gaseous fluxes from the plants) allows us to understand how the entire system works and depending on how frequent we measure the processes we may understand how the water cycle balances at a variety of temporal scales.

HUMAN INTERACTIONS WITH THE ENVIRONMENT

Environmental monitoring provides important support to sustainability science and engineering. As we mentioned in the introductory paragraph, by continuous and long-term measurement, we improve our chances of making good decisions about the environment. Monitoring helps design infrastructure adapted to environmental changes, improve agricultural production systems, and provide guidance for smart development.

For example, as we strive to develop renewable energy systems, it is imperative that we understand related environmental variables and their time variation. As green components become part of the electric power grid, there is a need to have real-time and long-term measurements about the weather, wind, and solar radiation, particularly for wind- and solar-based technologies used for power generation.

It is important to consider to what extent we affect the system we are trying to measure by our attempt to measure it. Several examples come to mind. At the sensor level, current flow in a thermistor increases temperature by self-heating and we must correct for this unintended rise in temperature. At the system level, a clearing in a forest for a tower may affect the canopy processes we want to monitor. For wildlife monitoring, depending upon size and other factors, a Global Positioning System (GPS) collar or a radio tag may affect the individual animal of a species we are monitoring.

CONTINUOUS REAL-TIME MONITORING

The development of electronic technologies has allowed us to collect environmental data as the processes unfold, that is, real time, and to repeat these measurements for long periods. In this context, real time does not necessarily mean measuring instantaneously at all times but rather that we can keep up with the rate of change of the process under measurement. What is important is that we can implement a consistent frequency of sampling that captures the dynamics of the process. By means of electronic devices, we can generate, store, and transmit environmental data. Principles of electronic technology for these purposes are the subject matter of Chapters 2–6, and specific examples of their applications are covered in Chapters 11–14.

Before these technologies were available, monitors used mechanical devices to sense environmental variables and to record these in a variety of manners. For example, air temperature changes were sensed by the differential thermal expansion of a bi-metallic strip and relative humidity by changes in length of human hair with humidity. The responses of these sensors were used to move a pen over a clock-driven rotating drum, thus producing a continuous record of temperature and relative humidity. These instruments are accurate but demand careful operation, maintenance, drum paper supplies and are difficult to deploy in harsh environments and remote locations. Electronic-based instruments are easier to operate, have reduced costs, and allow for deployment and long-term autonomous operation.

DATA MANAGEMENT AND THE WORLD WIDE WEB

As we collect environmental data continuously and over long term, the need for organizing, storing, and managing these data arises. Database design and management then becomes an essential tool that an environmental monitoring practitioner must know and understand. Integration of technologies from sensors to databases opens the possibility to use web-based frameworks for making data available for a variety of purposes. The result may be an environmental monitoring network, a real-time early warning system, a global monitoring network linked by satellites, and many other variations. What is common and critical in all these efforts is the concept of long-term, continuous, real-time measurements of environmental conditions and making them available to the public. In this book, we briefly cover some fundamental notions of database management and web-based technology in Chapter 10 and exercise practical aspects in Lab 10 of the companion Lab Manual (Acevedo 2024).

EXAMPLE: GLOBAL MONITORING

An excellent example of environmental monitoring is the measurement of atmospheric CO_2 concentrations recorded in Mauna Loa, Hawaii (Vaughan et al. 2001; Lovett et al. 2007). This long-term record helped to gain an increased understanding of global climate change, one environmental challenge we face today. A visit to the website of National Oceanic and Atmospheric Administration (NOAA)'s Global Monitoring Laboratory (NOAA 2021b) will inform us of recent values of monthly average of CO_2 concentration in parts per million (ppm). For example, the monthly average for December 2021 was 415.01 ppm, almost 2 ppm up from 413.12 ppm for December 2020.

In this case, ppm is a unit expressing dry air mole fraction defined as the number of CO_2 molecules divided by the number of all molecules in air, including CO_2 itself, after water vapor has been removed; for example, a value of 400 ppm represents a mole fraction of 0.000400 (NOAA 2021b).

Downloading the data files from this website and using the program R (R Project 2022), we can display the time series for CO_2 concentration. We will learn how to do this in Chapter 11 and Lab 11 of the companion Lab Manual (Acevedo 2024). In this chapter, we only show the results to provide an example. In Figure 1.2, we see two lines: the dashed line represents the monthly average values, centered in the middle of each month (NOAA 2021b). We clearly see that it swings up and down during the year according to the seasons. Removing the average of this seasonal cycle yields the solid line, which shows a clear increasing trend in the last ten years. Approximately, we see a change from 392 to 415 ppm or about 10 ppm in 10 years, resulting in about 2 ppm increase per year.

This brings up an important point. We often want to process the data acquired by an environmental monitoring program. In this case, we *filter* out fluctuations of the data in order to observe gradual or secular changes. In this example, the filter is implemented by a moving average of seven (an odd number) adjacent seasonal cycles centered on the month to be corrected (NOAA 2021b). Note that we need to make an exception for the first and last 3.5 years of the record because we cannot complete a full 7-year sequence. In these cases, we take an average of the seasonal cycle over the first and last seven years, respectively.

Figure 1.3 shows the same variables but over a longer period starting a couple of years before 1960, when monitoring commenced. We again see the clear seasonal fluctuation and that the trend (solid line after filtering out seasonality) displays an increase in the rate of change over time. We address how to estimate the rate of change in Lab 11 of the companion Lab Manual (Acevedo 2024) using the concepts of exponential and doubly exponential functions.

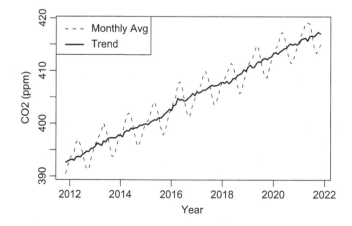

FIGURE 1.2 Monthly mean CO_2 at Mauna Loa during Nov 2011–Nov 2021. Plotted by R using data from NOAA, *Trends in Atmospheric Carbon Dioxide, Mauna Loa, Hawaii*. Retrieved from NOAA (2021b).

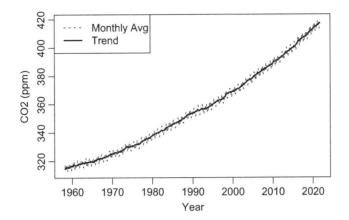

FIGURE 1.3 Monthly mean CO_2 at Mauna Loa, entire record (March 1958–Nov 2021). Plotted by R using data from NOAA, *Trends in Atmospheric Carbon Dioxide, Mauna Loa, Hawaii.* Retrieved from (NOAA 2021b).

STATISTICS AND DATA ANALYSIS

Describing data quantitatively, as well as subsequent rigorous analysis, requires the use of probability and statistics. In this chapter, we present some basic notions that we will exercise using the program R (R Project 2022) in Lab 1 of the Lab Manual companion to this textbook (Acevedo 2024).

RANDOM VARIABLES: DISTRIBUTIONS AND MOMENTS

A random variable (RV) is a rule or a map associating a probability to each event in a sample space. When the events are defined from intervals contained in a range of real values, we have a *continuous* RV. For example, concentration of a gas in the atmosphere or a chemical in water. A *probability density function* (PDF) $p(X)$ is based on intervals; the probability of the value being in an infinitesimal interval of X between x and $x + dx$ (Figure 1.4a) is given by

$$p(x)dx = P[x < X \leq x + dx] \tag{1.1}$$

here $p(x)$ is always positive or zero, that is $p(x) \geq 0$.

The probability of a value being in an interval of X between a and b can be found using the integral

$$P[a < X \leq b] = \int_{a}^{b} p(x)dx \tag{1.2}$$

which is the area under the curve $p(x)$ in each interval x and $x + dx$ from a to b (Figure 1.4b). When the interval is the whole range of values of X, then the value of the integral should be 1

$$\int_{-\infty}^{+\infty} p(x)dx = 1 \tag{1.3}$$

we have indicated the entire range of real values by selecting the limits from minus infinity ($-\infty$) to plus infinity ($+\infty$), or from a very large negative value to a very large positive value.

Consider, for example a uniform continuous RV. The density has the same value over the range $[a, b]$

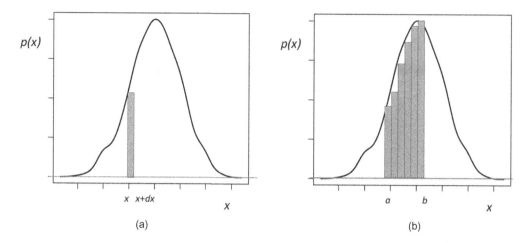

FIGURE 1.4 PDF of a continuous RV. Left: probability is the area under the curve between two values. Right: Probability of X having a value in between a and b.

$$U_{a,b}(x) = \begin{cases} \dfrac{1}{b-a} & \text{for } a \leq x \leq b \\ 0 & \text{otherwise} \end{cases} \tag{1.4}$$

as shown in the top graph of Figure 1.5.

The *cumulative distribution function* (CDF) at a given value is defined by summing or accumulating all probabilities up to that value.

$$F(x) = P[X \leq x] = \int_{-\infty}^{x} p(s)\,ds \tag{1.5}$$

Please note that the value at which we evaluate the CDF is the upper limit of the integral. Variable s is a dummy variable to avoid confusion with x. The CDF $F(x)$ at a value x is the area under the density curve up to that value. The value of the CDF for the largest value of X is equal to 1. For example, for the uniform continuous RV, $U_{a,b}(x)$ is a ramp with slope $1/(b-a)$. See the bottom graph in Figure 1.5.

The first moment of X is the *expected value* of X denoted by operator $E[X]$ applied to X, this is $E[X]$ or the *mean* of X.

$$\mu_X = E[X] = \int_{-\infty}^{+\infty} xp(x)\,dx \tag{1.6}$$

As an example, consider an RV uniformly distributed in [0,1]. In this case, $b=1$, $a=0$. We know that $p(x) = 1/(b-a) = 1$.

$$\mu_X = E[X] = \int_{0}^{1} x\,dx = \frac{1}{2}x^2 \Big|_{0}^{1} = \frac{1}{2} \tag{1.7}$$

The expected value or mean is a theoretical concept. To calculate it, we need the PDF of the RV. The mean is not the same as the statistic known as the *sample mean* which is the arithmetic average of n data values x_i comprising a sample

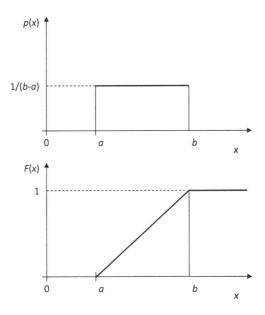

FIGURE 1.5 PDF and CDF for a uniform RV. Integration of a constant yields a linear increase (ramp function).

$$\bar{X} = \frac{1}{n} \sum_{i=1}^{n} x_i \tag{1.8}$$

Note that the mean (first moment) is denoted with the Greek letter μ, whereas the sample mean, or average is denoted with a bar on top of X, that is to say $\bar{X}$.

The second central (i.e., with respect to the mean) moment is the *variance* or the expected value of the square of the difference with respect to the mean

$$\sigma_X^2 = E[(X - \mu_X)^2] \tag{1.9}$$

Calculated by

$$\sigma_X^2 = E[(X - \mu_X)^2] = \int_{-\infty}^{+\infty} (x - \mu_X)^2 \, p(x) \, dx \tag{1.10}$$

The variance is a theoretical concept and to calculate it, we need the PDF of the RV. The *standard deviation* is the square root of the variance $\sigma_X = \sqrt{\sigma_X^2}$.

The variance or second central moment is not the same as the statistic known as the *sample variance*, which is the variability measured relative to the sample mean, i.e., the average of the square of the deviations from the sample mean

$$s_X^2 = \frac{1}{n-1} \sum_{i=1}^{n} (x_i - \bar{X})^2 \tag{1.11}$$

where $n-1$ is used instead of n to account for the fact that the sample mean was already estimated from the n values. We write s_X to denote the sample variance to differentiate from the variance σ_X^2 (Acevedo 2013). The *sample standard deviation* is the square root of the sample variance

$$s_X = \sqrt{\frac{1}{n-1} \sum_{i=1}^{n} (x_i - \bar{X})^2} \tag{1.12}$$

Variability can be expressed as a *coefficient of variation*, defined as the ratio of the sample standard deviation to the sample mean.

$$Cv = \frac{s_X}{\bar{X}} \tag{1.13}$$

The expected value of the sample mean $\bar{X}$ of X is equal to the mean μ_X of X, $E(\bar{X}) = \mu_X$ and the variance of the sample mean $\bar{X}$ of X is the variance of X divided by the sample size n $E[(\bar{X} - \mu_{\bar{X}})^2] = \sigma_X^2 / n$. The square root of the variance of the sample mean is the standard deviation of the sample mean or *standard error* of the estimate of the mean

$$\sigma_e = \sqrt{\frac{\sigma_X^2}{n}} = \frac{\sigma_X}{\sqrt{n}} \tag{1.14}$$

The expected value of the sample variance is $E(s_x^2) = (n - 1/n)\sigma_x^2$. For a large value of n, the sample mean approaches the mean, and the mean of the sample variance is approximately the same as the variance.

The first and second central moments (mean and variance) are also referred to as *parameters* of the RV distribution, and are different from the *statistics*, which are associated with the *sample* (Acevedo 2013). Another way of looking at this is to think of the PDF as a theoretical model expressing the underlying probability structure of the RV. These functions allow the calculation of the moments. However, the statistics are calculated from observed data and are used to estimate the moments.

Besides the mean and variance, to characterize a distribution we can use the *median*, which is the value that divides the area under the PDF curve into two equal parts or the value x at which the CDF attains $F(x) = 0.5$. We can further divide the area under the PDF into equal parts or *quantiles*. For example, if we divide it into four equal parts, we obtain *quartiles* and in this case $F(x_1) = 0.25$, $F(x_2) = 0.50$, $F(x_3) = 0.75$ where x_1, x_2, x_3 are the quartiles. Note that the inter-quartile interval x_3-x_2, or the difference between the third quartile and the first quartile has probability $= 0.75 - 0.25 = 0.5$. If we use 100 equal parts, we obtain *percentiles* $x_1, x_2, ..., x_{99}$ where $F(x_1) = 0.01$, $F(x_2) = 0.02$, ... and so on until $F(x_{99}) = 0.99$.

The ith-order statistic of a sample is equal to the ith smallest value and is denoted by $x_{(i)}$. The first-order statistic $x_{(1)}$ and the nth-order statistic $x_{(n)}$ are the minimum and maximum values in a sample of size n. The quantiles are equal to the order statistics when we calculate n quantiles in a sample of size n. In addition, when $n = 2m + 1$ then n is odd, and the median is the mth-order statistic $x_{(m+1)}$. However, when n is even, say $n = 2m$, then the median is the average of $x_{(m)}$ and $x_{(m+1)}$ but is not itself an order statistic.

NORMAL OR GAUSSIAN DISTRIBUTION

The *Normal* or *Gaussian* RV has a PDF

$$N_{\mu,\sigma}(x) = \frac{1}{\sqrt{2\pi}\sigma} \exp\left[-\frac{(x-\mu)^2}{2\sigma^2}\right] \text{ for } -\infty < x < +\infty \tag{1.15}$$

with mean μ and variance σ^2. This is a symmetrical PDF, i.e., the area under the curve left of the mean is the same as the area under the curve right of the mean. The CDF $F(x)$ gives the area under the PDF curve up to a value x. Thus, the area under the curve right of the point x is the same as $1-F(x)$. The area of the curve on both sides of the mean increases with distance: at one standard deviation on both sides $\mu \pm \sigma$ the area is 0.68, at two standard deviations $\mu \pm 2\sigma$ is 0.95, and at three standard deviations $\mu \pm 3\sigma$ is 0.99 (Acevedo 2013).

A normally distributed sample is symmetric. That is to say, the mean is equal to the median. Very commonly, the data are not symmetrical, i.e., there is higher frequency left or right of the mean. Two important cases can occur: positive (mean < median) or values are biased toward the right and negative (median < mean) or values are biased toward the left. As an example, consider a normal variable with mean $\mu = 1$ and variance $\sigma^2 = 0.25$ (the standard deviation is $\sigma = 0.5$). What is the probability of obtaining a value in between 0.5 and 1.5? This interval is one standard deviation away from the mean on each side. Therefore, the probability is 0.68.

A *Standard Normal* is a normal with zero mean $\mu = 0$ and unit variance $\sigma^2 = 1$. To obtain a standard normal from a normal, subtract the mean and divide by the standard deviation

$$Z = \frac{X - \mu_X}{\sigma_X} \tag{1.16}$$

the new variable Z is standard normal. Its mean is 0 and variance is 1.

All values to the left of the mean are negative ($z < 0$) and all values to the right of the mean are positive ($z > 0$). Because the normal is symmetric, calculating the area under the standard PDF curve from $-\infty$ up to a value $-z_0$ (left of the mean) is the same as calculating the area under the curve from that value $+z_0$ to $+\infty$ (right of the mean). See Figure 1.6.

The area under the curve right of the point z_0 is the same as $1 - F(z_0)$, where $F(z)$ is the CDF; important values are $F(0) = 0.5000$, $F(1) = 0.8413$, $F(2) = 0.9772$, and $F(3) = 0.9987$. See Figure 1.7. What is the probability that the variable is within the interval $+k\sigma$ around the mean? where $k = 1, 2, 3$? Or what is $P(|z| \le k\sigma)$ for $k = 1, 2, 3$? For $k = 1$, to obtain the probability of the variable taking values right of the mean and up to $+\sigma$ is $F(1) - F(0) = 0.8413 - 0.5 = 0.3413$. See Figure 1.7. Due to symmetry, this should be the same as left of the mean 0.3413, then $P(|z| \le \sigma) = 2[F(1) - F(0)] = 2 \times (0.8413 - 0.5000) = 0.682$. You can verify that for $k = 2$ $P(|z| \le 2\sigma) = 0.954$ and for $k = 3$, $P(|z| \le 3\sigma) = 0.997$. The values at the $0.1, 0.2, 0.3, \dots, 0.9$ quantiles or $10, 20, 30, 40, \dots, 90$ percentiles are also notable, approximately -1.28, $-0.84, -0.53, -0.25, 0.00, 0.25, 0.53, 0.84$, and 1.28.

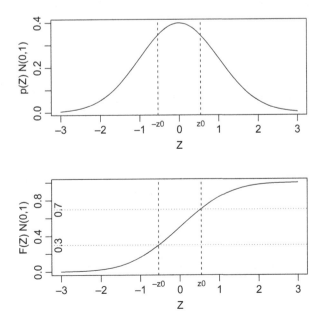

FIGURE 1.6 Standard normal PDF and CDF.

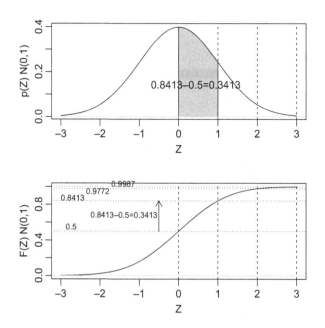

FIGURE 1.7 Standard normal distribution: probability values at 1, 2, and 3 standard deviation units (bottom). It also illustrates the equivalence between the area under the density curve between two points (top) and the difference in height under the cumulative curve (bottom).

COVARIANCE AND CORRELATION

In many cases, we are interested in how variables relate to each other; for example, the *bivariate* case of two RVs X and Y. An important concept is the joint variation or the expected value of the product of the two variables, where each one is centered at the mean. This is called the *covariance* and can be written as

$$\text{cov}(X, Y) = E[(X - \mu_X)(Y - \mu_Y)] \tag{1.17}$$

Please note that this is a theoretical concept since the expectation operator implies using the distribution of the product. Expanding Equation (1.17), and since the expectation of a constant is the same constant, we obtain

$$\text{cov}(X, Y) = E[XY] - \mu_X \mu_Y \tag{1.18}$$

A derived concept is the *correlation coefficient* obtained by scaling the covariance to values less or equal than 1, upon dividing by the product of the two independent standard deviations

$$\rho = \frac{\text{cov}(X, Y)}{\sigma_X \sigma_Y} \tag{1.19}$$

because this product is always larger than the expected value of the product then the ratio is always less than 1. This fact can also be seen by calculating the correlation coefficient for maximum covariance, which occurs when X and Y are identical

$$\rho = \frac{\text{cov}(X,Y)}{\sigma_X \sigma_Y} = \frac{\text{cov}(X,X)}{\sigma_X \sigma_X} = \frac{\sigma_X^2}{\sigma_X^2} = 1 \qquad (1.20)$$

The same idea can be applied to a sample to obtain the *sample correlation coefficient r* where the covariance is a *sample covariance*, and the denominator corresponds to the product of the sample standard deviations.

For illustration, we see two contrasting situations in Figure 1.8. These plots are called *scatter plots*; each pair of values x_i, y_i is marked with a symbol on the *x–y* plane. On the top panel, we have *y.u* vs. *x*, where *x.u* is a sample uncorrelated to *x*; the sample correlation coefficient is $r=0.123$. On the bottom panel, we have *y.r* vs. *x*, where *y.r* is correlated to *x*; the sample correlation coefficient is $r=0.979$, much closer to 1. In addition, we have drawn dotted lines at the mean of each sample.

Data are often collected in a time series or in a space sequence. For example, air temperature taken every hour, and vegetation cover taken every 10 m along a line or transect. A *time series* plot allows visualizing patterns in successive data values. A special application of the covariance and correlation concepts is to calculate the co-variation between pair of values separated by a lag *L* (either time or space) by means of the *autocovariance* function $C_X(L) = \text{cov}(x_{i+L}, x_i)$. We often work directly with the *autocorrelation* function, $\rho_X(L) = \text{cov}(x_{i+L}, x_i)/\sigma_X^2$ or the covariance scaled by the variance, which is equal to the maximum value of the auto-covariance function $C_X(0) = \text{cov}(x_{i+0}, x_i) = \text{cov}(x_i, x_i) = \sigma_X^2$. Note that for $L=0$, the autocorrelation will attain the maximum value of 1 $\rho_X(0) = (C_X(0)/\sigma_X^2) = (\text{cov}(x_i, x_i)/\sigma_X^2) = 1$.

As a specific example, consider the number of Atlantic hurricanes per season from 1944 to 1999 shown in Figure 1.9 (top). This is a *time series* plot; it allows seeing patterns in successive data values.

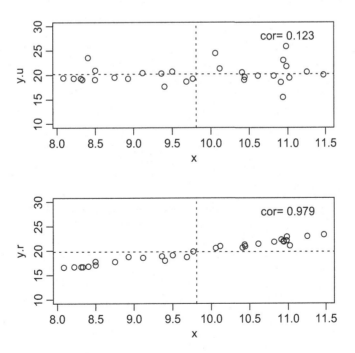

FIGURE 1.8   Scatter plots for two pairs of samples. Uncorrelated (top) and correlated (bottom).

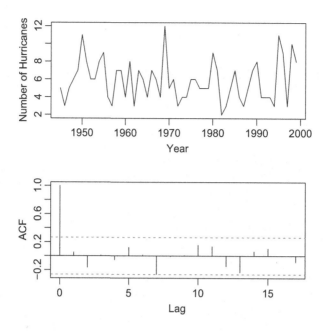

FIGURE 1.9 Number of Atlantic hurricanes in a season. Time series plot and autocorrelation plot.

EXPLORATORY DATA ANALYSIS

Classical parametric statistical inference, to be discussed in the next section, depends on outlier-free and Gaussian data. Before applying these inferential methods, it is a good idea to explore the data and see if they conform to the assumptions. One way of accomplishing this is by visual inspection of several plots; we can refer to this process as Exploratory Data Analysis (EDA).

In the *index plot*, observations are arranged serially according to the number of the observation allowing to see the variability of the data and identify potential outliers (Figure 1.10 left). For example, we can tell that there are potential outliers; three observations (10, 53, and 74) that have very low values and one (38) that has a very high value.

The *boxplot* or *box and whiskers* plot (Figure 1.10 right) shows the sample median (a line inside the box), the first and third quartiles or lower and upper *hinges* (edges of the box), and the minimum and maximum non-outlier values (the whiskers). These last two values are determined from the extremes of the range (or fence), which are the hinge (lower and upper, respectively) minus or plus a factor (e.g., 1.5) of the inter-quartile distance (iqd, for short). Values above or below the extremes of the range are outliers and identified as circles on the plot. For example, Figure 1.10 (right) shows the following values: lower hinge (first quartile)=38, upper hinge (third quartile)=54, and median= 46. In this case, the iqd is 54 – 38=16, and therefore using 1.5×16=24 for the range, we obtain 38 – 24=14 and 54+24=78 for the extremes of the range. The lowest value contained within the range is 30 (this sets the lower whisker) and the largest value is 75 (upper whisker). In this case, below 14 we have 3 values (7, 10, 13) and above 78 we have 1 value (96). All these four values are outliers and displayed as small circles (Figure 1.10 right-hand side).

A histogram is a graphical display of the distribution of a *sample*, representing the frequency with which you obtain a value falling in intervals (bins) of the range of the continuous variable. Given a large enough sample, a histogram can help to characterize the PDF of a variable, representing an approximation to the PDF (Figure 1.11 left) which can be further emphasized by a *density* approximation (Figure 1.11 right).

The empirical CDF or ECDF is constructed by sorting observations from smallest to largest to decide their position on the horizontal axes; these are the *i*th-order statistics. Once sorted, we rank the observations. Then, we divide these ranks by the number of observations to obtain

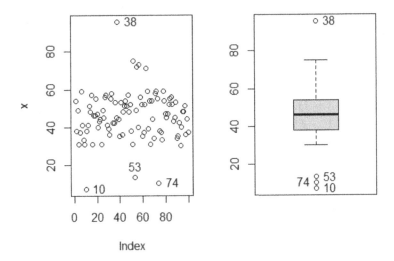

FIGURE 1.10 Index plot and boxplot to explore data showing identification of outliers.

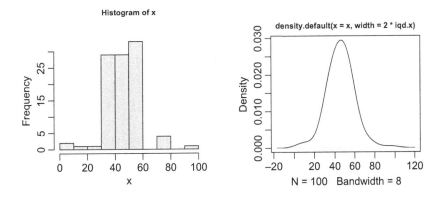

FIGURE 1.11 Histogram and density plots to explore data.

fractions of 1. Finally, these fractions go in the vertical axes. Naturally, 100 could multiply these fractions if we want the information in percentiles. We can explore normality by converting to Z scores, calculating the ECDF of the standardized sample, and comparing to the theoretical CDF of a standard normal. To obtain the z_i score for the data value x_i, we use

$$z_i = \frac{x_i - \overline{X}}{s_X} \tag{1.21}$$

where s_X is the sample standard deviation. In essence, we center the data at 0 by subtracting the sample mean (average), thus shifting the distribution, and then scaling up or down, according to s_X. For example, the ECDF of the standardized dataset of 100 numbers used in the boxplot above is shown in Figure 1.12 (left), together with the theoretical CDF of the standard normal. In this graph, we can see that the ECDF follows relatively well the expected or theoretical CDF for the normal distribution in the interval $(-1, 1)$ but departs significantly outside this interval.

Another tool to visualize whether the data follow a theoretical distribution is the *quantile-quantile* or *Q–Q plot*. For example, we can compare the quantiles of the data to the theoretical quantiles of a normal distribution. This is done using the standard normal Z: the z scores are ranked, and the ranks converted to fractions (values in between 0 and 1) or percentile values in 0–100 to obtain quantiles. Then these values are plotted in the vertical axis where the values of a standard normal

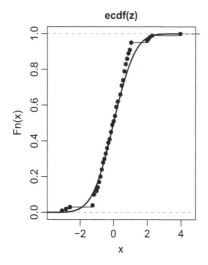 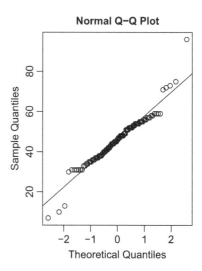

FIGURE 1.12   ECDF and *Q–Q* plots to explore data.

at the theoretical quantiles are used in the horizontal axis (Figure 1.12 right). If the data are normal, then the plot should follow the straight line drawn from corresponding quantile values. The outliers we saw in the boxplot are those observations that depart of the line.

STATISTICAL INFERENCE

The goal of *statistical inference* is to obtain probability statements about population parameters (mean, variance) from data. It involves two important concepts: *hypothesis testing* and *confidence intervals*. We seek to answer questions like these: is the sample drawn for a normal distribution? What is the uncertainty of an estimate of the population mean? For practicality, in the following, we will explain some simple methods of inference.

HYPOTHESIS TESTING

In the simplest approach, we work with the probability (α) of an unlikely event (interval called the *critical region*) that we set up from the PDF if the *null hypothesis* (H0) were to be correct. Once we analyze the data, if the result is within the critical region (i.e., the unlikely event), then we believe that it is unlikely that H0 is correct, and we reject it. On the contrary, if the experimental result is different from the improbable event, then there is no reason to reject H0.

Let us explain these ideas using a simple example. Suppose we want to know whether the sample mean departs significantly from the expected value of zero for a standard normal PDF $N(0,1)$. Setup H0: $\mu = 0$, select *type I error* $\alpha = 0.05$ or 5%, and identify the unlikely event dividing 0.05 into two equal parts of 0.025 because the sample mean departs either from above or below, i.e., *two-tail*, one on the left and one on the right. We know that for an $N(0,1)$, these values are approximately –2 and 2. Once we analyze the data and obtain more than 2 or less than –2, we reject H0 with at most a 5% probability of error ($\alpha = 0.05$). If the value obtained is larger than –2 and less than 2, then we cannot reject the null hypothesis.

The value α is the maximum error we are willing to accept. You should note that 5% is an arbitrary number that has remained popular in the literature as a convenient level of significance. However, we can calculate the *p-value* or probability that the measured outcome would occur and use it as the value at which the H0 can be rejected given the result of the experiment. Using the *p-value*, there is no need to establish a significance level a priori but just judge the magnitude of the *p*-value. If we believe that the *p*-value is small enough, then we can reject the null. For example, suppose we get a value of –3, we can calculate the probability that we get a value as low as –3 given that H0 is correct to be $(1-0.997) \times 2 = 0.0015$, meaning we can reject H0 with probability of error 0.15%.

To address the type II error of falsely failing to reject the H0, we set up an alternative hypothesis H1. The probability of error in rejecting H1 when it is true is the *type II error* (β). The power is $1 - \beta$ or probability of accepting H1 given that it is true. Then the total probability of error is the sum of the probability of false rejection of the null (false positive) and the probability of false rejection of the alternative (false negative)

$$P[\text{error}] = \alpha P(H0) + \beta[1 - P(H0)] \tag{1.22}$$

For a given α, we select the critical region in such a way that we minimize β or maximize power $1 - \beta$. In other words, we fix the type I error a priori and then design the test to minimize the type II error β given this type I error.

Expanding the previous example, suppose H0: $\mu < 0$ corresponds to a standard normal $N(0,1)$, select $\alpha = 0.025$, and consider only *one tail* for the critical region because we are interested only in the value being less than 0. Therefore, the unlikely event is that the value exceeds two. Setup H1: $\mu = 2$ corresponding to a normal $N(2,1)$. We know the area under the curve for $N(2,1)$ left of the critical value 2 is 0.5 and this determines the probability $\beta = 0.5$ that we accept H0 (reject H1) given that H1 is true. In this case, the type II error is high, and the power is low $1 - \beta = 1 - 0.5 = 0.5$. The difference between the two means $2 - 0 = 2$ is the *effect size*.

By using additional examples, we can determine that power increases with effect size. Intuitively, for the same α, the area under the H1 density curve left of the critical point decreases as we shift the H1 mean to the right. Consider H1: $\mu = 4$ for $N(4,1)$, in this case, the type II error of falsely rejecting H1 is $\beta = (1-0.9954)/2 = 0.023$, the power is $1 - 0.023 = 0.977$, and the effect size is 4. Suppose the sample value is one, we cannot reject H0 because the value is less than two and thus reject H1 with type II error of 2.3%. Suppose the sample value is 2.5, now we can reject H0 with type I error of 2.5% and accept H1.

CONFIDENCE INTERVALS

For large sample size n, the distribution of the sample mean $\overline{X}$ of X is approximately normal with mean equal to the expected value of $\overline{X}$ (which is equal to μ_X) and standard deviation equal to $\sigma_X / \sqrt{n}$ or standard error (see Equation (1.14)), that is to say $\overline{X} \sim N(\mu_X, \sigma_X / \sqrt{n})$.

Convert this distribution to a standard normal $N(0,1)$

$$Z = \frac{\overline{X} - \mu_X}{\dfrac{\sigma_X}{\sqrt{n}}} \tag{1.23}$$

Now we can say that 95% of the values of Z fall in the interval $[-2, 2]$. Thus, the probability that this interval contains the mean is 0.95. We define this as the 95% *confidence interval*. In general, we can use other quantiles of the standard normal. For quantile $z_{\alpha/2}$, we have the $100 \times (1-\alpha)\%$ confidence interval. Note that this is defined as an error rate α, i.e., the probability that an interval [a, b] fails to include the mean true value; common values for α are 0.01, 0.05, and 0.1 or 1%, 5%, and 10%.

PARAMETRIC METHODS

Classical inferential statistics proceeds as outlined so far in this chapter and it can be of two major types: parametric and nonparametric. In *parametric* methods, the PDF for the statistic is based on a parameter and assumes normal, outlier-free, non-serially correlated data; whereas *nonparametric* methods are independent of the distribution (normality not needed) and usually based on ranks of observations. In this section, we consider examples of parametric tests.

For example, the *t*-test is a parametric test that uses the sample to estimate unknown parameters and therefore it has reduced *degrees of freedom (df)* to compensate for the use of data in estimation of the parameters. In this case, we estimate the standard error from the sample standard deviation s_X. The number of degrees of freedom is df = number of observations in sample minus the number of parameters estimated from the sample.

The *t* distribution is similar to the normal but with heavier tails, which depend on size (*n*) of the sample or degrees of freedom $df = n-1$; the *t* statistic approaches the normal if the sample size is large

$$t = \frac{\bar{X} - \mu_X}{s_e} = \frac{\bar{X} - \mu_X}{\frac{s_X}{\sqrt{n}}} \tag{1.24}$$

This distribution can be used to test that the sample mean is equal to the true mean. The null H0: $\bar{X} = \mu_X$. Reject H0 if *t* is large for a given α. The number of degrees of freedom $df = n-1$, since s_X is estimated from the sample.

As an example, suppose we have $n = 20$, the sample mean is 0.68, and the sample standard deviation is 0.67. The H0 is that the mean $\mu_X =$ is 0. The critical values for $\alpha = 0.05$ are -2.09 and 2.09 for a t density with $df = 19$. Calculating *t* from Equation (1.24), we get $t = \frac{0.68 - 0}{0.67/20} = 4.52$; therefore, we can reject H0 at $\alpha = 0.05$. The probability of this t-value is 0.000116 and a *p*-value of 0.00023 for two-tails. With this low *p*-value, we would reject H0 given this outcome.

The *t*-test can also be used for one-sided situations; that is to say, to see if the sample mean is less than the true mean or to see if the sample mean is greater than the true mean. In addition, the *t* distribution can be used to test the equivalency of two samples (*t* statistic is redefined, s_e depends on n_1 and n_2). In this case, the null H0: $\mu 1 = \mu 2$. The degrees of freedom $df = n_1 + n_2$ 2, since two parameters are estimated from the sample (Acevedo 2013).

The *F* test is a parametric test for the equality of variances of two samples. Based on the *F* statistic given by ratio of two sample variances $F = s_1^2 / s_2^2$, the degrees of freedom are $df_1 = n_1 - 1$ and $df_2 = n_2 - 1$. The null H0: $\sigma_1^2 = \sigma_2^2$, meaning when we cannot reject H0, then there is no evidence to say that the variances are different. This test is performed prior to the *t*-test for two samples since the *t*-test assumes similar variances.

The parametric *correlation test*, Pearson's classical product-moment measure of correlation, allows establishing the statistical significance of the sample correlation coefficient *r*. The null hypothesis is that the samples are uncorrelated; this is to say H0: $\rho = 0$. Assume normality for both variables and use a *t* statistic of $t = \frac{r\sqrt{n-2}}{1-r^2}$ with $df = n-2$. We reject H0 if *t* is large for a given α; concluding that samples are correlated. However, if we cannot reject, then there is no evidence to say that the samples are correlated.

NONPARAMETRIC METHODS

Many nonparametric methods are based on a *rank* transformation of the data; values are sorted from smallest to largest. Thus, the actual values are no longer relevant, thereby reducing the impact of outliers. Let us denote $R(x_i)$ as the rank of observation x_i of the sample. There are several ways of resolving ties.

A nonparametric alternative to the *t*-test for equality of means of two samples is the *Mann-Whitney* or *Wilcoxon Rank Sum Test*. The ranks should spread out uniformly if it is true that the samples come from the same distribution. This test assumes that observations are independent and are based on ranks of the combined samples. The test statistic is the sum of ranks, and it can be approximated by a standard normal for large sample sizes.

An alternative to the *t*-test for one sample is the *Wilcoxon signed-rank test*, which is used when we want to see if the median is equal to a given value. We take the difference between the observations and the hypothetical median. Denote these differences by x_i and assign ranks $R(x_i)$ based on the absolute values $|x_i|$ of the differences x_i. However, to compute the test statistic, we only sum those ranks corresponding to positive differences $x_i > 0$, yielding an approximation to a standard normal for large sample size.

A popular method is the rank-based *Spearman's measure of correlation*. It is used to test for similarity between two sets of ranks, with a null hypothesis of zero correlation. Spearman's rank correlation coefficient is based on summing the square of differences in rank.

SIMPLE LINEAR REGRESSION

Throughout the book, we will encounter the concept of parameter estimation for a model or of sensor calibration, both of which require a basic understanding of regression. In this section, we review some basic notions of simple regression.

Let Y be an RV defined as the *dependent* or *response* variable, and X another RV defined as the *independent* or *explanatory* variable. This is a *bivariate* or two-variable situation. Assume that we have a joint sample x_i, y_i $i = 1, \ldots, n$ or measurements of X and Y. These data pairs display in the X–Y plane as a scatter plot. See the example in Figure 1.13.

Denote by $\hat{Y}$ the Linear Least Squares estimator of Y from X

$$\hat{Y} = b_0 + b_1 X \tag{1.25}$$

this is the equation of a straight line with *intercept* b_0 and *slope* b_1. For each data point i, we have the *estimated* value of Y at the specific points x_i

$$\widehat{y_i} = b_0 + b_1 x_i \tag{1.26}$$

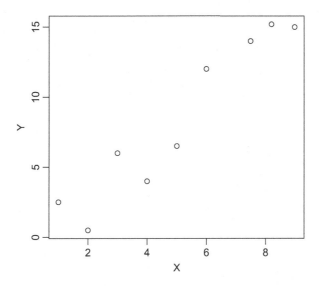

FIGURE 1.13 Scatter plot of X and Y.

the *error* or *residual* for data point i is

$$e_i = \widehat{y}_i - y_i \tag{1.27}$$

take the square and sum over all observations to obtain the total squared error

$$q = \sum_{i=1}^{n} e_i^2 = \sum_{i=1}^{n} (\widehat{y}_i - y_i)^2 = \sum_{i=1}^{n} (b_0 + b_1 x_i - y_i)^2 \tag{1.28}$$

We want to find the coefficients (intercept and slope) b_0, b_1 which minimize the sum of squared errors (over all $i = 1, ..., n$); to do this, we find the values of b_0 and b_1 that make the slope or derivative of q zero. A complete demonstration is given in Acevedo (2013) but in this book, we only state the results; for b_1

$$b_1 = \frac{\sum_{i=1}^{n} (x_i - \bar{X})(y_i - \bar{Y})}{\sum_{i=1}^{n} (x_i - \bar{X})^2} = \frac{s_{cov}(X,Y)}{s_x^2} \tag{1.29}$$

Here the numerator is the sample covariance of X and Y, whereas the denominator is the sample variance of X. For b_0 is

$$b_0 = \bar{Y} - b_1 \bar{X} \tag{1.30}$$

In summary, Equations (1.29) and (1.30) are used to calculate the coefficients b_0, b_1 which determine the regression line (Figure 1.14).

Rewriting Equation (1.30)

$$\bar{Y} = b_0 + b_1 \bar{X}$$

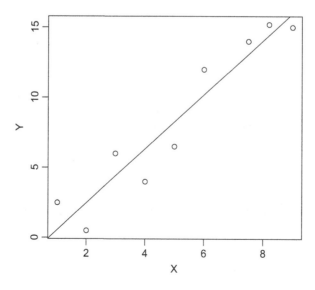

FIGURE 1.14   Regression line added to the scatter plot.

we note that the regression line goes through the sample means of X and Y. Using the correlation coefficient in Equation (1.29), we can rewrite as

$$b_1 = \frac{s_{\text{cov}}(X,Y)}{s_X^2} = \frac{rs_X s_Y}{s_X^2} = r\frac{s_Y}{s_X} \tag{1.31}$$

In other words, the slope is the sample correlation coefficient multiplied by the ratio of sample standard deviations of Y over X.

There are three important error terms in regression. In the following, SS denotes "sum of squares" and MS denotes "mean squares". The mean squares total error in Y is the sum of squared differences of sample points minus the sample mean divided by n, $MS_T = \left(\sum (y_i - \bar{Y})^2\right)/n = s_Y^2$ or sample variance of Y. The mean square "model" or "explained" error MS_M is the sum of squared differences of estimated points minus the sample mean $MS_M = \left(\sum (\hat{y}_i - \bar{Y})^2\right)/n$. The mean square "residual" or "unexplained" error SS_E is the sum of the squares of the difference between estimated and observations $MS_E = \left(\sum (\hat{y}_i - y_i)^2\right)/n$. Now the total error is the sum of model error and the residual, that is to say $s_Y^2 = MS_E + MS_M$.

A common measure of goodness-of-fit is the ratio of the model error to the total error

$$R^2 = \frac{MS_M}{MS_T} = \frac{MS_M}{MS_E + MS_M} = \frac{MS_M}{s_Y^2} \tag{1.32}$$

when MSE (which is minimized by the least squares procedure) is very small, then R^2 approaches 1. Note that R^2 is the fraction (or percent) of variance of Y explained by the regression model.

In addition, substituting the predictor equation, and recognizing the sample variance of X, we obtain

$$R^2 = \frac{\left(\frac{1}{n}\right)\sum b_1^2 \left(X - \bar{X}\right)^2}{s_Y^2} = \frac{b_1^2\left(\frac{1}{n}\right)\sum \left(X - \bar{X}\right)^2}{s_Y^2} = \frac{b_1^2 s_X^2}{s_Y^2} \tag{1.33}$$

and by recalling the expression for b_1

$$R^2 = \frac{\left(\frac{s_{\text{cov}}(X,Y)}{s_X^2}\right)^2 s_X^2}{s_Y^2} \tag{1.34}$$

therefore

$$R^2 = \left(\frac{s_{\text{cov}}(X,Y)}{s_X s_Y}\right)^2 = r^2 \tag{1.35}$$

The square root of R^2 is equal to r, which is the correlation coefficient.

The residual standard error is the standard deviation of the residuals

$$s_e = \sqrt{\frac{\sum (y_i - \hat{y}_i)^2}{n-2}} \tag{1.36}$$

We have used $n-2$ because two parameters (slope and intercept) were estimated.

We do not just look at the R^2 to evaluate a regression. It is necessary to check the assumptions of the method and to examine the significance of the regression, the confidence interval, and the patterns of the unexplained or residual error. For example, we examine the following. Consider the p-value of a test to check the statistical significance of the trend, i.e., whether the slope is non-zero. Look at the scatter plot of Y versus X to see how good the linear assumption is. If the y_i points seem to follow a definite non-straight pattern or curve, then linearity is suspicious even when getting a good R^2. Consider the random residual error by looking at a plot of the residuals as a function of the estimated or predicted y. The residuals should be scattered up and down around zero (i.e., "just noise"), telling us that the error is independent of the position in the regression line.

FROM MEASURING TO KNOWING, ANALYSIS, AND MODELING

A model is a simplified representation of reality based on concepts, hypotheses, and theories of how a real system works. Some models represent reality as a set of mathematical equations based on the *processes* at work (Acevedo 2012). For example, a differential equation representing tree growth over time based on increment of its diameter. For this purpose, we use the concept that diameter increases faster when the tree is smaller and that growth decreases when the tree is large. This process-based or mechanistic method is in contrast to *empirical* models that build a quantitative relationship between variables based on data without explicit consideration of the process yielding that relation (Acevedo 2013). For example, using regression, we can derive a predictor of tree height as a function of tree diameter based on measured data from many trees of different heights and diameters.

However, we use empirical models to estimate parameters of the process-based models based on data from field and laboratory experiments, as well as monitoring programs. For example, we can use a mechanistic model to calculate the flow of a stream using water velocity and cross-sectional area, but we estimate velocity using an empirical relation of velocity to water depth. In addition, we can use empirical models to convert output variables of process-based models to other variables. For example, we can predict tree diameter increment from a process-based model of tree growth and then convert diameter to height using an empirical relation of height versus diameter.

Temporal dynamics and spatial gradients make the concept of *rate of change* have paramount importance in environmental monitoring and modeling. Therefore, one interesting application of environmental monitoring results is to analyze the *dynamics* of the environmental systems, that is, changes over time. Moreover, we can integrate the results with mathematical and simulation models to predict future behavior of the environmental system. This book does not emphasize the mathematical fundamentals to understand and analyze models nor the methodology to simulate models. You can use Acevedo (2012) for this purpose.

We use datasets created from monitoring results to inform, calibrate, and evaluate models. There are models readily available today through various government agency and university websites such as the U.S. Environmental Protection Agency (USEPA) and the U.S. Geological Survey (USGS).

Interdisciplinary

Environmental monitoring requires interdisciplinary work among scientists and engineers with various backgrounds and training. Electrical engineers would be very familiar with the electronic technology underlying sensors and systems but monitoring also requires knowledge of materials as well as mechanical and chemical engineering disciplines. Computer science and engineering provide important tools for programming algorithms and database management.

In most cases, an engineer would benefit from an understanding of the monitored natural system. Such an expertise is often provided by scientists in various disciplines that converge in environmental sciences. For example, principles of biology, chemistry, and physics are very helpful to understand the basic interactions of a sensor with its environment. Mathematical and statistical skills are needed for data processing and analysis and for the fundamental nature of data quality assurance and control.

SCALES AND RESOLUTION

Like in many other fields, two important scales play a role in a monitoring program: *spatial* and *temporal*. Under spatial considerations, we include whether we want measurements at a single point, or one- (e.g., transect), two- (e.g., horizontal grid), or three-dimensional (3D) arrays (e.g., horizontal grid plus height). For example, we may want to measure soil moisture near the soil surface at a series of points representing topographic position to study the effects of terrain elevation on surface soil moisture; or more complicatedly, measure the air temperature in a horizontal grid at various height levels of a forest. For all these arrangements, we would have to consider *resolution*, such as detailed requirements in terms of grid cell size, number of vertical strata measured, and *extent*, such as area covered.

Timescale is characterized by similar considerations, resolution by time step or interval of measurements, and time step of data reporting and storage. For example, data could be measured every second but averaged every 10 minutes for storage. Similarly, depending on the purpose of the monitoring program, we may further process the stored data. For example, for regulatory purposes or compliance with standards, we may calculate averages every hour, reporting 24 values in a day, and then calculate the maximum in an 8-hour period, such as calculating 8-hour ground-level ozone (McCluney 2007; ASL & Associates 2023; USEPA 2021a, b). Data may be reported and analyzed weekly, monthly, seasonally, and annually, depending on the objectives of the program. Many long-term modern monitoring programs do not limit the temporal extent to a period or horizon of measurements. However, naturally, shorter studies may establish a limited period, such as a year or a decade.

A related concept to resolution is used in analytical chemistry, the *lower limit of detection*, or limit of detection, is the lowest quantity of a substance that can be distinguished from the absence of that substance (a blank value) within a specified confidence limit (e.g., 1%).

We often must make decisions about how to report and store the results of measurements. For example, even though we may have many decimal places for a number, these will not make sense in terms of the process and devices employed to obtain the value. Let us recall a simple rule to round numbers. If the number is greater than 0.5, then round up; for example, round 38.6 up to 39.0. If the number is less than 0.5, then round down; for example, round 38.4 down to 38.0. If the number is equal to 0.5, then break the tie, which is round to the nearest even number or to the nearest odd number. For example, round 36.5 down to 36, the nearest even number, and round 37.5 up to 38. This rule results in rounding up or down in equal proportions. Together with rounding, recall simple rules of significant figures: zeros to the left do not count but zeros to the right do count.

PRECISION AND ACCURACY

Precision is the variation in measured values as we repeat the measurements. High precision corresponds to small variation in measured values upon repeated measurements, whereas low precision implies larger variations for the same conditions (Artiola and Warrick 2004). It is similar to the concept of *noise* or variability and to the concept of *random error*. A useful measure of precision is the sample standard deviation defined by Equation (1.12) (ISO 2023; Bell 1999). In some instances, outliers can be detected and removed when taking the sample mean.

Accuracy is the difference between the measured value and its true or reference value, expressed often as the maximum error one can expect in the measurement (Artiola and Warrick 2004). It is often related to the concept of *bias* or *systematic error*. A useful measure of accuracy is the difference between the sample mean and the true value, and whether that difference is greater than the standard deviation, i.e., not just within the noise or random error.

Often, in colloquial terms, many people use precision and accuracy to mean the same thing, but in measurement, they have distinct meanings (Figure 1.15). Precision refers only to the variability of measurements and does not indicate whether the readings are correct (Artiola and Warrick 2004). An instrument can be accurate but not precise, or precise but not accurate, or both precise and accurate, or neither.

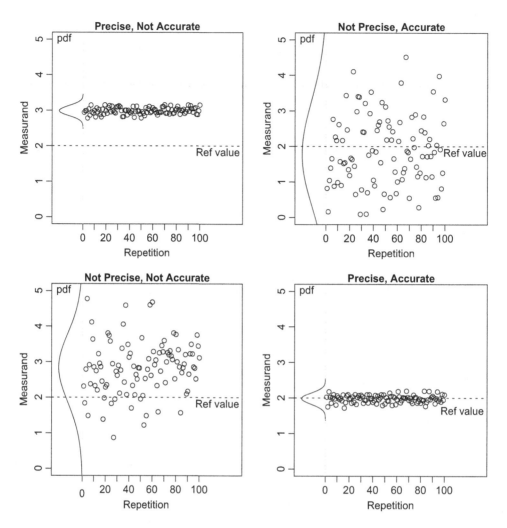

FIGURE 1.15 Precision and accuracy are distinct concepts. At the left of each graph, we illustrate the empirical probability density of the sample. Higher precision occurs for less spread or variance of this probability density. The larger the distance between the mean of this probability density and the reference value, the lower the accuracy. Four distinct cases are illustrated.

The terminology related to these concepts can be more complex. For example, the International Vocabulary of Basic and General Terms in Metrology (ISO 2023) recommends to use the terms *repeatability* and *reproducibility* to relate to how close are the results of successive measurements; the first term is closeness under the same conditions, whereas the second are under changed conditions.

REMOTE SENSING: AIRBORNE AND SPACEBORNE PLATFORMS

There is great variety of airborne and spaceborne platforms and instruments to monitor environmental systems remotely, i.e., the sensors are located at a distance above the ground that allows covering a broader spatial range (Huete 2004). *Remote sensing* includes taking images of the land or ocean for specific purposes as needed, for example by airplanes and unmanned aircraft systems or *drones*, as well as repetitive collection of imagery of the same area, for example by satellites orbiting the Earth. Remote sensing of the environment is an extensive topic and there is a wealth of information, books, and journals on these monitoring approaches.

In this book, we touch on the use of remotely placed sensors (air- or spaceborne) responding to different electromagnetic (EM) reflection and absorption of sunlight by land, soil, vegetation, and surface water. Atmospheric effects, due to the reflection and scattering of particulate matter or absorption by atmospheric gases, modify the reflected signal from the ground, which can be considered to correct for these effects. We also discuss the use of spectrometers at ground level to acquire data that can serve to ground-truth the remote sensors. Drones for environmental parameters sensing is in the middle between ground data collection and remote sensing from space (Wallerman et al. 2018).

Another class of remote sensors relies on the response to an EM wave signal shot from the platform to the ground, rather than sunlight. These include LiDAR (Light Detection And Ranging) and RADAR (RAdio Detection And Ranging). LiDAR uses laser pulses to map objects or the ground in 3D, for example digital elevation models that can be used to monitor land erosion. Radar uses EM in the radio and microwave part of the spectrum, and the backscatter can be used to monitor surface water, forest biomass, and many other environmental systems. Both LiDAR and Radar are also used at ground level to monitor a variety of processes.

The concept of resolution applies to remote sensing imagery in terms of spatial, spectral, temporal, and radiometric resolution. We will discuss these concepts in Chapter 8. Analysis of remote sensing images requires techniques of image processing (Haneberg 2004), and there is a trend to seek solutions of data classification problems and biophysical parameter estimation by machine learning techniques, which we will cover in Chapter 9.

MORE ON APPLICATIONS

Monitoring applications are so diverse that it is difficult to prescribe generalized guidance on how to structure a monitoring program. This means that the intended application should play an important role in designing the monitoring program.

Environmental monitoring contributes to the development of environmental impact assessments, as well as in many circumstances in which human activities carry a risk of harmful effects on the natural environment (Artiola et al. 2004a). Often, we design monitoring activities to establish the baseline, status, and trends of environmental variables. In these cases, the results of monitoring are subject to scrutiny and discussion, and in some cases litigation.

Other examples include environmental remediation and restoration, and biological conservation. In remediation and restoration, a monitoring program helps to follow-up, assess how successful the implemented solutions are over time, as well as help design and develop solutions and corrections. In biological conservation, we can detect changes in organism and ecosystem responses to evaluate the success of the conservation effort.

Monitoring helps to build understanding for scientific and engineering purposes, and many monitoring networks have been formed for this purpose. Other applications include informing the public. For example, environmental and ecological observatories collect real-time information on environmental conditions using ground-based network of stations and remote sensing to make data readily available to the public and amenable to modeling, analysis, and synthesis (Figure 1.16). For example, during the early 2000s, at the University of North Texas, in collaboration with the City of Denton, we operated a Texas Environmental Observatory (TEO) that provided data on ultraviolet radiation, total column ozone, soil moisture, and water quality.

As we progress from sensors to systems and realize the importance of real-time continuous monitoring with very long-term horizons, we will realize that this integration provides us a way of moving forward toward a better future. We can take better care of planet Earth, because we have a better understanding of the complicated dynamics of the planet.

Whether you are involved in making policy, regulatory actions, decision-making, education, or public outreach, monitoring helps you to have the information and tools to better orient your efforts. Many existing programs offer opportunities for training in monitoring activities. For example, we developed workshops based on TEO for local governments and teachers. Attendees learned about

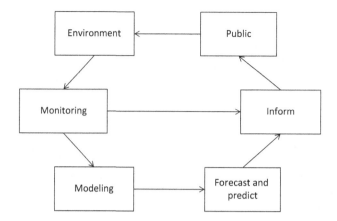

FIGURE 1.16 Simple representation of an environmental observatory paradigm, which consists of integrated modeling and monitoring and emphasizes feedback.

various tools and methods of data collection, including wireless soil moisture sensors (Chapter 6), surface water multiple parameter probes (Chapter 12), database management and web services (Chapter 10), and environmental models. Attendees left with an understanding of the benefits of cyberinfrastructure and knowledge of how real-time environmental data are being collected and utilized in their local area (TEO 2015).

EXAMPLES OF PROGRAMS AND AGENCIES

In the United States, we have excellent examples of systems developed by government agencies. The USGS has a variety of systems such as the Water Quality Watch of surface water (USGS 2021d), the National Water Information System (USGS 2021c), earthquake monitoring (USGS 2021b), and volcanoes monitoring (USGS 2021a). From NOAA, examples are the weather observation and forecast system (NOAA 2021a), climate monitoring (NOAA 2021c), and ocean and coast monitoring (NOAA 2021d). Most National Aeronautics and Space Administration (NASA) efforts on Earth observation are spaceborne and airborne, but several programs are ground-based. The U.S. Department of Agriculture through the Natural Resources Conservation Service has a National Environmental Monitoring Initiative (USDA 2021). The National Science Foundation has started the National Ecological Observatory Network (NEON 2021). Unfortunately, some efforts have stopped, such as the Environmental Monitoring and Assessment Program of the U.S. EPA and the Ecological Monitoring and Assessment Network of Canada.

Many countries have monitoring programs; for example, in Sweden, the Swedish Environmental Protection Agency has a national monitoring program for contaminants and other environmental stressors (SEPA 2021).

At a global scale, there are programs of the United Nations, such as the Global Atmosphere Watch (WMO 2021a) conducted by the World Meteorological Organization (WMO 2021b), the Global Climate Observing System (GCOS 2021), the World Conservation Monitoring Centre (UNEP-WCMC 2021), and the Global Environment Monitoring System (UNEP 2021) conducted by the UN Environment Programme.

ENVIRONMENTAL MONITORING BOOKS

Several books on environmental monitoring provide additional resources. Artiola et al. (2004b) have a broad coverage, including environmental characterization, sampling, statistics, geographic information systems, remote sensing, and automated data acquisition, among many topics.

The book edited by Wiersma (2004) is a comprehensive reference book treating monitoring of air, water, and land; it includes integrated monitoring at the landscape level, as well as case studies of existing monitoring programs. Kim et al. (2009) cover many aspects of biological, chemical, and atmospheric environmental monitoring. Cole (2017) is a guide to the theory and practice of environmental monitoring that can be used to prepare environmental impact assessments and covers air, water quality, and soils. Kazantsev (2017) describes the processes and activities required to monitor and assess the quality of the environment and natural resources, including air, water, soil, and biota. Campbell (1997) covers sensor technologies, including optics, electrochemical, gas sensors and analyzers, piezoelectric, and biosensors.

EXERCISES

Exercise 1.1

Define an RV from the outcome of soil moisture measurements in the range of 20%–40% in volume. Give an example of an event. Assuming that it can take values in [20,40] uniformly, plot PDF and CDF, and calculate the mean and variance.

Exercise 1.2

At a site, monthly air temperature is normally distributed. It averages to 20°C with standard deviation of 4°C. What is the probability that a value of air temperature in a given month exceeds 24°C? What is the probability that it is below 16°C or above 24°C?

Exercise 1.3

Suppose we have collected 50 values for a sample of ozone and the average is 2.00, with standard deviation of 0.5. What would be the value of t when testing that the mean is equal to 2.5?

Exercise 1.4

Monthly rainfall at a site is classified into two groups: one group for El Niño months and the other for La Niña months (defined according to sea surface temperature in the Pacific Ocean). We have 100 months for each group. The variance of each group is the same. Is it true that rainfall during El Niño is different to that during La Niña? What type of test would you run? What is H0? Suppose you get a p-value=0.045. What is the conclusion of the study?

Exercise 1.5

Consider the following six measurements of concentration X in ppm of a water constituent: 1.1, 1.0, 1.3, 1.1, 1.2, and 1.1. Suppose the true value is 1.5. State the precision and accuracy of these values. Hint: calculate average and standard deviation then use these to look at the difference between the average and the true value and variability.

Exercise 1.6

We are designing a monitoring program for a rectangular area of 10 km^2. Which option gives the lowest spatial resolution if we install a monitoring station in each one of the resulting cells? Assume all cells have the same size for each design option. Option A: divide the area in ten rows of five cells each. Option B: divide the area in five rows of ten cells each. Option C: divide the area in ten rows of ten cells each. Option D: divide the area in five rows of five cells each.

Exercise 1.7

Using NOAA's Earth System Research Laboratory, Global Monitoring Division website for CO_2 measured at Mauna Loa (NOAA 2021d), calculate average growth rate in ppm/year for the 1960–1969 decade and for the 2000–2009 decade, and compare. Hint: use annual mean data. Subtract first CO_2 ppm value of the decade from last CO_2 ppm value of the decade and divide by the number of years. Alternatively, average the rate ppm/year values given in the website for each decade.

REFERENCES

Acevedo, M. F. 2012. *Simulation of Ecological and Environmental Models*. Boca Raton, FL: CRC Press, Taylor & Francis Group. 464 pp.

Acevedo, M. F. 2013. *Data Analysis and Statistics for Geography, Environmental Science & Engineering. Applications to Sustainability*. Boca Raton, FL: CRC Press, Taylor & Francis Group. 535 pp.

Acevedo, M. F. 2024. *Real-Time Environmental Monitoring: Sensors and Systems, Second Edition – Lab Manual*. Boca Raton, FL: CRC Press, Taylor & Francis Group. 463 pp.

Artiola, J. F., I. L. Pepper, and M. L. Brusseau. 2004a. "Monitoring and Characterization of the Environment." In *Environmental Monitoring and Characterization*, edited by J. F. Artiola, I. L. Pepper and M. L. Brusseau. Burlington, MA: Elsevier Academic Press. pp. 1–9.

Artiola, J. F., I. L. Pepper, and M. L. Brusseau, eds. 2004b. *Environmental Monitoring and Characterization*. Burlington, MA: Elsevier Academic Press. 410 pp.

Artiola, J. F., and A. W. Warrick. 2004. "Sampling and Data Quality Objectives for Environmental Monitoring." In *Environmental Monitoring and Characterization*, edited by J. F. Artiola, I. L. Pepper and M. L. Brusseau. Burlington, MA: Elsevier Academic Press. pp. 11–27.

ASL & Associates. 2023. *Calculating the 8-Hour Ozone Standard*. Accessed May 2023. https://www.asl-associates.com/cal8hr.htm.

Bell, S. 1999. A Beginner's Guide to Uncertainty of Measurement. *Measurement Good Practice* 11 (2):34.

Campbell, M., ed. 1997. *Sensor Systems for Environmental Monitoring: Sensor Technologies*. Dordrecht: Springer. 310 pp.

Cole, H., ed. 2017. *Environmental Monitoring*. New York: Larsen and Keller Education, 324 pp.

GCOS. 2021. *Global Climate Observing System (GCOS)*. Accessed December 2021. https://gcos.wmo.int/.

Haneberg, W. 2004. *Computational Geosciences with Mathematica*. Berlin, Heidelberg: Springer. 381 pp.

Huete, A. R. 2004. Remote Sensing for Environmental Monitoring. In *Environmental Monitoring and Characterization*, edited by J. F. Artiola, I. L. Pepper and M. L. Brusseau. Burlington, MA: Elsevier Academic Press. pp. 183–206.

ISO. 2023. *ISO/IEC Guide 99:2007, International vocabulary of metrology — Basic and general concepts and associated terms (VIM)*. accessed May 2023. https://www.iso.org/standard/45324.html.

Kazantsev, T., ed. 2017. *Fundamentals of Environmental Monitoring*. Burlington ON: Arcler Education Inc., 276 pp.

Kim, Y. J., M. G. Ulrich Platt, and I. Hitoshi, eds. 2009. *Atmospheric and Biological Environmental Monitoring*. Dordrecht: Springer. 311 pp.

Lovett, G. M., D. A. Burns, C. T. Driscoll, J. C. Jenkins, M. J. Mitchells, L. Rustad, J. B. Shanley, G. E. Likens, and R. Haeuber. 2007. Who Needs Environmental Monitoring? *Frontiers in Ecology and the Environment* 5 (5):253–260.

McCluney, L. O. 2007. *Calculating Design Values*. Accessed 2014. http://www.epa.gov/ttn/airs/airsaqs/conference/AQS2007/Session%20Handouts/Calculating%20Design%20Values%20updated%202007.pdf.

NEON. 2021. *National Ecological Observatory Network*. Accessed December 2021. https://www.neonscience.org/.

NOAA. 2021a. *National Weather Service*. Accessed December 2021. https://www.weather.gov.

NOAA. 2021b. *Trends in Atmospheric Carbon Dioxide, Mauna Loa, Hawaii*. Accessed December 2021. https://gml.noaa.gov/ccgg/trends/.

NOAA, Climate Prediction Center. 2021c. *Monitoring and Data*. Accessed December 2021. http://www.cpc.ncep.noaa.gov/products/MD_index.shtml.

NOAA, National Ocean Service. 2021d. *Monitoring Oceans and Coasts*. Accessed December 2021. http://oceanservice.noaa.gov/observations/monitoring/.

R Project. 2022. *The R Project for Statistical Computing*. Accessed January 2022. https://www.r-project.org/.

SEPA. 2021. *Swedish Environmental Protection Agency*. Accessed December 2021. https://www.naturvards-verket.se/en.

UNEP. 2021. *GEMSTAT – Global Environment Monitoring System*. Accessed December 2021. http://www.gemstat.org/.

UNEP-WCMC. 2021. *United Nations Environment Programme – World Conservation Monitoring Centre*. Accessed March 2015. http://www.unep-wcmc.org/.

USDA, Natural Resources Conservation Service. 2021. *National Water and Climate Center*. Accessed December 2021. https://www.nrcs.usda.gov/wps/portal/wcc/home/.

USEPA. 2021a. *NAAQS Table*. Accessed December 2021. https://www.epa.gov/criteria-air-pollutants/naaqs-table.

USEPA. 2021b. *Timeline of Ozone National Ambient Air Quality Standards (NAAQS)*. Accessed December 2021. https://www.epa.gov/ground-level-ozone-pollution/timeline-ozone-national-ambient-air-quality-standards-naaqs.

USGS. 2021a. *Comprehensive Monitoring Provides Timely Warnings of Volcano Reawakening*. Accessed December 2021. https://www.usgs.gov/programs/VHP/comprehensive-monitoring-provides-timely-warnings-volcano-reawakening.

USGS. 2021b. *Earthquake Hazards Program*. Accessed December 2021. http://earthquake.usgs.gov/.

USGS. 2021c. *National Water Information System: Web Interface*. Accessed December 2021. http://waterdata.usgs.gov/nwis.

USGS. 2021d. *WaterQualityWatch – Continuous Real-Time Water Quality of Surface Water in the United States*. Accessed December 2021. http://waterwatch.usgs.gov/wqwatch/.

Vaughan, H., T. Brydges, A. Fenech, and A. Lumb. 2001. Monitoring Long-Term Ecological Changes through the Ecological Monitoring and Assessment Network: Science-Based and Policy Relevant. *Environmental Monitoring and Assessment* 67:3–28.

Wallerman, J., J. Bohlin, M. B. Nilsson, and J. E. S. Franssen. 2018. Drone-Based Forest Variables Mapping of ICOS Tower Surroundings. In *IGARSS 2018 IEEE International Geoscience and Remote Sensing Symposium*, Valencia: IEEE 22–27 July 2018.

Wiersma, G. B., ed. 2004. *Environmental Monitoring*. Boca Raton, FL: CRC Press. 792 pp.

WMO. 2021a. *Global Atmosphere Watch Programme*. Accessed December 2021. https://public.wmo.int/en/programmes/global-atmosphere-watch-programme.

WMO. 2021b. *World Meteorological Organization – Weather – Climate –Water – Environment*. Accessed December 2021. https://public.wmo.int/en.

2 Programming and Single-Board Computers

INTRODUCTION

Computers and programming have become of paramount importance for environmental monitoring and are now part of sensor systems as embedded systems, dataloggers, and sensor networks. Working with computers in the field of monitoring requires knowledge of iterative calculation and measurement loops, serial communication, analog to digital and digital to analog conversion, networks, database management, and web applications. This chapter emphasizes single board computers (SBCs) and microcontrollers (MCUs), provides basic concepts of computer organization and architecture, and focuses on their application for environmental monitoring. For this purpose, we describe the Raspberry Pi and Arduino, and basic concepts of programming in Python, Arduino, Hyper Text Markup Language (HTML), Hypertext Preprocessor (PHP), Java Script (JS), and Structural Query Language (SQL). We will employ these devices and programming languages later in Chapters 3–6 when we study sensors, data acquisition, dataloggers, and sensor networks.

COMPUTER ORGANIZATION AND ARCHITECTURE

In this section, we review terminology and basicw concepts of computer organization and architecture (Null and Lobur 2012). Basic computer organization includes a CPU (*Central Processing Unit*), *memory*, and I/O (*input/output*) devices. Exchange of data among these components occurs over a *bus*. The CPU performs computations, decides logical flow, and processes I/O requests, while memory is used to store data processed by the CPU, and I/O peripherals allow interaction with the computer.

Within the CPU, arithmetic operations are conducted by the ALU (*Arithmetic/Logic Unit*), whereas a *Control Unit* decodes and executes instructions, which are a set of bits specifying operations. A set of registers serve as program counter that points to the memory address containing the instruction to be executed. Then the control unit fetches the instruction and executes it; these are the *fetch/execute* steps.

Two major types of memory are RAM (random access memory) and ROM (read-only memory) with variants for each. For example, SRAM (static RAM) provides data permanence while powered, whereas DRAM (dynamic RAM) requires periodic refreshing. In general, SRAM is volatile because data are lost after turning power off, however, there are special types of non-volatile SRAM that have applications in datalogging.

There are several types of DRAMs, such as synchronous DRAM (SDRAM) and double data rate SDRAM (DDR SDRAM). The latter has had several generations, DDR1, DDR2, DDR3, and culminating with DDR4, which is currently in use. DDR1 writes two words of data per clock cycle, DDR2 doubles it by writing four words of data per cycle, and DDR3 doubles it again to eight words of data per cycle.

ROM can be programmable (PROM) for installing programs and erasable (EPROM). We will expand on these in a later section on MCUs.

SINGLE BOARD COMPUTERS

A *Single Board Computer* (SBC) is a microcomputer on a single printed circuit board, including CPU, memory, and I/O ports (Figure 2.1). SBCs were originally intended for use in embedded

DOI: 10.1201/9781003425496-2

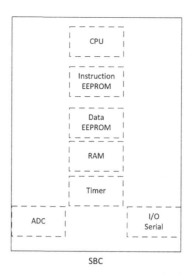

FIGURE 2.1 SBC components.

applications, i.e., provide computing capabilities as part of a larger equipment or system (Null and Lobur 2012). Embedded systems typically did not include peripherals for human interaction; they were connected to a general-purpose computer to load a program or to retrieve stored data.

As such, in contrast to computers, SBC functions were specialized to the needs of the larger system instead of providing a flexible computer platform. This situation has changed as SBCs became more powerful and low-cost and are able to provide a flexible computer platform. Such is the case of the *Raspberry Pi*, which has evolved to provide a Linux computing platform. SBCs employ a variety of microprocessors and MCUs as CPU or complete systems. There are also applicable as *Edge* devices or devices close to the process or data generating sources, in contrast to devices remote from the process (e.g., on the *cloud*).

A/D AND D/A CHANNELS

In addition to digital pins or lines, and because of their use in embedded systems, some SBCs include A/D pins (ADC) for measurement and D/A pins for control. We will discuss A/D and D/A conversion in detail in Chapter 4. For now, let us just motivate their use by pointing out that the output of a *transducer* (which will be covered in detail in Chapter 3) is an analog voltage or current output signal that is proportional to the continuous measurand variable. Therefore, we need to convert to digital in binary form to be able to interface with MCUs and computers (Kester, Bryant, et al. 2015; Kester, Sheingold, et al. 2015).

The quantization or discretization provided by an A/D converter depends on its number of bits. With a *resolution* of n bits, we can divide the maximum analog signal range in 2^n levels. This means $2^n - 1$ steps between levels, and thus a voltage in the range of a step is represented as a digital number stored as binary. Therefore, for a maximum analog voltage of V_{AD}, the *voltage resolution*, i.e., the smallest increase in voltage to produce an increase in digital numbers is

$$V_{res} = \frac{V_{AD}}{2^n - 1} \tag{2.1}$$

The voltage resolution corresponds to the voltage eliciting the smallest change in digital value or the LSB (*least significant bit*).

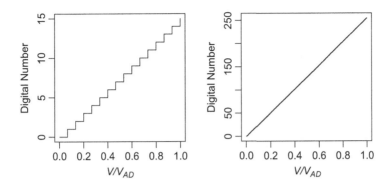

FIGURE 2.2 A/D output as a function of analog input: four bits (left) and eight bits (right).

For example, with $n=4$ bits, we have $2^4 = 16$ levels and $2^4 - 1 = 15$ steps that can be assigned to digital numbers 0–15, or 0000–1111 in binary.

```
0 0 0 0
0 0 0 1
0 0 1 0
0 0 1 1
0 1 0 0
0 1 0 1
0 1 1 0
0 1 1 1
1 0 0 0
1 0 0 1
1 0 1 0
1 0 1 1
1 1 0 0
1 1 0 1
1 1 1 0
1 1 1 1
```

This is illustrated in Figure 2.2 (left), where the horizontal axis is the voltage V scaled as a fraction of voltage V_{AD}. In this case, the voltage resolution is $V_{res} = \dfrac{V_{AD}}{2^4 - 1} = \dfrac{V_{AD}}{15} = 0.0666 \times V_{AD}$ or 6.66% of V_{AD}. For instance, given $V_{AD} = 5V$ the voltage resolution is $V_{res} = \dfrac{5V}{15} = 0.333V$. In contrast, the same $V_{AD} = 5V$, discretized using $n=8$ bits, yields $2^8 = 256$ levels and $2^8 - 1 = 255$ steps or voltage intervals (Figure 2.2 right). Then 5 V / 255 steps = 0.0196 V / step, which means the voltage resolution is 19.6 mV.

SERIAL COMMUNICATIONS

Typically, collected monitoring data are stored in files, which are retrieved locally or remotely. Users copy data files to other devices but not move them; this way, files are preserved and available for retrieval by multiple users. Useful access to data files is via serial port connections, such as RS-232 and Universal Serial Bus (USB), or by Ethernet connections using appropriate protocols, such as Transmission Control Protocol/Internet protocol (TCP/IP) (we will discuss networks in the next section). Other interfaces and protocols to exchange data among devices will be covered in Chapter 5.

The symbol rate (in *baud*) is the number of symbols per second or waveform changes per second. This rate expresses itself differently according to the modulation method employed. When we have two levels per symbol, e.g., binary 0 and 1, then each symbol carries one bit of data, and thus the bit rate (in bits per sec) is the same as the symbol rate (in baud). Although in principle one can have any baud value, only a set of values is commonly used; for example, doubling from 300, say 1200, 2400, 4800, 9600 at the slow range, or 57,600, 115,200, 230,400 at a faster range.

Environmental sensors measuring a variable intermittently (at low frequency) generate small amount of data and thus typically require only low speed, say 9600 baud. However, as we increase the amount of data collected, as it occurs in more demanding high meteorological frequency applications, or multimedia data, we require higher baud rates.

As part of the serial communication protocol, besides those bits strictly coding the *data* values, there are other bits, such as *start*, *parity*, and *stop*. A common setting is one byte or eight bits of data and ordered to send the LSB first. The parity bit is used to detect transmission error, by sending always an even or odd number of binary 1s, and checking the number of 1s in a character allows to detect an odd number of errors. However, it is common to omit the parity bit. The stop bit at the end of each character allows to synchronize the data transmission. A very common setup is 8/N/1 or 8 data bits, none for parity, and 1 stop. In summary, a data packet is formed by one start bit, a data frame of five to nine bits, none or one parity bit, and one to two stop bits (Figure 2.3).

UNIVERSAL ASYNCHRONOUS RECEIVER TRANSMITTER

Universal Asynchronous Receiver Transmitter (UART) is a common serial communication interface between two devices using two lines, transmit Tx and receive Rx (Figure 2.4). It is *asynchronous*, i.e., there is not a clock signal controlling the communication, and data rate or speed is configurable; both devices must have the same speed since there is no clock. An inactive line is at

Packet			
Start	Data Frame	Parity	Stop
1 bit	5-9 bits	0-1 bits	1-2 bits

FIGURE 2.3 Serial communication data packet.

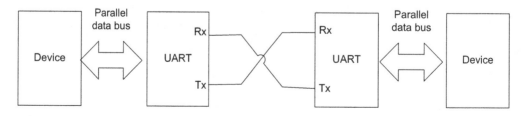

FIGURE 2.4 UART lines and arrangement for the communication of two devices.

logical 1 or high voltage level (e.g., 3.3 V, 5 V); a UART device initiates transmission by setting a logical 0 or low voltage level. Transmission ends setting the line at logical 1.

As we will discuss in forthcoming sections, MCUs and SBCs include UART functions and provide Tx and Rx lines at their I/O pins. For example, an Arduino UNO has Rx and Tx lines at digital pins 0 and 1, respectively, and a Raspberry Pi Zero has Tx and Rx lines at GPIO (General Purpose I/O) pins 14 and 15.

RS-232

The RS-232 is a standard defining signals between two devices, particularly the voltage levels corresponding to logical 0 and 1, the purpose of the signal, and the connectors. It was originally defined as a Recommended Standard (hence the RS preceding the number 232) by the Electronic Industries Association. In essence, the standard defined the exchange between a DTE (*data terminal equipment*, e.g., a terminal) and a DCE (*data circuit-termination equipment*, e.g., a modem). Since RS-232 specifies large voltage levels, e.g., 15 V, drivers are required to adapt it to the UART low voltage levels. The RS-232 standard was initiated in the 1960s to interconnect devices but later PCs started employing RS-232 to connect peripherals. The USB, a standard developed in the 1990s, has substituted RS-232 for these PC applications. However, many environmental monitoring dataloggers and other scientific and industrial systems still use RS-232 for serial communications.

Originally, the RS-232 standard specified a D-25 or 25-pin D-subminiature connector, which is actually a large size connector when compared to today's commonly used ones. Most popular had been to implement the RS-232 using a DE-9 or 9-pin D-subminiature connector (Figure 2.5 and Table 2.1). To connect the RS-232 of a datalogger (a DTE) to a PC (a DTE), a null-modem

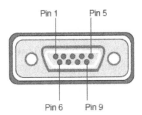

FIGURE 2.5   RS-232 using a D-9 or a 9-pin D-subminiature connector: (a) male and female; (b) pinout.

TABLE 2.1
RS-232 Pinout at a D-9 Connector

Pin of DE-9	Signal	Signal Abbreviation
1	Carrier detect	DCD
2	Received data	RXD
3	Transmitted data	TXD
4	Data terminal ready	DTR
5	Signal ground	GND
6	Data set ready	DSR
7	Request to send	RTS
8	Clear to send	CTS
9	Ring indicator	RI

TABLE 2.2
Null-Modem Connection

Pin	Signal	Pin	Signal
1	DCD	4	DTR
2	RXD	3	TXD
3	TXD	2	RXD
4	DTR	1	DCD
4	DTR	6	DSR
5	GND	5	GND
6	DSR	4	DTR
7	RTS	8	CTS
8	CTS	7	RTS
9	RI (not connected)	9	RI (not connected)

connection simulate the RS-232 handshake between a DCE and a DTE (Table 2.2). This cable permits two DTEs to communicate with each other without DTE mediation and can be terminated by DE-9 at both ends wired as shown in Table 2.2. Nowadays, laptops and PCs do not include RS-232 but USB, and therefore a very useful accessory in environmental monitoring work is an RS-232 to USB adapter to interface legacy monitoring equipment to laptops and PCs (Figure 2.6).

USB

The USB standard specifies cables, connectors, power, and protocols for communication and power supply between computers and peripherals, offering the great advantage of not requiring prior configuration. USB has continuously evolved through four generations and more than a dozen types of connectors, including standard, mini, and micro sizes and all have been deprecated except the most recent *USB Type-C* or *USB-C*.

Basically, the USB interface consists of two devices, the host and the peripheral, and uses a *master/slave* protocol. However, one host can connect to many peripherals (up to 127) by an extension hub following a tree network topology. Peripheral devices cannot interact with one another except via the host, and two hosts cannot communicate over their USB ports directly.

FIGURE 2.6   Examples of RS-232 to USB cable for compatibility with legacy systems.

In 1996, *USB 1.0* specified a data transfer rate of 12 Mbps, whereas now with *USB4*, the data transfer rate is 40 Gbps. Still, this data transfer rate is lower than the Ethernet 100 Gbps interface, which we will discuss later in the chapter.

A USB *data packet*, which is the unit of data transfer, consists of several fields starting with *Sync*, which synchronizes the clock of the receiver and transmitter, and PID (Packet ID), which consists of eight bits where the upper four indicate the packet type (to be described shortly) and the lower four bits are for error detection. Depending on the type of packet, the remaining fields can be ADDR for destination address (made of 7 bits, thus supporting 127 devices), ENDP for endpoints, CRC for cyclic redundancy check, DATA for payload, and EOP for end of packet.

The four types of data packets are *token packets*, *data packets*, *handshake packets*, and *start of frame packets*. A token data packet indicates the type of message and consists of Sync, PID, ADDR, ENDP, CRC, and EOP fields (Figure 2.7). Data packets carry the *payload* data, consisting of Sync, PID, Data, CRC, and EOP fields. Handshake packets recognize data packets and report errors, consisting simply of Sync, PID, and EOP. Start of frame packets flag the start of a new frame of data, consisting of Sync, PID, Frame Number, ENDP, CRC, and EOP. A typical transaction includes several packets given in sequence: for example, token, data, and handshake.

Furthermore, there are four types of data transfer: *Control*, *Bulk*, *Interrupt*, and *Isochronous*, which serve different purposes and may include one or several transactions. The control type is used to configure and control a device, the bulk type is convenient for transferring a large amount of data (e.g., stored data in a pen drive to a folder on a PC), the interrupt type is convenient for sending small amounts of data (e.g., mice or keyboards), and the isochronous type helps to transmit real-time information like audio and video data at a constant rate.

In addition to data transfer, USB is used for powering devices. For this purpose, a USB cable has two sets of wires, one set for DC power (positive and ground) and the other set for data. Configuration of these wires has evolved through a variety of connectors, such as Type-A, Type-B, and micro. The most recent is USB-Type C, which is a 24-pin connector. In the companion lab guides, we will work with a variety of USB cables both for data transfer and to power devices.

Token Packet					
Sync	PID	Address	Endpoint	CRC	EOP
8 bytes	8 bits	7 bits	4 bits	5 bits	3 bits

Data Packet				
Sync	PID	Data	CRC	EOP
8 bytes	8 bits	0-1023 bytes	16 bits	3 bits

Handshake Packet		
Sync	PID	EOP
8 bytes	8 bits	3 bits

Start of Frame					
Sync	PID	Frame number	Endpoint	CRC	EOP
8 bytes	8 bits	7 bits	4 bits	5 bits	3 bits

FIGURE 2.7 USB data packets.

NETWORKS

This section intends to provide information about networks that we need to understand the laboratory activities to be conducted in lab session 2 described in the companion lab manual. We will discuss other interfaces and associated protocols in Chapter 5.

OPEN SYSTEMS INTERCONNECTION (OSI) MODEL

The OSI model has seven logical layers (Table 2.3), tiered hierarchically such that a layer serves the layer above it while being served by the layer below it. These layers are abstractions and can contain sub-layers to perform their functions. The *Media Access Control* (MAC) data communication protocol is part or a sub-layer of layer 2 or data link layer. The MAC sub-layer provides addressing and access of multiple devices to a shared medium. A medium is a general term that refers to a communication link, for example *Ethernet*. The MAC is an interface between layer 1 (Physical layer) and the logical link control sub-layer of layer 2 and emulates a full-duplex channel in a multi-point network.

MAC addresses are 48-bit (6-byte) binary addresses represented in 12 hexadecimal digits. Each digit is 48/12=4 bits or a *nibble*. The MAC address corresponds to the network card or adapter of the device assigned when it is manufactured and therefore it is a permanent address. For example, the MAC address B8:27:EB:7B:55:E4 has 6 bytes or 12 nibbles, E4 is the least significant byte and B8 is the most significant byte. The first three bytes is the OUI (*Organizational Unique Identifier*) that is assigned by the Institute of Electrical and Electronics Engineers (IEEE) to the manufacturer. In this example, the OUI is B8:27:EB corresponds to Raspberry Pi. The last three bytes 7B:55:E4 constitute the *Network Interface Controller* specific number and it is assigned by the manufacturer.

TABLE 2.3
OSI Seven-Layer Model

Type	Data Unit	Layer	Function
Host	Data	7 Application	Convert to application
Host	Data	6 Presentation	Data representation, encryption, convert machine dependent to independent
Host	Data	5 Session	Inter-host communication managing sessions
Host	Segments	4 Transport	Reliable addressing, routing, and delivery of messages between points
Media	Packet	3 Network	Addressing, routing, and delivery of messages between points
Media	Bit/Frame	2 Data link	Reliable direct point-to-point connection
Media	Bit	1 Physical	Direct point-to-point connection

TCP/IP

The five-layer TCP/IP model is like the OSI model; layers 1–4 are the same and then the fifth TCP/IP layer takes the role of OSI layers 5–7. In other words, functions of the three top OSI layers (Application layer, Presentation layer, and Session layer) are merged to a single fifth TCP/IP layer named Application layer.

Layer 4, transport layer, of the TCP/IP includes the *TCP* and *User Datagram Protocol* as well as address for different applications called *port numbers*. Layer 3, Network layer, includes the IP that uses IP addresses to identify source and destination of packets and routing functions. At Layer 2, the data link layer, we have the Local Area Network (LAN) standards such as Ethernet. In this layer, the MAC address is used to identify source and destination of packets.

IPv4 addresses are 32-bit binary addresses, divided into 4 octets or sets of 8 bits (given in decimal, e.g., 192.168.10.80) used by the IP for delivering packets to another device. Being 8 bits, each octet can range from 0 to 255. An IPv4 address has two parts: the network part and the host part. A network has two reserved addresses; one is the first (the network address) and the other one is the last (the broadcast address). The Domain Name System (DNS) translates the domain name to the IP address.

There are several classes of IPv4 address sets. Class C is for small networks up to 256 IP addresses, e.g., 192.168.1.xxx, where xxx is 0–255. Class B is for intermediate networks up to 65,536 IP addresses (e.g., 126.168.xxx.xxx). Class A is for large networks up to 16,777,216 IP addresses (e.g., 10.xxx.xxx.xxx).

In this book, we will focus on Class C networks. The three most significant bits are reserved to be 110. The first three octets identify the network and the fourth octet the host. The minimum value that the first octet has is 110–00000, which is 192 in decimal; with a maximum value of 110–111111 or decimal 223. Networks 192.168.0.0 to 192.168.255.0 are reserved for private use.

The purpose of the *subnet mask* is to identify the network part of the IP address. All the bits of the subnet mask for the network address are set at 1 and all the bits of the host part are set at 0. The decimal equivalent of an octet with all bits set at 1 is 255. For example, the subnet mask 255.255.255.0 identifies the first three octets as the network address and the last octet as the host address.

ETHERNET

Introduced in the 1980s, and covered under IEEE 802.3, Ethernet has evolved to high data rates and remains the primary element for wired LAN, defining the physical layer and MAC sub-layer of the data link layer for wired networks. Nowadays, Ethernet cable is twisted pair and fiber optics (for large networks). In terms of the OSI model, Ethernet provides support at Layers 1 and 2 and uses the IP (Layer 3). To avoid collisions using the same cable, each device is connected to a network switch.

Ethernet Frame						
Preamble	SFD	Destination MAC address	Source MAC address	Length	Payload	FCS(CRC)

FIGURE 2.8 Ethernet data frame.

An Ethernet frame has several fields (Figure 2.8). *Preamble*: provides alert and timing pulse for transmission; *Start Frame delimiter*; *Destination MAC Address*: physical address of destination stations; *Source MAC Address*: physical address of the sending station; *Length*: number of bytes in the data field; *Data*: This is a variable sized field that carries the data or payload from the upper layers; *Padding*: This is added to the data to bring its length to the minimum requirement of 46 bytes; and *Frame Check Sequence*: this is CRC, containing the error detection information.

WIRELESS FIDELITY (WI-FI)

In Chapter 6, we will cover wireless technology and relationships with environmental monitoring, with emphasis on applications of Wi-Fi to telemetry and wireless sensor networks. For now, we briefly explain some notions of Wi-Fi needed for lab session 2 described in the companion lab manual. Wi-Fi stands for *Wireless Fidelity*, utilizes an unlicensed part of the spectrum, and is a widely used wireless technology certified under the IEEE 802.11 standard. Wi-Fi technology and its associated standards evolved since 1999 when 802.11a and 802.11b specified the 5-GHz band (with 54-Mbps data rate) and the 2.4-GHz band (with data rate of 11 Mbps), respectively. In two decades, the standard went through 802.11g (54 Mbps for 2.4 GHz), 802.11n (600 Mbps and, including both 2.4 and 5 GHz), and 802.11ac (1.3 Gbps for 5 GHz). In 2020, 802.11ax specified much higher data rates for 2.4 GHz and 5 GHz, and an additional 6-GHz band. Wi-Fi is embraced by a worldwide network of companies (Wi-Fi Alliance 2022) issuing its own certification, and identifying Wi-Fi generations that map to the IEEE 802.11. For example, the Wi-Fi 4 and 5 generations correspond to IEEE 802.11n and IEEE 802.11ac, respectively, Wi-Fi 6 corresponds to 802.11ax (2.4/5 GHz), while Wi-Fi 6E corresponds also to 802.11ax but includes a 6-GHz band.

Using 2.4 GHz allows for longer distances, however the bandwidth for 2.4 GHz is about 100 MHz, which divided into 20-MHz channels makes most of them overlapping, which leads to interference (only three of the channels are non-overlapping). For 5 GHz, the bandwidth is larger allowing for higher data rates and more non-overlapping channels (reducing interference). However, 5 GHz has shorter range.

INTERNET AND WORLD WIDE WEB (WWW)

TCP/IP is used to exchange information on the Internet, the global network of computers and devices that started in 1969 funded by the U. S. government and made available for commercial applications in 1995. We already discussed the method of defining IP addresses for devices connected to a network. Considering the maximum value of each octet, IPv4 addresses range from 000.000.000.000 to 255.255.255.255; that is $2^8 \times 4$ or 4,294,967,296 IP addresses, or approximately 4.3 billion addresses. These are not enough for devices connected now to the Internet around the world; therefore, many devices now use IPv6 addresses. We will not discuss IPv6 in this book.

An information exchange service on the Internet is the WWW (or the web for short) using HTML and other web programming languages such as PHP, as the basis to display documents. We will learn fundamentals of these two languages later in this chapter and when using the lab guides in the companion lab manual. A *web server* provides web pages via files that are processed by *web clients* representing the user or device side of the web, that use *web browsers* to display documents or search on the Internet.

INTERNET OF THINGS (IoT)

As more devices on the Internet interconnect with sensors and actuators (physical objects or "things"), the concept of the IoT, combined with control and data analysis, is used in many industrial and commercial applications, such as home automation, remote monitoring, autonomous vehicles, predictive maintenance, manufacturing, and facilities management. Such an approach and the IoT concept are applicable to the framework for environmental monitoring covered in Chapter 1, which builds up from continuous and long-term data collection to data analysis and modeling. In Chapter 6, we will cover devices that can be employed in IoT.

SBC, SYSTEM ON A CHIP (SoC), AND MCU

SoC

An SoC is an integrated circuit (IC) that includes a complete computer; in a manner similar to MCU, they can be used in embedded systems (Null and Lobur 2012). Compared to MCU, SoCs tend to have more on-board memory and have more powerful processors. These added capabilities allow SoCs to run full operating systems and exploit additional external memory and peripherals.

ARM ARCHITECTURES

ARM computers are based on reduced instruction set computing (RISC). The name ARM derives from the developers ARM Holdings, a British company. Because of the reduced instruction set, ARM processors require fewer transistors than complete instruction set processors such as ×86, and therefore reduce cost and power consumption. Thus, ARM processors are common in mobile devices. Chip manufacturers, e.g., Atmel, Texas Instruments (TI) fabricate ARM architecture processors.

An example of SBC based on ARM is the Technologic Systems TS-7800, which is approximately 12 cm × 10 cm. The TS-7800 is based on a Marvell MV88F5182 ARM9 CPU at 500 MHz. It has a 32-bit PCI bus, Ethernet, USB 2.0, RS-232, a Real-Time Clock (RTC), and five-channel ten-bit ADC. The TS-7800 has 128 MB of DDR SDRAM (actually first generation or DDR1) memory and 512 MB of flash memory (Technologic 2021). The board is powered by 5 V DC.

SBC EXAMPLE: RASPBERRY PI

The Raspberry Pi is a small low-cost SBC created in the UK to promote teaching of computer science in schools (Figure 2.9). It runs Linux, Debian, and Arch Linux ARM, as well as other OS. Other components of the SoC are GPU, SDRAM, and USB port. At its inception as Raspberry Pi 1, it included an SoC (Broadcom BCM2835) 700-MHz ARM processor (ARM11, which is a family of 32-bit RISC microprocessors, using ARMv6 instruction set). The Raspberry Pi has been evolving and now comprises a variety of devices, such as the Raspberry Pi 3 Model B+, Raspberry Pi 4 Model B, and Raspberry Pi Zero W. The latter is currently the smallest and lowest cost Raspberry Pi SBC.

We will employ the Raspberry Pi Zero W in the lab sessions described in the lab manual. The Raspberry Pi Zero W includes 802.11b/g/n Wi-Fi using 2.4 GHz, Bluetooth 4.1, Bluetooth Low Energy, 1 GHz, single-core CPU, 512-MB RAM, Mini HDMI port and micro USB On-The-Go port, Micro USB power, and HAT-compatible 40-pin header (Raspberry Pi 2022b).

MICROCONTROLLERS

An MCU is a microcomputer on a single IC, including CPU, memory, and I/O ports (Figure 2.10). There are several alternative abbreviations for MCU such as µC and uC. In this book, we will use MCU. In a similar fashion as SBC, MCUs are intended for use in embedded applications,

FIGURE 2.9   Raspberry Pi 4 Model B and Raspberry Pi Zero W.

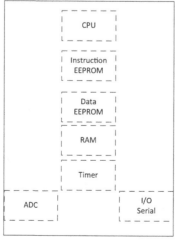

FIGURE 2.10   MCU components.

i.e., provide computing capabilities as part of a larger equipment or system. In contrast to microprocessors, its functions are specialized to the needs of the larger system instead of providing a flexible computer platform. In addition, embedded systems typically do not have peripherals for human interaction; we connect them to a computer to load a program or to retrieve stored data. In addition to digital ports, MCUs input capabilities include A/D for measurement and in some cases, the output includes D/A for control. MCUs tend to be low-cost, low power, and low speed.

MCUs have been used for many dedicated applications since when were developed in the 1970s. The EPROM technology is allowed for installing programs in non-volatile memory and is re-programmed by erasing the PROM using UV light. In the 1990s, low-cost Electrically EPROM (EEPROM) technology (memory erasable by using electrical signals) expanded the applications of MCUs by allowing flexible prototype development. MCU's popularity was further expanded by including *flash memory* (which is a special type of EEPROM) and facilitating development and use. Low-cost MCUs are eight-bit. There are currently 16-bit and 32-bit MCUs available. Applications must fit in small-capacity memory. In environmental monitoring, MCUs allow important applications; for example, using an MCU to monitor solar radiation (Mukaro and Carelse 1999).

MCU EXAMPLE

As an example, we present the Atmel ATmega 128. This is a low-power eight-bit AVR RISC-based MCU that combines 128 KB of programmable flash memory, 4-KB SRAM, a 4-KB EEPROM, an eight-channel ten-bit A/D converter, and a Joint Test Action Group (JTAG) interface for on-chip debugging (Atmel 2011). The device supports throughput of 16 MIPS at 16 MHz and operates between 4.5 and 5.5 V.

The Atmel® AVR® core RISC architecture includes 32 general-purpose working registers directly connected to the ALU, allowing two independent registers to be accessed in one single instruction executed in one clock cycle (Atmel 2011).

IN-CIRCUIT SERIAL PROGRAMMING (ICSP)

For specialized applications, manufacturers program the MCU before installing it on a circuit board. However, for flexible applications, it is best to program the MCU after placing it in a circuit board. ICSP is a particular instance of this approach, where the programming computer communicates serially with the MCU using I/O pins (Microchip Technology 2003). This approach offers many advantages in prototyping and producing MCU-based systems. For example, adding calibration parameters and ID codes to the MCU memory, and upgrading the system in the field after deployment. ICSP uses an interface between the MCU and the programming device or programmer (Figure 2.11).

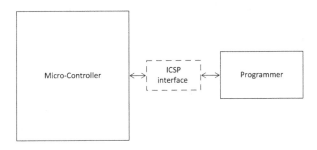

FIGURE 2.11 ICSP: MCU and programmer device.

MCU-Based SBC Example: Arduino

Arduinos have become very popular in a variety of fields since their hardware and software are open source and have been designed to facilitate its use by non-experts in electronics, engineering, or programming (Gertz and Di Justo 2012). It is easy to program using a cross-platform that runs on Windows, macOS, and Linux.

The Arduino UNO is an MCU board based on the ATmega328, has 14 digital I/O pins, 6 analog inputs (10-bit ADC), a 16-MHz ceramic resonator, a USB connection, a power jack (input 7–12 V), an ICSP header, and a reset button. It will derive power when connected to a computer with a USB cable or from a DC power source, AC-to-DC adapter, or battery (Arduino 2014).

The ATmega328 has a clock speed of 16 MHz, 32 Kb of flash memory, 2 Kb of SRAM, and 1 KB of EEPROM which can be read and written with functions from the EEPROM library. The ATmega328 on the Arduino UNO comes with a boot loader that allows to upload new code without the use of external hardware programmer.

The board can be powered from the DC power jack (7–12 V), the USB connector (5 V), or the VIN pin of the board (7–12 V). There are several pins associated to power: (1) VIN to provide an input voltage (7–12 V) to the Arduino from an external power source, (2 and 3) Ground (GND), (4) 5 V, provides a regulated 5V from the regulator on the board, (5) 3.3 V, provides a 3.3 volt supply generated by regulator with maximum current draw of 50 mA, (6) RESET, to reset the board, and (7) IOREF, which provides an indication reference of the MCU operating voltage. Other ports include USB, and six basic pins used for ICSP. These lines include Serial Peripheral Interface (SPI) pins and RST, VTG, and GND. The SPI interface will be defined and discussed in Chapter 5.

Arduinos have add-on boards to complement their basic capabilities called "shields" because they typically fit on top of the Arduino resembling a protective shield. There are hundreds of shield makers and shields available (Shieldlist 2014), which enables an Arduino to perform a variety of functions. For example, an Ethernet connection shield allows an Arduino to connect to the Internet; a GPS shield allows it to obtain location and datalogger shield which is relevant to environmental monitoring.

We have seen three main examples of SBC: the TS-7400, the Raspberry Pi, and the Arduino UNO. Their sizes are compared in Figure 2.12. More comparison features are given as an exercise at the end of this chapter.

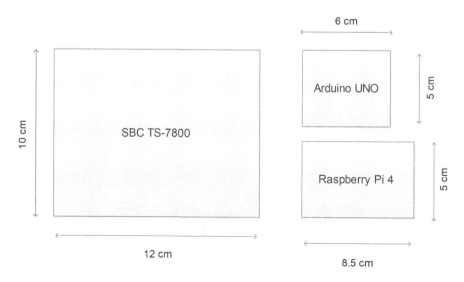

FIGURE 2.12   Comparing size of SBCs: TS-7400, Raspberry Pi, and Arduino UNO.

Example MCU-Based SBC

One additional system we will use as example is TI MSP430 LaunchPad. This SBC can be programmed as easily as an Arduino using an Integrated Development Environment (IDE). It has 20 pins to interconnect it to other circuits. There are now "booster packs" to connect to these pins, in the same manner as Arduino shields. It has a general-purpose switch, a reset switch, and two LEDs indicators. The LaunchPad is based on the MSP430G2452 MCU. Which has speed 16 MHz, 16-KB Flash, 512-B RAM, 8-channel 10-bit ADC, Comparator, two 16-bit Timers, up to one Inter Integrated Circuit (I2C) interface, two SPI, and one UART. With low power operation is suitable for battery-operated applications. Open-source software is available from several communities, such as Energia (Energia 2014), which is similar to Arduino. We will discuss the I2C interface in Chapter 5.

CONCEPTS OF PROGRAMMING

In lab session 1 of the lab manual companion to this book (Acevedo 2024), we learned how to code simple programs in R and some terms applicable to programming languages in general. In this chapter, and corresponding lab session 2 of the lab manual, we will continue learning programming in R as well as include Python, Arduino code (which is based on C++), and programs to display data and graphs on the web using HTML, Cascading Style Sheets (CSS), PHP, and JS. We cover more details of Python, Arduino, HTML, CSS, PHP, and JS programming in lab guide 2 and subsequent sessions of the lab manual.

The required setup for programming includes a text editor to create computer programs, a compiler (such as in C++) to convert the code into binary format, or an interpreter to execute the programs directly (as in R and Python). *Syntax* is a collection of rules on how to write the code, and it varies from language to language.

Variables are identifiers pointing to a memory location containing a value for the variable; for example, `calc = 5.7`. In many languages, you must specify the type of variable, for example integer (`int number = 2`), floating point (`float voltage = 3.7`), and string (`string name = "channel"`).

Control structures allow changing the flow of execution of instructions. A basic one, *if-else*, allows to perform a function if a condition is met or an alternative function if the condition is not met.

```
if (measurement >500){
        calibration();
} else {
        measure();
}
```

The syntax of how we write these lines of code varies according to the language. Here we are using curly braces for the functions to be executed by the "if" part and the "else" part. The lines of code stating the function end in semicolon.

Other important control structure allows to create *loops* such as a control structure based on *for* loop

```
for (int i = 0; i<=100; i++) {
        a = 2*i;
}
```

Where we increment integer i from 0 to 100 and make variable a equals to 2 times i. Likewise, the *while* control structure

```
while (int i<100) {
      a = 2*i;
}
```

Also of great importance in programming are *data structures*, which provide ways of organizing information. We encountered the array data structure and data frame when we worked with R in lab session 1. In R, there are several other structures such as *lists* that allow combining a variety of other objects.

PYTHON

Python is an interpreted general-purpose programming language that has become very common in data science, as well as interfacing systems and data acquisition. Interpreter and libraries are freely available from the Python Web site (Python 2022a) as well as other contributors. A tutorial is also available from the Python Web site (Python 2022b) and the Raspberry Pi documentation website (Raspberry Pi 2022a)

Python does not require a character to terminate statements, and blocks are specified using indentation. Statements to be followed by an indentation level end with a colon. Values are assigned using the "=" sign. You can decrement or increment values with the operators += or –= with the amount on the right-hand side.

For example, in this control structure

```
for i in range(5):
  print(i)
```

the first line ends in ":" and the second line is indented. The response would be the numbers, 0, 1, 2, 3, 4. Note that the sequence starts at 0 and ends at the range minus 1.

Data structures in Python include *lists*, *dictionaries*, and *tuples*. Lists are similar to one-dimensional arrays, but they are more general because you can also have lists of other lists. Dictionaries are essentially *associative arrays* or hash tables, and tuples are one-dimensional arrays. Python arrays can be of any type. The type of variable does not have to be declared, it suffices to give the name and assign a value. A list is created by comma-separated items between square brackets. For example, a list of squares of numbers 1, 2, and 3.

```
x =[1**2,2**2,3**2]
x
[1, 4, 9]
```

ARDUINO

The Arduino UNO is programmed using the Atmel IDE. Once installed, you run the Arduino software and will see the script editor (scripts in Arduino are called sketches). After variable declaration, an Arduino program has two functions: setup() and loop(). In the setup() function, we specify I/O pins, communication, and other configurations. In the loop() function, we perform instructions to be executed in a repeatedly manner. In Arduino programming, each line of code ends with semicolon ";".

For example, in setup(), we could start the serial port at 9600 baud

```
void setup() {
   Serial.begin(9600);
   }
```

and in loop(), we can program a control structure (for loop) that doubles the value of integers from 0 to 100 and prints the integers and the calculated value. Function delay() pauses execution for a given number of milliseconds; 1000 ms in this case (which is 1 second).

```
void loop() {
   int a;
   for (int i = 0; i<=100; i++) {
     a = 2*i;
     Serial.print(i);
     Serial.println(a);
     delay(1000);
   }
}
```

HTML

HTML elements inform a web browser how to display the contents of web pages, by specifying headings, paragraphs, images, tables, links, and other characteristics of the contents. It does this by defining tags enclosed in symbols < >, using a "tagname" at the start of an element and slash "/tagname" to define the end of an element. For example, <head> will be the start of element heading and </head> is the end of that element. Some tags called empty, have no ending tag; for example, <br> that inserts line breaks in text. At the time of this writing, the latest version of HTML was HTML5, and HTML6 was expected to be released soon.

Consider the following HTML document and its display as a web page (Figure 2.13) to explain some of the tags.

```
<!DOCTYPE html>
<html>
<head>
<title>Real-Time Environmental Monitoring</title>
</head>
<body>
<h1>RTEM Test Page</h1>
<h2>Learning HTML</h2>
<p>Examples of URL links: external web page and a local file</p>
<p><a href=https://github.com/mfacevedol/rtem>RTEM GitHub Repository</a></p>
<p><a href="data/datasonde.csv">Datasonde File</a></p>
</body>
</html>
```

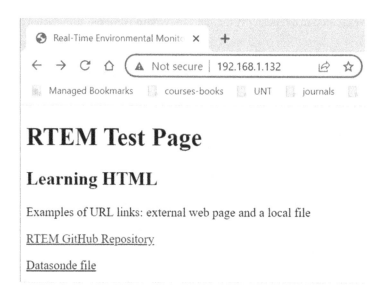

FIGURE 2.13   Web page displayed by the HTML example.

Here, `<!DOCTYPE html>` declares that the content is an HTML5 document. Next, `<html>` is the root element of an HTML page endingin `</html>`. The `<head>` element closed with `</head>` contains information about the HTML page; in this case, `<title>` and `</title>` specifies a title for the HTML page (shown in the tab for the page, in Figure 2.13). Now, `<body>` and `</body>` delimit the container for the visible content, such as headings, paragraphs, images, hyperlinks, tables, and lists. In the example, the `<h1>` and `<h2>` and elements define a heading, and each `<p>` ending in `</p>` element defines a paragraph.

Nested within the second and third `<p>` tags in this example, we see another tag `<a>` defining a hyperlink to an external website and to a local file available for download. Here `"href="` is the attribute specifying the Uniform Resource Locator (URL), for the site or for the file to be linked. An URL can be defined in a variety of ways. In the first case, we use the address for the site `https://github.com/mfacevedol/rtem`. In the second case, we are using a relative URL to a comma separated values (CSV) file contained in a subdirectory `"data"` of the directory housing the HTML document. Before closing with `</a>`, we define the label to be shown on the page (Figure 2.13).

CSS

CSS is used to style HTML documents, that is how the elements of an HTML document are displayed on the web page and presented to the user. It is very useful because it separates content from format and can control the style of multiple HTML documents from a common set of stylesheets. Stylesheets can be contained in `<style>` `</style>` tags within the HTML document. The syntax of a stylesheet is to declare the HTML element that will be styled (the selector), then enclosed in curly braces, the properties, and their values separated by colon. For example,

```
<style>
p {
  font-family: Arial;
  font-size: 20px;
}
</style>
```

Here the selector is the paragraph tag and properties are the font family and size. Stylesheets can be included within the <head> elements in the HTML document or in a separate *.css file called from the HTML document. In this case, the external file is referenced using a link tag

```
<link rel="stylesheet" href="example.css">
```

The example.css file should not contain HTML tags; thus, the stylesheet will not be enclosed in <style> tag.

PHP

PHP is a programming language employed at the web server side or embedded in HTML to create dynamic web pages. A semicolon ";" is used to terminate statements. Language keywords are such as if, else, null, echo, foreach are not case-sensitive, but it is a good practice to be consistent and using all the same, for example, all lower case. However, constants and variables are case-sensitive. Variables are defined preceding by the dollar sign $. Comments and remarks can be in one line preceded with double slash // or pound # symbols, and multiple line using /* to start and */ to end. Currently, the latest version of PHP is 7.4.

Consider the following example and Figure 2.14 to introduce PHP scripts. In this example, we extract each record of a csv datalog file as $line and use echo to display $line on a web page terminating with HTML tag <br> at the end of the line. The period in echo between $line and the br tag is string concatenation.

```
<?php
$lines = file("data/datasonde.csv");
foreach ($lines as $line_num=> $line) {
    echo $line . "<br>";
}
?>
```

We can insert this piece of code within the body tags of the example HTML document discussed above, after the <a href line, and preceding it with a <p> tag announcing the file content as shown below.

When executed, the results will be as shown in Figure 2.14 where we can see the lines of the datasonde.csv file.

```
<!DOCTYPE html>
<html>
<head>
<title>Real-Time Environmental Monitoring</title>
</head>
<body>
<h1>RTEM Test Page</h1>
<h2>Learning HTML</h2>
<p>Examples of URL links: external web page and a local file</p>
<p><a href=https://github.com/mfacevedol/rtem>RTEM GitHub Repository</a></p>
<p><a href="data/datasonde.csv">Datasonde Data File</a></p>
<?php
$lines = file("data/datasonde.csv");
foreach ($lines as $line_num=> $line) {
echo $line . "<br>";
}
?>
</body>
</html>
```

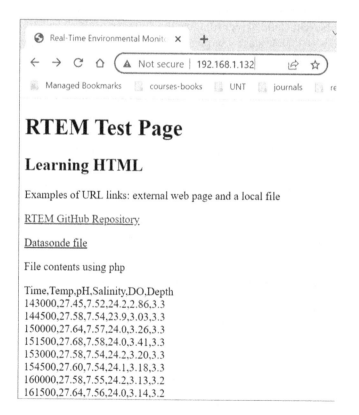

FIGURE 2.14   Web page displaying the datasonde file contents by the PHP example.

JS

JS is commonly used for development of web sites since it can change the behavior of elements of HTML documents and therefore make a web page dynamic. For environmental monitoring applications, JS can be used to interact with PHP results, read data, plot graphs, and a variety of tasks. JS code is enclosed within `<script>` `</script>` tags and can contain functions that are called by other elements of the HTML document. For example,

```
<script>
Function test() {
  Code here …}
</script>
```

Then function `test()` can be called somewhere else in the HTML code; for example, a button that can be clicked to execute it.

```
<button type ="button" on click="test()">Test</button>
```

There are many JS libraries available for a variety of purposes such as data manipulation, visualization, forms, and animations. An interesting JS library is Data-Driven Documents (D3) or `D3.js` which is designed to manipulate HTML elements based on data. We will use it in the lab sessions for several purposes and particularly to plot graphs. One graph type is Scalable Vector Graphics (SVG) that can be defined from HTML code. Related to JS, a common file format to store and transport data is JS Object Notation (JSON) that uses JS code to specify the contents and it can be read by other programming languages.

SQL

SQL is a language to retrieve and manage databases. It is used, for example to create a database and add new tables to the database, as well as query a database, retrieve, insert, update, and delete records. We postpone discussing SQL until Chapter 10 and Lab 10 when we cover databases.

EXERCISES

Exercise 2.1

Consider a ten-bit ADC. Calculate the number of steps and the voltage resolution for a maximum voltage of 5 V.

Exercise 2.2

Consider the following control structure

```
for (int k = 0; k<=10; k++) {
b = 10-k;
}
```

What would happen with variable b during execution?

Exercise 2.3

Consider the following Python code

```python
for i in range(5):
    print(i)
```

What would be the interpreter response?

Exercise 2.4

Consider the following Arduino code

```arduino
void setup() {
  Serial.begin(9600);
  }

void loop() {
  int a;
  for (int i = 0; i<=100; i++) {
    a = 2*i;
    Serial.print(i);
    Serial.println(a);
    delay(1000);
  }

}
```

What would be the response on the serial monitor?

Exercise 2.5

Consider the following HTML code

```html
<!DOCTYPE html>
<html>
<head>
<title>Real-Time Environmental Monitoring</title>
</head>
<body>
<h1>RTEM Exercise Page</h1>
<h2>Learning HTML</h2>
<p>Examples of URL links: external web page and a local file</p>
<p><a href=https://github.com/mfacevedol/rtem>RTEM GitHub Repository</a></p>
<p><a href="data/salinity.csv">Salinity Data File</a></p>
</body>
</html>
```

What would be the display on a web browser?

REFERENCES

Acevedo, M. F. 2024. *Real-Time Environmental Monitoring: Sensors and Systems*, Second edition Lab Manual. Boca Raton, FL: CRC Press, Taylor & Francis Group. 463 pp.

Arduino. 2014. *Arduino Uno*. Accessed October. http://arduino.cc/en/Main/ArduinoBoardUno.

Atmel. 2011. 8-bit Atmel Microcontroller with 128Kbytes. In *System Programmable Flash ATmega128 and ATmega128L,* edited by Atmel, J. San Jose, CA: Atmel Corporation.

Energia. 2014. *Energia*. Accessed October 2014. http://energia.nu/.

Gertz, E., and P. Di Justo. 2012. *Environmental Monitoring with Arduino: Building Simple Devices to Collect Data About the World Around Us*. Sebastopol, CA: O'Reilly Media Inc.

Kester, W., J. Bryant, and J. Buxton. 2015. *Practical Design Techniques for Sensor Signal Conditioning, Section 8 ADCs for Signal Conditioning*. Accessed March 2015. http://www.analog.com/media/en/training-seminars/design-handbooks/732529616sscsect8.PDF.

Kester, W., D. Sheingold, and J. Bryant. 2015. *Chapter 2 Fundamentals of sampled data systems*. Accessed March 2015. http://www.analog.com/en/education.html.

Microchip Technology. 2003. *In-Circuit Serial Programming™*. Chandler, AZ: Microchip.

Mukaro, R., and X. F. Carelse. 1999. A Microcontroller-Based Data Acquisition System for Solar Radiation and Environmental Monitoring. *IEEE Transactions on Instrumentation and Measurement* 48 (6):1232–1238.

Null, L., and J. Lobur. 2012. *The Essentials of Computer Organization and Architecture*. Sudbury, MA: Jones & Bartlett Learning.

Python. 2022a. *Python*. Accessed January 2022. https://www.python.org/.

Python. 2022b. *The Python Tutorial*. Accessed January 2022. https://docs.python.org/3/tutorial/.

Raspberry Pi. 2022a. *Raspberry Pi Documentation*. Accessed January 2022. https://www.raspberrypi.com/documentation/.

Raspberry Pi. 2022b. *Raspberry Pi Zero W*. Accessed January 2022. https://www.raspberrypi.com/products/raspberry-pi-zero-w/.

Shieldlist. 2014. *Arduino Shield List*. Accessed October. http://shieldlist.org/.

Technologic. 2021. *TS-7800: Technologic Systems*. Accessed December. https://www.embeddedarm.com/products/TS-7800.

Wi-Fi Alliance. 2022. *The Worldwide Network of Companies that Brings You Wi-Fi®*. Accessed September 2022. https://www.wi-fi.org/.

3 Sensors and Transducers
Basic Circuits

INTRODUCTION

This chapter starts with a very basic and quick review of concepts of electric circuits that are needed to understand sensors and transducers. We discuss sensors defined as elements that respond to a change in conditions by changing its properties. An electrical transducer converts variations of one form of energy to variation in electrical energy. For example, a voltage divider circuit with a resistive sensor of temperature produces an output voltage related to resistance. We discuss passive (e.g., thermocouples) and active sensors (e.g., thermistors) using temperature as an example. This chapter also provides an overview of parameter estimation by fitting to non-linear models by transformation. These techniques are useful in design and calibration of sensors.

PRINCIPLES OF ELECTRICAL QUANTITIES

This section provides a very basic and quick review of some concepts of electric circuits that we need to understand sensors and transducers. Several textbooks provide an introductory treatment of circuit analysis, for example Irwin and Nelms (2011). Electrical *charge* is a fundamental property of matter that interacts with electromagnetic fields. Charge can be positive or negative; at the subatomic level, protons represent positive charge, whereas electrons have negative charge. The unit of charge is the Coulomb or C, where 1 C is the equivalent charge of 6.2×10^{18} electrons. In a conductor, free electrons can contribute to a change of negative charge, which for practical purposes is thought of "flow of charge", although it is not really flow but a rate of change.

Voltage is the potential energy difference between two points in an electric field, measured per unit charge. Being potential energy means it is available to perform work; in this case to move a unit charge against an electric field. Intuitively, voltage is the energy available to cause electron flow in a conductor. Its unit is Volt or V, which is defined as Joule/Coulomb, and is named Volt in honor of Alessandro Volta. In general, work and charge vary with time. Denoting charge by q, voltage by v, and work by w, the definition of voltage is the rate of change of work with respect to charge

$$v(t) = \frac{dw}{dq} \tag{3.1}$$

Current is charge (electrons) flow rate through a material (e.g., a conductor) or in other words, the rate of electric charge change through a conductor. Its unit is the Ampere, in honor of André M. Ampère, or "Amp" or A for short, which is one Coulomb of change of charge in 1 second of time, i.e., 1 A is 1C/s. Denoting current by $i(t)$, we can use the derivative of charge with respect to time

$$i(t) = \frac{dq}{dt} = \dot{q} \tag{3.2}$$

As we can see, voltage and current are related by the fundamental notion of rate of change of charge and work required to produce it. Multiplying Equations (3.1) and (3.2), we obtain a very interesting result

$$v(t)i(t) = \frac{dw}{dq} \times \frac{dq}{dt} = \frac{dw}{dt} = p(t) \tag{3.3}$$

DOI: 10.1201/9781003425496-3

In short, multiplying voltage and current we obtain power $p(t)$ which is rate of change of work. Dimensionally, $V \times A$ or $\dfrac{J}{C} \times \dfrac{C}{s}$ = Watt or W. In other words, electric power is the rate of change of work required to move charge at a given rate. This relation is a fundamental equation

$$p(t) = v(t)i(t) \tag{3.4}$$

Thus, high power can occur due to high voltage and low current or to low voltage and high current. Neither voltage nor current alone constitute power; we need both. If $v(t) = 0$, power will be zero (no potential to do work), and if $i(t) = 0$, power will be zero (no rate of change of charge).

It is customary to use lower case letters to denote time-varying electrical quantities and t to denote time. Therefore, $p(t) = v(t)i(t)$ is the instantaneous value of power at instant t. It is also common to denote constant values by capital letters; thus for constant voltage and current, power (P) is equal to current (I) multiplied by voltage (V), that is to say

$$P = V \times I \tag{3.5}$$

Suppose voltage is 12 V constant for a 1-day (24 hours) interval during which current drawn is 1 A for the first half day and 0 A later. What is the power drawn (consumed) during the day? What is the energy consumed at the end of the day? Answer: During the first half day, power is $12\,V \times 1A = 12$ W. During the second half day, the power is $12\ V \times 0A = 0$ W. Therefore,

$$p(t) = \begin{cases} 12\ \text{W in } 0 \le t \le 12 \text{ hours} \\ 0\ \text{W in } 12\,h < t \le 24 \text{ hours} \end{cases}$$

The energy consumed, or work done by the voltage source is

$$e(t) = w(t) = \int_{0}^{24\,h} p(x)\,dx = 12 \int_{0}^{12\,h} dx + 0 \int_{12\,h}^{24\,h} dx = 12 \times 12 \text{ Wh} = 144 \text{ Wh}.$$ It is common to express electrical energy in Watt-hour or Wh for short.

Voltage is analogous to the work done to lift a weight against the pull of gravity; current is analogous to the movement of that weight. Similarly, voltage is analogous to water stored in an elevated tank; current is analogous to water flow in a pipe draining the tank.

By convention, polarity of current in a circuit is opposite to electron flow, and polarity of voltage is a drop as it goes from + to – in the circuit, whereas it is a rise going from – to + (Figure 3.1).

We generate voltage using a variety of processes and devices. Some important ones are chemical reactions (e.g., batteries, fuel cells), radiant energy (e.g., solar cells), and using interaction of magnetic and electric field (e.g., alternator).

Circuits extract power from the electrical power supply, which is often a voltage source, e.g., a battery. A *load*, a circuit performing a function, consumes power supplied by the source. The function performed by the load varies. Resistors convert current to heat dissipation, capacitors store charge as potential energy. An integrated circuit (IC) uses the power for its resistors, capacitors, and transistors, internal to the IC.

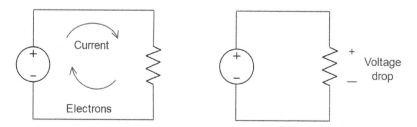

FIGURE 3.1 Conventions. A voltage source is represented by a circle with + and - symbols.

Direct current (DC) refers to the case where voltage, current, or power remain constant during a period in consideration, and there is no change of polarity; for example, the power provided by a battery. This type of electricity contrasts with a time-varying voltage, current, or power defined as *alternating current* (AC) that refers to a periodic variation of these quantities. The most common form is that of a sinusoidal variation for which polarity changes in a cyclical fashion. This type of power is supplied by an AC generator.

Ohm's law is an important relationship between current and voltage for a circuit element and based on the concept of *resistance R* (Figure 3.2)

$$R = \frac{V}{I} \tag{3.6}$$

Resistance's unit is Ohm defined as Volt/Amp and given by Greek letter omega Ω. When the V/I ratio remains constant for all values of V, the relationship is *linear* (Figure 3.3a) and we say that the element follows Ohm's law (it is an ohmic element). In other words, voltage is proportional to current. The trace in the current – voltage (I–V) plane is a straight line, with slope equal to the inverse of resistance.

Combining Equations (3.5) and (3.6), we can derive alternative expressions of power for an ohmic element

$$P = VI = (I \times R) \times I = I^2 R$$

or $\qquad\qquad\qquad\qquad\qquad\qquad\qquad\qquad\qquad$ (3.7)

$$P = VI = V \times \frac{V}{R} = \frac{V^2}{R}$$

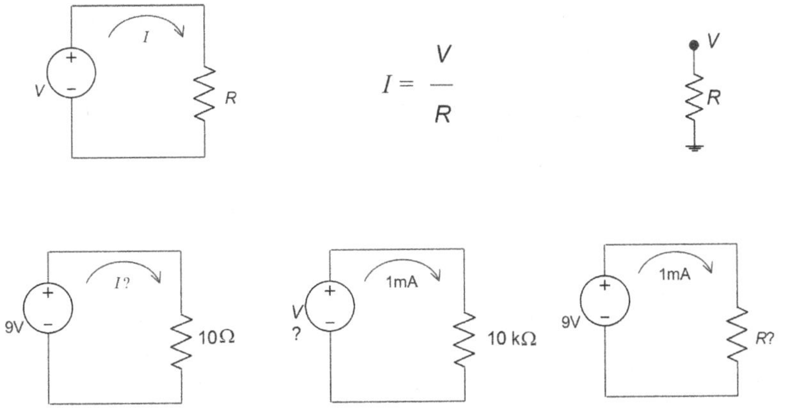

FIGURE 3.2   Ohm's law.

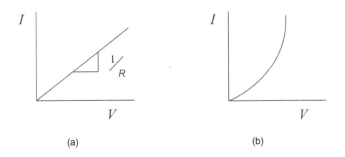

(a) $\qquad\qquad\qquad\qquad\qquad\qquad\qquad$ (b)

FIGURE 3.3   Voltage–Current (V–I) plane (a) Ohmic element and (b) Non-ohmic element.

When we have time-varying current or voltage, and the element is ohmic, it is still valid to use

$$R = \frac{v(t)}{i(t)}$$

and therefore, we can derive an expression for instantaneous power

$$p(t) = v(t)i(t) = i(t)^2 R = \frac{v(t)^2}{R}$$

Conductance G is the inverse of resistance $G = 1/R$ and therefore an equivalent statement of Ohm's law is that conductance G is the ratio of current to voltage $G = I/V$. Conductance has units of siemens, abbreviated S, where S = 1 Amp/Volt.

The *Joule heating* effect, also called Joule's first law, implies that power absorbed by a resistor or a conductor of resistance R produces heat flow. This heat is proportional to power and thus proportional to I^2 or the square of the current; also, this heat flow is independent of the direction of the current. As we will see later in this chapter, Joule heating has important implications in the design of sensors particularly those sensors designed to measure temperature.

When the relationship between V and I is non-linear, the element is *non-ohmic* and Equation (3.6) does not hold. Instead, current is a non-linear function of voltage, say

$$I = f(V)$$

or   (3.8)

$$i(t) = f(v(t))$$

for constant and time-varying quantities. In other words, the proportion between I and V changes with V (Figure 3.3b). An example of non-linear element is a diode (Figure 3.4) that shows a sharp increase of current with voltage once voltage overcomes a threshold.

Resistivity is a property of a conductive material and determines the resistance of an element with certain geometry. Resistivity is denoted by the Greek letter rho (ρ) and its units are ohms-m (Ω-m). For instance, take a wire of length l, and cross-sectional area A. We calculate resistance by

$$R = \rho \frac{l}{A}$$ (3.9)

Resistance increases for longer and thinner wires, whereas it decreases for shorter and thicker wires. For example, take a copper wire of 10-m long and 2-mm diameter. The resistivity of copper is 1.72×10^{-8} Ωm. Therefore, using Equation (3.9), we obtain a resistance of

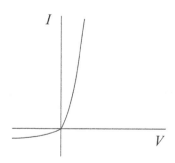

FIGURE 3.4   Example: a diode.

$$R = 1.72 \times 10^{-8} \frac{10}{\left(\frac{\pi}{4}\right)(2 \times 10^{-3})^2} = \frac{1.72 \times 0.1}{\pi} = 0.055\,\Omega$$

In AC circuits, we employ the concept of *impedance* to extend the concept of resistance. The impedance concept includes capacitors and inductors; it is defined similarly to DC using the ratio of voltage to current as in Ohm's law.

CIRCUITS: NODES AND LOOPS

We form circuits by connecting circuit elements in a network. The most useful concepts are those of *nodes* and *loops* (Figure 3.5). A node is a point at a distinct potential whereas a loop is formed by tracing the circuit starting at a given node and returning to it. *Ground* or common is a very important node defined to have zero potential or zero volts. All other nodes have potentials that are measured with respect to ground. Points directly connected to ground have the same voltage (0 V).

Kirchhoff's voltage law states that the sum of voltage around a circuit loop adds to zero as illustrated in the simple circuit of Figure 3.6 composed of one voltage source V and two resistors R_1 and R_2. All the voltages using the corresponding sign, positive for the source and negative for the drops V_1 and V_2, add to zero, that is to say $V - V_1 - V_2 = 0$ which is equivalent to $V = V_1 + V_2$. In other words, the sum of all voltage rises is equal to the sum of all voltage drops.

The equivalent resistance of resistances connected in series is calculated as the sum of all resistances. For example, two resistances R_1 and R_2 connected in series (Figure 3.7) yield an equivalent resistance R given by

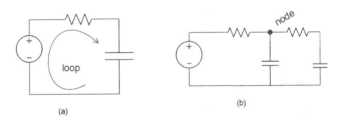

FIGURE 3.5 (a) Loops and (b) Nodes.

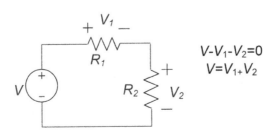

FIGURE 3.6 Kirchhoff's voltage law.

$$R = R_1 + R_2 \qquad (3.10)$$

We form a *voltage divider* using two resistances in series. The voltage across one resistance is proportional to the input voltage multiplied by the fraction of that resistance to the total equivalent resistance. For example, the voltage V_1 across resistance R_1 in Figure 3.8 is

$$V_1 = V \times \frac{R_1}{R_1 + R_2} \qquad (3.11)$$

This is an important concept to develop transducers from sensors, as we will see later in this chapter. A variable resistor or potentiometer is in essence a voltage divider; the position of the cursor determines the proportions R_1 and $R_2 = R - R_1$ of the total potentiometer resistance R. Therefore, the voltage across R_1 is given by Equation (3.11) at any cursor position.

Kirchhoff's current law states that the sum of currents in and out of a circuit node adds to zero, as shown in Figure 3.9 for two resistors $I - I_1 - I_2 = 0$. This is equivalent to say that the sum of all currents leaving a node must equal the sum of all currents entering a node $I = I_1 + I_2$.

The inverse of the equivalent resistance of resistances connected in parallel is calculated as the sum of inverses of all resistances. For example, two resistances R_1 and R_2 connected in parallel (Figure 3.10) yield an equivalent resistance R given by

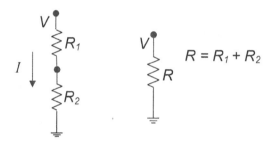

FIGURE 3.7   Resistances in series.

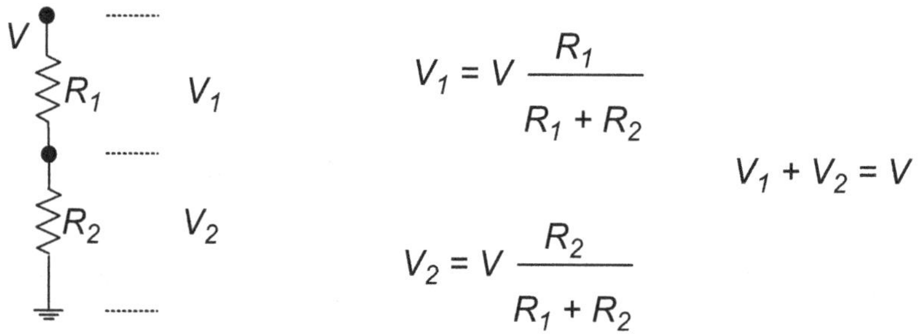

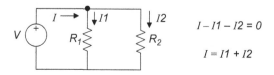

FIGURE 3.8   Voltage divider.

FIGURE 3.9   Kirchhoff's current law.

$$\frac{1}{R} = \frac{1}{R_1} + \frac{1}{R_2}$$ (3.12)

Or equivalently

$$R = \frac{R_1 R_2}{R_1 + R_2}$$ (3.13)

Recall that conductance G is the inverse of resistance $G = 1/R$ and therefore a more practical interpretation of Equation (3.13) is that the equivalent conductance of a parallel combination G is the sum of individual conductance values G_1 and G_2.

We form a *current divider* using two resistances in parallel. The current through one resistance is proportional to the input voltage multiplied by the fraction of the other resistance to the total equivalent resistance. For example, the voltage I_1 through resistance R_1 in Figure 3.11 is

$$I_1 = I \times \frac{R_2}{R_1 + R_2}$$ (3.14)

For many purposes, we interpret a circuit as performing a function or producing an *output* for a given *input*. We typically conceptualize this as the relationship between an input voltage and an output voltage. A linear circuit produces an output voltage proportional to the input voltage over a range (Figure 3.12 left). A linear amplifier is an example of this type of relationship. However, we also have non-linear circuits for which the output voltage does not have the same proportion over a range. A simple example is a voltage comparator, which produces a completely different value once the input goes over a threshold (Figure 3.12 right).

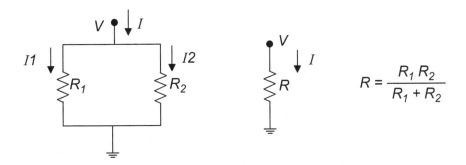

FIGURE 3.10   Resistances in parallel.

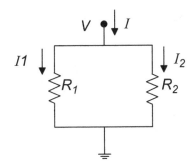

FIGURE 3.11   Current divider.

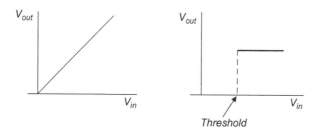

FIGURE 3.12   Response function: example of linear response (a) and non-linear response (b).

The relationship of Equation (3.11) can be interpreted as input-output if we think of V as the input voltage V_{in} and V_1 as the output voltage V_{out}. Renaming and rearranging, we obtain

$$\frac{V_{out}}{V_{in}} = \frac{R_1}{R_1 + R_2} \tag{3.15}$$

MEASURING VOLTAGES, CURRENTS, AND RESISTANCES

A *voltmeter* measures voltage (AC or DC), an *ammeter* measures current (AC or DC), and an *ohmmeter* measures resistance. A *multimeter* combines all these functions in one instrument. Voltmeters and ammeters derive their power from the circuit under measurement, whereas an ohmmeter requires a battery; it measures current and indirectly measures resistance by using Ohm's law. Therefore, do not use an ohmmeter while a circuit is hot or powered on. Many multimeters denote the resistance function by the symbol "Ω" or the word "ohms". The ohmmeter is also useful for a continuity test, i.e., detecting whether there is a continuous electrical connection from one point to another by indicating zero resistance or short circuit. Or equivalently indicating lack of continuity by showing very large resistance or open circuit.

Analog meters use a pointer (needle) that moves over a printed scale, and polarity given by the direction of movement of needle. The resistance scale is logarithmic; zero resistance (short circuit) occurs at one extreme of the scale, whereas infinite resistance (open circuit) swings to the opposite end.

Digital multimeters (DMM) convert the analog voltage, current, or ohms to a digital number and then display it on a numerical readout; polarity given by the sign on the readout. We have discussed *analog-to-digital conversion* (ADC) in Chapter 2. In the DMM context, the number of digits is an important specification of the display; for example, a three and 1/2 (3 ½) digit digital display would show $\pm 1XXX$, where X denotes a full digit (0–9). Seven short segments make up a digit. The leftmost digit is the leading digit or most significant digit (Figure 3.13). It is the "1/2 digit" and can indicate 1 as a maximum. The range is 0–1999, positive or negative. Therefore, we also call these meters a "2000-count" meter. Similarly, a four and 1/2 (4 ½) digit would show $\pm 1XXXX$ and therefore has a range of 0–19,999, positive or negative (Figure 3.13). In resistance mode, these meters indicate non-continuity by a non-numerical code on the display (e.g., "OL" for Open-Loop, or "–").

Voltmeters select the measuring range by switching resistances in a voltage divider circuit (resistances in series), whereas ammeters select range by changing resistance in a current divider circuit (resistances in parallel). A DMM may select the range automatically (auto ranging) or manually (the operator moves a switch to find the best range). The hands-on exercises in the Lab Manual companion to this book will often require using a DMM to measure voltage, current, and resistance.

The maximum value for a particular scale determines the *full scale* (*FS*). DMMs use a percentage of this *FS* value and the least significant digit (LSD) to specify their accuracy. Recall from Chapter 1 that accuracy is the difference between the measured value and the true value. More specifically, we specify DMM accuracy by a percentage of *FS* plus number of LSD. For example, consider a 3½ DMM with $\pm 2\%$ of *FS* and three LSD in the 20 V range. When reading

FIGURE 3.13 Displays: 3½ digits and 4½ digits.

FIGURE 3.14 A V DC reading on 200 V, 20 V, and 2 V scale.

a value in this range, we will see a reading of *XXX* (see Figure 3.14 for an example); note that three LSD is 0.01 and therefore the accuracy would be $\pm\left(20\times0.2/100+0.03\right)=\pm0.07$ V. If the measured value is 1.5, it means that the true value can be anywhere between $1.5-0.07=1.43$ V and $1.5+0.07=1.57$ V. Were we to use the 2-V range, we would observe a reading of 1*XXX* (see Figure 3.14 for an example), and now the three LSD contribution is only 0.003. Therefore, accuracy has improved to $\pm\left(2\times0.2/100+0.003\right)=\pm0.007$ V. Thus, if we measure 1500 V, the true value can be in between 1493 and 1507 V.

As an example, these are specifications of a low-cost 3 1/2 digit DMM. DC Volts: ranges 2–20–200 V, accuracy ±0.8% of *FS*, ±1 LSD; AC Volts: ranges 200–500 V (60 Hz), accuracy ±1.5% of *FS*, ±5 LSD; DC Current: ranges 2–20–200 mA, accuracy ±2.0% of *FS*, ±1 LSD; Resistance: ranges 200–2K–20K–200K–2M Ω, accuracy ±2.0% of *FS*, ±3 LSD.

For measuring voltage, select AC or DC and use correct polarity; typically connect the red test lead to positive (+) and the black test lead to negative (–). Select the range: start with the highest range; if the reading is small, set to the next lower range. The reading should be larger now; you would iterate if the reading continued to be small. Use the lowest range setting that does not "over-range" the meter. An over-ranged analog meter needle goes all the way to the side of the scale, past the full-range scale value. An over-ranged digital meter displays the letters "OL", or a series of dashed lines, or some other symbol (this indication is manufacturer-specific).

For measuring resistance, start with a simple test of continuity: set the meter to its highest resistance range and touch the two test probes and check for 0 ohms (short circuit). An analog multimeter has a potentiometer to calibrate it for "zero" ohms. To measure a resistance value: connect the test probes across the resistor and obtain a reading. If the reading is close to zero, select a lower resistance range on the meter and iterate until you use the appropriate range.

For measuring current, open the circuit and insert an "ammeter" in series with the circuit so that all current flowing through the circuit also must go through the meter. Measuring current in this manner makes the meter part of the circuit. Ideally should not cause voltage drop, assuming it has very little internal resistance. Therefore, the ammeter will act as a short circuit if placed in parallel to a source of voltage, causing a high current and potentially damaging the meter.

An *oscilloscope* allows us to see time-varying voltages by sweeping horizontally at a speed such that the horizontal deviation is proportional to real time. The proportionality is a timescale. On the scope screen, we then see waveforms or AC voltages such that vertical deviations correspond to voltage values and horizontal deviations correspond to time. When measuring DC, the vertical deviation is constant, and we just see a horizontal line. When measuring transient voltages, we may see a horizontal line moving down or up (if the sweep is too fast compared to the transient), or a trace describing the transient (if the sweep speed is commensurate with the transient). Adjust the Volts/division knob on the oscilloscope until the voltage appears on the screen. Estimate the voltage by counting the divisions and multiplying by the number of volts per division.

SENSORS

A sensor is an element that changes a property when its environment changes. In other words, it responds to a change in conditions by changing its properties. Commonly, we say that produces a *response* or *output* to a *signal* or *input*. Some examples are: a thermistor, which changes resistance with changes of temperature; a thermocouple, which produces a voltage due to temperature changes; a light-dependent resistor (LDR) that changes resistance with changes in light intensity, and a photovoltaic (PV) cell that produces a current when illuminated by solar radiation.

Active and *passive* sensors are distinguished according to their requirement for power (Brown and Musil 2004; Sheingold 1980). An active sensor requires external energy to generate the response signal; typically, it produces a signal using a power source. Examples are a thermistor and a LDR. In contrast, a passive sensor does not require external power; examples are a thermocouple and a PV cell.

FROM SENSORS TO TRANSDUCERS

An electrical *transducer* converts variations of one form of energy to variation in electrical energy. For example, we can place an active sensor of temperature (say a thermistor), in a voltage divider circuit, to get an output voltage related to resistance. This output voltage depends on the circuit voltage supply and on temperature since the resistance changes with temperature. The circuit is now a transducer converting thermal energy to electrical energy (Brown and Musil 2004). This electrical energy comes from the circuit power supply.

Passive sensors are themselves transducers since they convert variations of one form of energy to variations in electrical energy. For example, a PV cell converts variations in solar energy to voltage variations; a thermocouple converts thermal energy to electrical energy.

Although technically a sensor is not necessarily the same as a transducer, in colloquial terms it is customary to refer to a transducer as a sensor.

SENSOR SPECIFICATIONS: STATIC

We can talk about three major types of specifications: static, dynamic, and environmental (Brown and Musil 2004). In this chapter, we introduce some of the *static specifications*, i.e., when the sensor is at steady-state conditions, which means when the measurand is not changing.

- *Accuracy*: difference between the transducer measurement with respect to true and reference value (see Chapter 1). It is customarily given as the maximum expected error, expressed as percentage of *FS* output (example 1% of 50°C or 0.5°C).
- *Precision*: how well the transducer yields repeated values of the same value; or variation of output for the same input (see Chapter 1).
- *Resolution*: the smallest change in measurand detected by the transducer.
- *Sensitivity*: change of output for a given change in input, or magnitude of the response to a unit change in input; for example, 20 μV/°C.
- *Linearity error*: an indication of linearity of the response with respect to changes in measurand; it is based on deviation of the output with respect to the ideal linear response or straight line over the range of the measurand. Pallás-Areny and Webster (2001) define several ways of defining the straight line; we will focus on two of these. For *least squares linearity*, the straight line is defined by least-squares regression, thus equalizing maximum negative and positive errors. In *end-points linearity*, the straight line goes from the minimum to the maximum output. The linearity error can be summarized by various metrics or statistics of the deviation values. We focus on two of these: *maximum deviation* and *root mean square* (*RMS*) of the deviations; both expressed as a percentage of *FS*.
- *Hysteresis*: represents the difference in output between an increasing input and a decreasing input.

We will further discuss sensitivity and linearity as we present the following examples.

RESISTIVE SENSORS

Many physical variables affect resistance of a device and therefore it is common to use sensors based on the variation of resistance. For example, position may vary the resistance of a potentiometer. We use a resistive sensor in a circuit designed to produce a voltage related to the change in resistance of the sensor. This voltage is the signal indicating the physical quantity that we want to measure. Such a circuit must be powered and therefore these resistive sensors are *active*, i.e., require energy to operate.

Examples of resistive sensors are potentiometers, resistive temperature detectors, LDR, thermistors, liquid level sensors, strain gages, resistive gas sensors, liquid conductivity sensors, and resistive hygrometers.

THERMISTORS: TEMPERATURE RESPONSE

Thermistors can be of two types: NTC (negative temperature coefficient) and PTC (positive temperature coefficient). Those of NTC type are made from semiconductor material and resistance decreases gradually with temperature. In opposite fashion, a PTC, made from ceramic, will have a resistance that increases quickly with temperature.

We will focus on NTC thermistors for which the resistance decreases non-linearly with temperature. A general model is the B parameter equation

$$R = R_0 \exp\left(B\left(T^{-1} - T_0^{-1} \right) \right) \qquad (3.16)$$

where T is the temperature in K, R is the thermistor resistance in Ω, T_0 is the nominal value of T (25°C = 298 K), R_0 is the nominal value of R at T_0, and B is a parameter in K (Rudtsch and von Rohden 2015). Figure 3.15 shows an example of the model response for $B = 4100$ K, $T_0 = 25$°C, and $R_0 = 10$ kΩ.

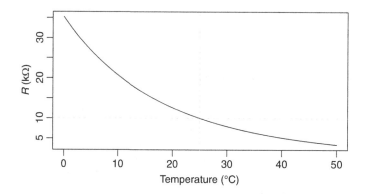

FIGURE 3.15   Thermistor resistance vs. temperature using the B parameter model.

The B parameter equation can be inverted to calculate temperature given resistance by rearranging terms and taking logarithm of both sides of the equation

$$\frac{1}{T} = \frac{1}{T_0} + \frac{1}{B}\ln\left(\frac{R}{R_0}\right) \tag{3.17}$$

Denoting $a_0 = 1/T_0$ and $a_1 = 1/B$, we obtain

$$\frac{1}{T} = a_0 + a_1 \ln\left(\frac{R}{R_0}\right) \tag{3.18}$$

This equation can be considered the first-order approximation $n = 1$, of a series

$$\frac{1}{T} = \sum_{i=0}^{n} a_i \left(\ln(R/R_0)\right)^i \tag{3.19}$$

With known nominal T_0 and R_0, parameter B can be estimated from data of R and T using linear regression. For this purpose, denote $x = \frac{1}{T} - \frac{1}{T_0}$ as independent variable and $y = \ln\left(\frac{R}{R_0}\right)$ as dependent variable, then $y = Bx$ and therefore B can be estimated by linear regression though the origin (Figure 3.16). We exercise this concept in Lab 3 of the companion Lab Manual (Acevedo 2024).

Equation (3.17) can be rewritten as

$$\frac{1}{T} = A_0 + A_1 \ln(R) \tag{3.20}$$

where $A_0 = \frac{1}{T_0} - \frac{1}{B}\ln(R_0)$ and $A_1 = \frac{1}{B}$, which can in turn be considered the first order, $n = 1$, of the series

$$\frac{1}{T} = \sum_{i=0}^{n} A_i \left(\ln(R)\right)^i \tag{3.21}$$

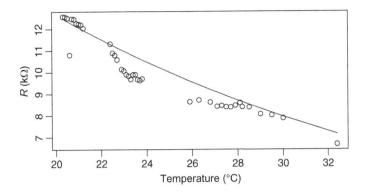

FIGURE 3.16 Measured temperature and resistance together with B parameter model calibration by regression.

Equation (3.20) can also be derived from solid-state physics principles by writing the conductance as $G = 1/R = F \exp(-\Delta E/2kT)$ where F is a proportionality constant, k is Boltzmann's constant, and ΔE is activation energy. Using logarithm, we obtain Equation (3.20) with $A_0 = \dfrac{2k \ln(F)}{\Delta E}$ and $A_1 = \dfrac{2k}{\Delta E}$ (Rachakonda et al. 2014). However, due to large variations in these terms, the coefficients are still estimated by regression.

A derived model is the Steinhart-Hart equation (Steinhart and Hart 1968; McGee 1988) that uses $n = 3$ and assumes $A_2 \simeq 0$

$$\frac{1}{T} = a + b \ln(R) + c \left(\ln(R) \right)^3 \tag{3.22}$$

where a, b, c are the coefficients A_0, A_1, and A_3 and can be estimated by *polynomial regression*. In Figure 3.17, we illustrate a graph for this model, for $a = 1.4 \times 10^{-3}$, $b = 2.37 \times 10^{-4}$, and $c = 9.90 \times 10^{-8}$.

Briefly, in polynomial regression, the predictor is a linear combination of increasing powers of X. In this case, we formulate the non-linear relation as a polynomial instead of a functional relationship, which is useful when you do not know what model to apply. Although a solution is always found, we may not know the meaning or interpretation attached to coefficients. Changing variables $Y = 1/T$ and $X = \ln(R)$, we write the predictor as

$$Y = \sum_{i=0}^{n} A_i X^i \tag{3.23}$$

which yields a good fit. However, because we do not know the meaning of the coefficients, we cannot claim that we have a better understanding of a generic response of temperature and resistance. One practical application of polynomial regression is the calibration of sensors and instruments. By transforming each power of X into a variable, the polynomial regression is converted to a multivariable linear regression problem. We will use this method in Lab 3 of the Lab Manual (Acevedo 2024).

For more precise measurements, some authors recommend to use all terms, including A_2, and expand to $n = 4$ (Rudtsch and von Rohden 2015). For simplicity, we will use Equations (3.16) and (3.22) in the experiments described in Lab 3 of the Lab Manual.

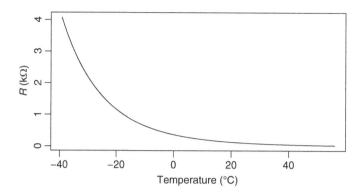

FIGURE 3.17   Steinhart-Hart model: resistance vs. temperature.

EXAMPLE: FROM THERMISTOR TO TEMPERATURE TRANSDUCER

As a specific example, we will consider a thermistor and design a circuit to make a temperature transducer. We can use a voltage divider circuit to make a simple transducer with an NTC thermistor, where the thermistor is connected to a voltage source and a fixed resistor R_f (Figure 3.18). In this circuit, R_t is the thermistor resistance. The transducer output voltage $V_{out} = V_f$ is equal to the input voltage source V_s times the ratio of resistances

$$V_{out} = V_f = V_s \frac{R_f}{R_t + R_f} \tag{3.24}$$

where R_t is the thermistor resistance. We can combine the circuit Equation (3.24) with the B parameter Equation (3.16)

$$V_{out} = V_f = V_s \frac{R_f}{R_t + R_f} = V_s \frac{R_f}{R_0 \exp\left(B\left(T^{-1} - T_0^{-1}\right)\right) + R_f} \tag{3.25}$$

To see the effect of temperature on the circuit response, we can obtain a chart of the behavior of V_{out} as a function of temperature for several values of the fixed resistor R_f, yielding the result shown in Figure 3.19 for $V_s = 12$ V, $B = 4100$ K, $T_0 = 25°C$, and $R_0 = 10$ kΩ. Note that linearity of the response is a function of R_f. Qualitatively, for $R_f = 10$ kΩ, the response is reasonably linear over this range of temperature.

CALCULATING SENSITIVITY AND LINEARITY ERROR

Range is the span between minimum and maximum value of the measurand (in this case temperature) (for example 0°C–50°C in Figure 3.19), whereas *FS* is the span between the maximum and minimum output voltage obtained in the range (say for example ~2.5–~9 V for $R_f = 10$ kΩ in Figure 3.19). In this case, the measurand is variable T and the minimum and maximum values are T_{min}, T_{max} the range is $\Delta T = T_{max} - T_{min}$. Assume the output is $V_{out}(T_{min})$ and $V_{out}(T_{max})$ for T_{min}, T_{max}, then the *FS* is

$$FS = \Delta V_{out} = V_{out}(T_{max}) - V_{out}(T_{min}) \tag{3.26}$$

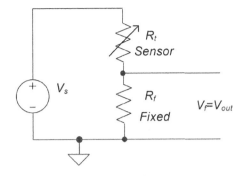

FIGURE 3.18 Simple transducer based on a voltage divider.

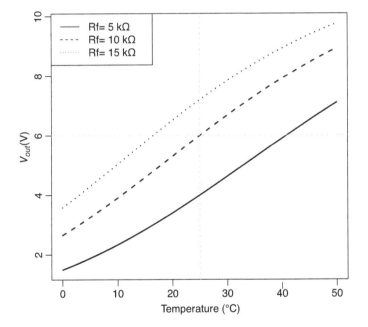

FIGURE 3.19 Response of voltage divider circuit for several values of R_f.

In the example for $R_f = 10$ kΩ of Figure 3.19, the range is 50°C, and $FS = 9 - 2.5 = 6.5$ V.

Sensitivity is calculated by the change of output for a given change in input, or magnitude of the response for a unit change in input; that is to say, *FS* divided by the range

$$Sens = \frac{FS}{\Delta T} \tag{3.27}$$

In other words, we can calculate sensitivity as the slope of the straight line joining the *end-points* or extremes of the range, i.e., the value of V_{out} at minimum value of T and the highest value of V_{out} or the one at the maximum value of T. In the example for $R_f = 10$ kΩ of Figure 3.19, $Sens = \dfrac{6.5 \text{ V}}{50°\text{C}} = 130 \text{ mV}/°\text{C}$. We will do this calculation in Exercise 3.1.

As discussed under the static specifications section, there are several ways to define the straight line used to calculate linearity error. The *end-points* method is based on the linear response passing through V_{min} and V_{max}, which is defined by

$$V_{lin}(T) = Sens \times (T - T_{min}) + V_{min} \tag{3.28}$$

Calculating the *least squares* linearity error requires performing linear regression based on values V_{out} and T to obtain the coefficients, intercept b_0 and slope b_1, which define the straight line

$$V_{lin}(T) = b_1 \times T + b_0 \tag{3.29}$$

To demonstrate, assume that we have a set of n values of temperature T_i where $I = 1, ..., n$, corresponding values $V_{out}(T_i)$, as well as calculated values $V_{out}(T_i)$ from Equations (3.28) for the end-point method, and (3.29) for the least squares method. Then for each i, the deviation of the actual response with respect to the linear would be

$$z_i = V_{out}(T_i) - V_{lin}(T_i) \tag{3.30}$$

Applying the maximum of the absolute values of the deviation, and dividing by FS, we calculate linearity error a a percentage of FS

$$L_{max} = \dfrac{\max(z_i, i = 1, ..., n)}{FS} \times 100 \tag{3.31}$$

We will practice this calculation in Exercise 3.2. An alternative metric, as discussed earlier, is to calculate the RMS of the deviations and divide by FS

$$L_{rms} = \dfrac{\left(\dfrac{1}{n} \displaystyle\sum_0^n z_i^2 \right)^{\frac{1}{2}}}{FS} \times 100 \tag{3.32}$$

In Lab 3 of the Lab Manual, we will learn how to implement these calculations using a function in R. For example, for $V_s = 12$ V, we obtain Figure 3.20 that is annotated to have $FS = 6270$ mV, sensitivity of 125.4 mV/°C, end-points (Ends) and maximum linearity error of 4.46% of FS, least-squares (LSQ) and maximum linearity error of 4.11% of FS, end-points RMS linearity error of 2.74% of FS, and least-squares RMS linearity error of 1.32% of FS. The gray segments indicate where the maximum deviation occurs; for end-points at about 35°C, whereas for least-squares at the upper extreme of the range 50°C.

READING OUTPUT VOLTAGE WITH A DIGITAL DEVICE

As pointed out in Chapter 2, it is practical to read the analog output voltage V_{out} from the transducer using a microcontroller or digital device equipped with an ADC. We know from Chapter 2 that for n bits and maximum voltage V_{AD}, the voltage resolution is

$$V_{res} = \dfrac{V_{AD}}{2^n - 1} \tag{3.33}$$

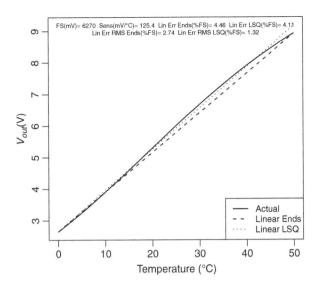

FIGURE 3.20   Response of voltage divider circuit for $R_f = 10\,\text{k}\Omega$ with calculated values of *FS*, sensitivity, and linearity error by two methods and two metrics.

The digital number corresponding to a V_{out} value is the truncation of $V_{\text{out}}/V_{\text{res}}$

$$DN = trunc\left(\frac{V_{\text{out}}}{V_{\text{res}}}\right) \tag{3.34}$$

For example, a ten-bit ADC with 5 V will have a resolution of $V_{\text{res}} = \dfrac{5}{2^{10}-1} = \dfrac{5}{1024-1} = \dfrac{5}{1023} = 4.888$ mV. When the transducer produces $V_{\text{out}} = 2.00$ V, the digital value is $DN = trunc\left(\dfrac{2.00}{\frac{5.00}{1023}}\right) = 409$ and when V_{out} increases by the resolution to reach $V_{\text{out}} = 2.00488$ V, DN increases by one, i.e., $DN = trunc\left(\dfrac{2.00488}{\frac{5.00}{1023}}\right) = 410$. Note when $V_{\text{out}} = 5.00$ V, then $DN = 1023$.

INVERTING THE TRANSDUCER OUTPUT TO OBTAIN TEMPERATURE

Once we measure the transducer output voltage, we need to calculate temperature by inverting Equation (3.25), or calculate T from V_{out}. The thermistor resistance R_t is solved from the voltage divider equation

$$R_t = V_s \frac{R_f}{V_{\text{out}}} - R_f \tag{3.35}$$

and the B model is $\exp\left(B\left(T^{-1} - T_0^{-1}\right)\right) = \dfrac{R_t}{R_0}$, taking logarithms $B\left(T^{-1} - T_0^{-1}\right) = \ln\left(\dfrac{R_t}{R_0}\right)$, and solving for T

$$T = \left(T_0^{-1} + B^{-1} \ln\left(\frac{R_t}{R_0}\right)\right)^{-1} \tag{3.36}$$

In other words, we calculate R_t using Equation (3.35) given the output voltage and circuit parameters V_s and R_f. Then calculate temperature using Equation (3.36) given the prior calculation of R_t and the thermistor parameters T_0, B, and R_0.

When the transducer output is read using an ADC, the digital number is first converted to voltage units, and then we calculate measurand from voltage. Suppose we read digital number DN, the corresponding voltage is calculated from the inverse of equation (3.34). This is to say

$$V_{\text{out}} = DN \times V_{\text{res}} \tag{3.37}$$

For instance, when ADC is 10 bit and 5 V, a DN of 409 calculates out to $V_{\text{out}} = 409 \times 5/1023 = 1.999$ and a DN of 410 calculates out to $V_{\text{out}} = 410 \times 5/1023 = 2.00391$. At this point, you would realize that we could not recover the output voltage without an error, i.e., the *quantization error* because all voltage values within an interval of 0.004888 V were converted to the same digital number.

SELF-HEATING EFFECT

Thermistors are affected by *self-heating* problems because of Joule heating produced by current flowing in the thermistor; power absorption raises temperature above the intended measurement proportionally to power according to

$$T_a = T_t - \frac{\text{Power}}{k} = T_t - \frac{\dfrac{V_t^2}{R_t(T_t)}}{k} \tag{3.38}$$

Here T_a is the ambient temperature being measured, T_t is the temperature of the sensor (thermistor), V_t is the voltage across the thermistor, k is the thermal coefficient (in mW/°C), and R_t is the resistance of the thermistor (which in turn is a function of temperature T_t). The thermal coefficient k varies with ambient temperature being measured, e.g., 1.5 mW/°C in still air.

Therefore, we may need to correct for self-heating. Figure 3.21 illustrates this concept using a voltage source of $V_s = 12$ V and an $R_f = 10$ kΩ. The top figure shows how the actual ambient temperature T_a will be consistently underestimated because the temperature of the thermistor is higher. Similarly, the bottom graph shows the difference between T_a and T_t as a function of V_f. In this case, we can see the gap between T_a and T_t showing again the underestimation of T_a. We learn how to produce these graphs in Lab 3 of the Lab Manual.

To diminish the effect of self-heating, we can lower the voltage by decreasing V_s and therefore power consumption. Suppose we use $V_s = 3.3$ V. Recalculating the effect of self-heating, we would see that the temperature lines are indistinguishable from each other. Of course, we now have a much lower sensitivity in terms of mV/°C; being reduced by a factor 3.3 V/12 V = 0.275.

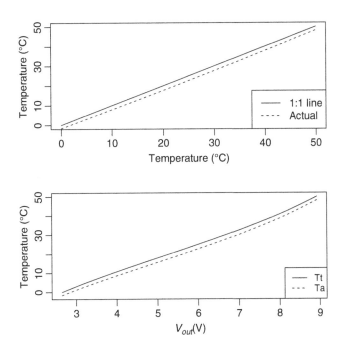

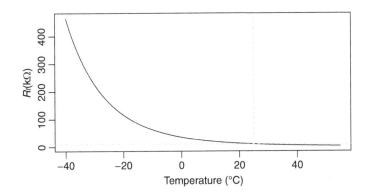

FIGURE 3.21 Self-heating effect for $Rf = 10\,k\Omega$ and $V_s = 12\,V$.

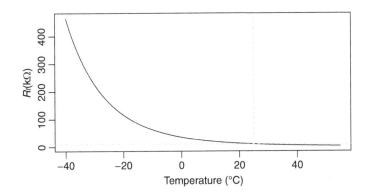

FIGURE 3.22 B parameter model resistance vs. temperature for a wider range of temperature.

EXAMPLE: WIDER TEMPERATURE RANGE

For a wider temperature range, say from $-40°C$ to $55°C$ as shown in Figure 3.22 (produced with $B = 4100\,K$, $R_0 = 10\,k\Omega$), the thermistor resistance at the low part of the range ($\sim460\,k\Omega$) is much higher than the resistance at the high part of the range ($\sim2.8\,k\Omega$). Such a large difference ($460/2.8 \sim 160$ times) may enhance linearity error, particularly in the low part of the range.

To compensate for such a large difference, we can add one resistor R_1 to the voltage divider as shown in Figure 3.23. Now the difference is reduced, for example for $R_1 = 10\,k\Omega$, we have $(460 + 10)/(2.8 + 10) \sim 36$ times. The transducer output voltage V_f is now related to the input voltage source V_s by

$$V_{out} = V_f = V_s \frac{R_f}{R_t + R_1 + R_f} \qquad (3.39)$$

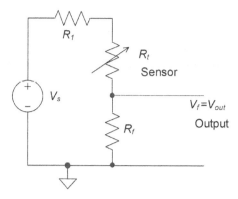

FIGURE 3.23 Voltage divider circuit with sensor and additional resistor R1 compensating for broad temperature range.

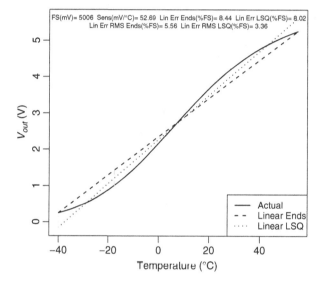

FIGURE 3.24 Response of voltage divider circuit for $R_f = 10\ k\Omega$, $R_1 = 10\ k\Omega$, $V_s = 12\ V$ with calculated values of FS, sensitivity, and linearity error.

To illustrate the effect of this modified divider, we repeat the calculations of output voltage, sensitivity and linearity using $B = 4100$, $R_f = 10$ kΩ, $R_1 = 10$ kΩ, and $V_s = 12$ V as an example (Figure 3.24). We can see how the non-linearity has spread to both, the low part of the range and high part of the range, which means a trade off in sensitivity and linearity. Sensitivity is reduced to 52.69 mV/°C, end-points and least-squares maximum linearity error increase to 8.44% of FS and 8.02% of FS, respectively, while end-points and least-squares RMS increase to 5.56% and 3.36% of FS.

When using this voltage divider, the inverse calculation of temperature from output voltage changes in the following manner: the calculation of thermistor resistance by Equation (3.35) needs to be changed to

$$R_t = V_s \frac{R_f}{V_{\text{out}}} - R_f - R_1 \tag{3.40}$$

prior to calculating temperature using Equation (3.36) derived from the thermistor B model.

EXAMPLE: A TEMPERATURE TRANSDUCER FOR AIR, SOIL, AND WATER

To illustrate the practical application of the above concepts, we will look at a commercial temperature transducer, the "107 Temperature Probe", manufactured by Campbell Scientific (Campbell Scientific Inc. 2014). It measures temperature of air, soil, or water in environmental monitoring stations, such as a weather station. Its range of measurement is from −35°C to 50°C.

The 107 probe is based on a thermistor with the parameters $a = 8.27 \times 10^{-4}$; $b = 2.08 \times 10^{-4}$; and $c = 8.06 \times 10^{-8}$ and connected in a voltage divider circuit. The output voltage corresponds to the drop across $R_f = 1$ kΩ resistor. In series with the thermistor, the 107 has $R_1 = 249$ kΩ resistor, which compensates for the large difference across the low part of the range and the high part of the range. For a source voltage V_s, the output voltage is given by Equation (3.39). Assuming $V_s = 2.5$ V, we can plot V_{out} vs. temperature (Figure 3.25), along with calculating sensitivity and linearity error. Sensitivity is 0.09 mV/°C, end-points and least-squares maximum linearity error are 9.59% of *FS* and 87.89% of *FS*, respectively, while end-points and least squares RMS are 4.28% and 2.83% of *FS*. Using the R_1 resistor, the transducer has spread the non-linearity to both parts of the range. For measurements in air, a radiation shield protects the transducer from heating by direct sun exposure (Figure 3.26).

EXAMPLE: THERMOCOUPLES

Thermocouples sense temperature as the difference in potential created when joining two metals of different characteristics. This joint does not require external energy to act as a sensor, thus a thermocouple is a passive sensor. We can better understand the working principles of thermocouples by looking at two additional effects: *Peltier* and *Thomson* effects. The Peltier effect is described for a thermocouple circuit consisting of two junctions (Figure 3.27), each one at a different temperature (Pallás-Areny 2000). A current I through such a thermocouple circuit generates a heat flow dQ/dt from one junction to another. Direction of heat flow depends on direction of the current flow and is proportional to current

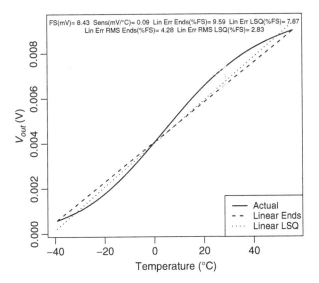

FIGURE 3.25 Transducer gain vs. temperature.

FIGURE 3.26   Radiation shield to protect the temperature transducer in air temperature measurement.

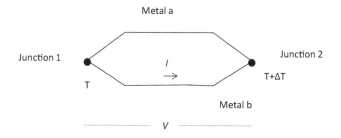

FIGURE 3.27   Thermocouple circuit.

$$\frac{dQ_{12}}{dt} = \pi_{12}(T + \Delta T)I \quad \text{and} \quad \frac{dQ_{21}}{dt} = -\pi_{12}(T)I \tag{3.41}$$

where $\pi_{12}(\cdot)$ is the Peltier coefficient (in units of V) which is a non-linear function of temperature.

The Thomson effect consists of heat dissipation or absorption in a conductor with a temperature gradient due to current flow. Heat flow is proportional to current (not to power, and thus not to the square of current) and therefore it is affected by the direction of current flow. Heat flow is also proportional to the temperature gradient. In each one of the conductors, of metal a and b, ΔT is the net difference in temperature

$$\frac{dQ_{12}}{dt} = \sigma_b I \Delta T \quad \text{and} \quad \frac{dQ_{21}}{dt} = -\sigma_a I \Delta T \tag{3.42}$$

where the coefficients σ_a and σ_b have units of V/°C. In Equation (3.42), we have assumed that the current is low enough to be able to neglect Joule heating. Heat is released going from junction 2 to junction 1 along wire of metal a, and heat is absorbed when going from junction 1 to 2 along wire of metal b.

Both, Peltier and Thomson, effects combined explain the *Seebeck* effect, which consists of the generation of a voltage V or electromotive force (emf, E) when the two junctions are at different temperatures. To see this, denote by dQ/dt the net heat flow and write the heat balance using Equations (3.41) and (3.42) neglecting Joule heating

$$\frac{dQ}{dt} = \pi_{12}(T + \Delta T)I - \pi_{12}(T)I + (\sigma_b - \sigma_a)I\Delta T \tag{3.43}$$

Divide both sides by I, and we get both sides in terms of voltage. The left side $(dQ/dt)I$ is equivalent to power over current; in other words, the net heat flow is converted to electrical potential difference V between the two junctions, which we can rewrite as $(dV/dT)\Delta T$; this term explicitly uses dV/dT known as the Seebeck coefficient. In this manner, we rewrite Equation (3.43) as

$$\frac{dV}{dT}\Delta T = \pi_{12}(T + \Delta T) - \pi_{12}(T) + (\sigma_b - \sigma_a)\Delta T \tag{3.44}$$

Now divide both sides by ΔT

$$\frac{dV}{dT} = \frac{\pi_{12}(T + \Delta T) - \pi_{12}(T)}{\Delta T} + (\sigma_b - \sigma_a) = \frac{d\pi_{12}}{dT} + \sigma_b - \sigma_a \tag{3.45}$$

moreover, use the definition of derivative for very small ΔT

$$\frac{dV}{dT} = \frac{d\pi_{12}}{dT} + \sigma_b - \sigma_a \tag{3.46}$$

which gives the Seebeck coefficient (change of voltage with temperature) as a sum of two terms: the derivative of the Peltier coefficient and the difference in Thomson coefficients. This expression holds for small variations of temperature between the junctions. Equation (3.46) is the basic *thermo-electric equation*. Note that this is non-linear because of the non-linearity of the Peltier coefficient.

Thermocouple types are defined according to what metals or alloys are joined together; one is designated positive and the other negative (Table 3.1). For example, Type J thermocouples are low-cost and have adequate sensitivity. There are standards for thermocouples of different types; for instance, the National Institute of Standards and Technology (NIST) Standard Reference Database or NIST ITS-90 Thermocouple database (Croarkin et al. 1993). The standards specify the relationship between voltage and temperature for a given thermocouple by tables or by fitted polynomials that approximate the non-linear relationship by increasing powers of V or E (inverse polynomial) or of voltage as a function of increasing powers of T (direct polynomial). See also OMEGA (2014).

An inverse polynomial is of the kind

$$T = a_0 + a_1 V + a_2 V^2 + a_3 V^3 + \dots \tag{3.47}$$

For example, Figure 3.28 shows a text file downloaded from NIST (1995) for a type J thermocouple that contains the inverse polynomial coefficients. In Lab 3, we will learn to read the coefficients from this file using R and obtain the plot shown in Figure 3.29. For the sake of illustration, assume the temperature range 0°C–760°C and that the non-zero coefficients from Figure 3.28 are $a_1, a_2, \dots,$ a_7. Take for example $V = 5\,\text{mV}$, applying Equation (3.47) we get 95.0655°C.

TABLE 3.1

Some Common Thermocouple Types

Type	Positive Metal/ Alloy	Negative Metal/Alloy	Temperature Range (°C) (Approximate)	Voltage Range (mV)	Error (°C)
T	Copper	Constantan	–200 to 350	26	±1
J	Iron	Constantan	180– to 750	43	±2.2
K	Chromel	Alumel	–200 to +1250	56	±2.2
E	Chromel	Constantan	–200 to +900	75	±1.0

Source: Adapted from Maxim (2007); Pallás-Areny (2000).

```
thermocoupleJ-inverse-coeff.txt
  1   Inverse coefficients for type J:
  2
  3   Temperature   -210.              0.              760.
  4     Range:         0.            760.             1200.
  5
  6     Voltage      -8.095           0.000            42.919
  7     Range:        0.000          42.919            69.553
  8
  9            0.0000000E+00   0.000000E+00  -3.11358187E+03
 10            1.9528268E+01   1.978425E+01   3.00543684E+02
 11           -1.2286185E+00  -2.001204E-01  -9.94773230E+00
 12           -1.0752178E+00   1.036969E-02   1.70276630E-01
 13           -5.9086933E-01  -2.549687E-04  -1.43033468E-03
 14           -1.7256713E-01   3.585153E-06   4.73886084E-06
 15           -2.8131513E-02  -5.344285E-08   0.00000000E+00
 16           -2.3963370E-03   5.099890E-10   0.00000000E+00
 17           -8.3823321E-05   0.000000E+00   0.00000000E+00
 18
 19   Error         -0.05           -0.04            -0.04
 20   Range:         0.03            0.04             0.03
 21
```

FIGURE 3.28   Inverse polynomial coefficients. Data downloaded from NIST (1995).

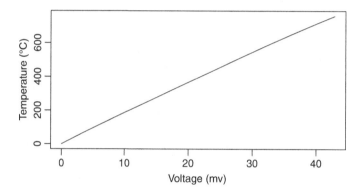

FIGURE 3.29   Thermocouple J, temperature calculated from mV for the range 0°C–760°C using NSIT inverse polynomial.

To use a thermocouple circuit to measure temperature, we use one junction at the temperature T that we wish to measure and the other junction at a reference temperature T_r as shown in Figure 3.30. It is imperative to minimize the current to avoid Joule heating as well as unintentional heating the environment surrounding the junctions. Many tables use 0°C as reference temperature. However, this is not a practical reference for many purposes, because it would involve an ice bath. A more practical approach is to use a reference voltage of equal magnitude of the emf that the reference junction would have provided, or to use a bridge circuit to provide this reference. We will study such circuits in the next chapter when we cover bridge circuits and as cold-junction compensation in the context of signal conditioning. Thermocouples are often used to measure soil temperature (Faucett et al. 2006; Sackett and Haase 1992). We will see an example in the next section.

EXAMPLES: USING THERMOCOUPLES

One example of application of thermocouples in environmental monitoring is the Campbell Scientific Averaging Soil Thermocouple (TCAV) probe, which can be used to measure soil temperature at various soil depths. The TCAV probe utilizes four or two pairs of type-E (chromel-constantan)

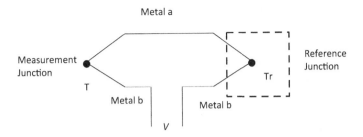

FIGURE 3.30   Thermocouple circuit for temperature measurement.

thermocouples. Each member of a thermocouple pair is buried at a different depth. The two pairs are separated at up to 1 m. These four soil temperatures are averaged, thereby providing an average soil temperature in the upper layer of soil for heat flux and energy balance calculations.

Another example of thermocouple application is measuring the average tree-trunk temperature to perform energy balance and from this obtain an estimate sap flow. The principle of operation is convection of heat by the flow of sap. We will discuss this further in Chapter 13.

EXERCISES

Exercise 3.1

Consider a thermistor in a voltage divider circuit (Figure 3.18). Assume $V_s = 12$ V, $R_f = 10$ kΩ, and a thermistor with $B = 4100$ K, $T_0 = 25°C$, and $R_0 = 10$ kΩ. Calculate V_{out} for the end-points and sensitivity in temperature range of 0°C–50°C.

Exercise 3.2

Consider the same temperature range, thermistor, and circuit of Exercise 3.1. Assume you know that the maximum deviation with respect to the end-points straight line occurs at 34.3°C. Calculate end-points maximum linearity error as a percentage of *FS*.

Exercise 3.3

Assume a temperature transducer built as a voltage divider as shown in Figure 3.23 with $R_f = R_1 = 10$ kΩ and $V_s = 5$ V using a thermistor with $B = 4100$ K, $T_0 = 25°C$, and $R_0 = 10$ kΩ. What is the temperature when the output is 2.00 V?

Exercise 3.4

Consider the temperature transducer of Exercise 3.3. The output voltage is read by an ADC of 10 bit and 5 V. Calculate the temperature when the digital value is 409.

Exercise 3.5

Consider the linearity error of Equation (3.47) relating temperature T to thermocouple voltage V for the range 0–760°C and voltage is 0–42.919 mV as shown in Figure 3.29. The exact calculation of the inverse polynomial yields temperature of 95.0655°C for 5 mV. Calculate sensitivity in mV/°C assuming a linear response and calculate temperature error due to end-points linearity for 5 mV as a percentage of FS.

REFERENCES

Acevedo, M. F. 2024. *Real-Time Environmental Monitoring: Sensors and Systems*. Second edition –Lab Manual. Boca Raton, FL: CRC Press, Taylor & Francis Group. 463 pp.

Artiola, J. F., I. L. Pepper, and M. L. Brusseau, eds. 2004. *Environmental Monitoring and Characterization*. Burlington, MA: Elsevier Academic Press.

Brown, P., and S. A. Musil. 2004. "Automated Data Acquisition and Processing." In *Environmental Monitoring and Characterization*, edited by J. F. Artiola, I. L. Pepper and M. L. Brusseau, 49–67. Burlington, MA: Academic Press.

Campbell Scientific Inc. 2014. *Model 107 Temperature Probe*. Logan, UT: Campbell Scientific Inc.

Croarkin, M. C., W. F. Guthrie, G. W. Burns, M. Kaeser, and G. F. Strouse. 1993. *Temperature-Electromotive Force Reference Functions and Tables for the Letter-Designated Thermocouple Types Based on the ITS-90*. Gaithersburg, MD: National Institute of Standards and Technology (NIST).

Faucett, R. P, E. C. Brevik, and S. Crow. 2006. Design of an Inexpensive Thermocouple-Based Soil Thermometer. *Soil Survey Horizons* 47:71–73.

Irwin, J. D., and R. M. Nelms. 2011. *Basic Engineering Circuit Analysis*. Hoboken, NJ: Wiley.

Maxim. 2007. *Application Note 4026. Implementing Cold-Junction Compensation in Thermocouple Applications*. Accessed September 2014. http://www.maximintegrated.com/en/app-notes/index.mvp/id/4026.

McGee, T. 1988. *Principles and Methods of Temperature Measurement*. New York: John Wiley & Sons.

NIST. 1995. *Tables of Thermoelectric Voltages and Coefficients for Download*. Accessed 2014. http://srdata.nist.gov/its90/download/download.html.

OMEGA. 2014. *ITS-90 Thermocouple Direct and Inverse Polynomials*. Accessed 2014. http://www.omega.com/temperature/z/pdf/z198-201.pdf.

Pallás-Areny, R. 2000. *Amplifiers and Signal Conditioners*. Boca Raton, FL: CRC Press.

Pallás-Areny, R., and J. G. Webster. 2001. *Sensors and Signal Conditioning*. New York: Wiley.

Rachakonda, P., D. Sawyer, B. Muralikrishnan, C. Blackburn, C. Shakarji, G. Strouse, and S. Phillips. 2014. *In-situ Temperature Calibration Capability for Dimensional Metrology*. Accessed December 2021. https://tsapps.nist.gov/publication/get_pdf.cfm?pub_id=915733

Rudtsch, S., and C. von Rohden. 2015. Calibration and Self-Validation of Thermistors for High-Precision Temperature Measurements. *Measurement* 76:1–6. https://doi.org/10.1016/j.measurement.2015.07.028.

Sackett, S., and M. Haase. 1992. *Measuring Soil and Tree Temperatures During Prescribed Fires with Thermocouple Probes*. Albany, CA: Pacific Southwest Research Station.

Sheingold, D., ed. 1980. *Transducer Interfacing Handbook*. Norwood, MA: Analog Devices, Inc.

Steinhart, J., and S. Hart. 1968. Calibration Curves for Thermistors. *Deep Sea Research and Oceanographic* 15 (4):497–503.

4 Bridge Circuits and Signal Conditioning

INTRODUCTION

In this chapter, we focus on relatively small departures of the measurand with respect to its nominal value and the corresponding linearized response of the sensor. We start by looking at a linearized thermistor and the response of a voltage divider when we use the approximate linear thermistor response. An important part of the chapter is designing transducer circuits based on bridge circuits, starting with the balanced source divider, and continuing with Wheatstone bridges, including quarter-bridge, half-bridge, and full-bridge. The latter is illustrated using strain gages and their application to soil tensiometers. This chapter discusses more details on transducer specifications, namely dynamic and environmental specifications. We then cover electrochemical sensors, piezoelectric sensors, and applications to environmental monitoring. The last part of this chapter is devoted to signal conditioning, and a detailed explanation of operational amplifiers and their use to linearize bridge circuits, as well as to amplify transducer output. Several other signal-conditioning topics included are common-mode rejection, noise, shielding, and current loop methods. These circuits and topics are expanded by computer and hands-on exercises in Lab 4 of the companion Lab Manual (Acevedo 2024).

LINEARIZED THERMISTOR: SMALL VARIATION ANALYSIS

When variations of the measurand are small, it is possible to simplify the analysis of the sensor response by assuming a linear response around the operating value of the measurand (Figure 4.1). For instance, if a thermistor has small variations around the nominal value, we could use a linearized resistance response around the nominal value as an approximation

$$R_t = R_0 + dR \approx R_0 + S dT \tag{4.1}$$

where S is the slope of the resistance response around the nominal temperature

$$S = \left. \frac{dR}{dT} \right|_{T=T_0} \tag{4.2}$$

Recall the B parameter equation from Chapter 3

$$R = R_0 \exp\left(B\left(T^{-1} - T_0^{-1} \right) \right) \tag{4.3}$$

where T is the temperature in K, R is the thermistor resistance in Ω, T_0 is the nominal value of T ($25°C = 298$ K), R_0 is the nominal value of R at T_0, and B is a parameter in K. The derivative $\frac{dR}{dT}$ is

$$\frac{dR}{dT} = R_0 B \exp\left(B\left(T^{-1} - T_0^{-1} \right) \right) \times \left(-T^{-2} \right) \tag{4.4}$$

evaluating at $T = T_0$

$$S = \left. \frac{dR}{dT} \right|_{T=T_0} = R_0 B \times \left(-T_0^{-2} \right) \tag{4.5}$$

DOI: 10.1201/9781003425496-4

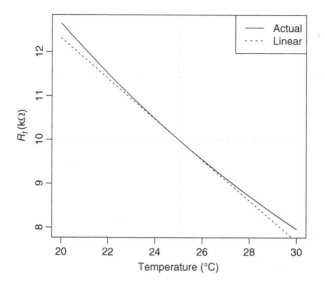

FIGURE 4.1   Linearized thermistor response.

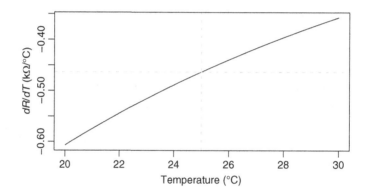

FIGURE 4.2   Derivative of the B model.

For example, when $B = 4100$ K, $T_0 = 25°C = 298$ K, and $R_0 = 10$ kΩ, the derivative over the range 20–30°C is negative and increases with temperature as shown Figure 4.2. At the nominal value, we obtain $S = \dfrac{dR}{dT}\Big|_{T=T_0} = 10 \text{ k}\Omega \times 4100 \text{ K} \times \left(-(298 \text{ K})^{-2}\right) = -0.4635 \text{ k}\Omega/\text{K}$. Note that it is equivalent to express this as a change per °C.

VOLTAGE DIVIDER WITH LINEARIZED THERMISTOR

Recall from Chapter 3 the expression of the output for a voltage divider transducer given by the circuit in Figure 3.23

$$V_{\text{out}} = V_s \frac{R_f}{R_t + R_1 + R_f} \tag{4.6}$$

We could simplify the nonlinear B model for R_t to a linearized resistance as in Equation (4.1). Assume $R_f = R_0$ and denote by V_0 the transducer voltage for nominal T_0, we can derive the change of the transducer output voltage due to the change of temperature as

$$dV = V_s \frac{R_f}{R_f + dR + R_1 + R_f} - V_0 \text{ which reduces to } dV = V_s \frac{R_f}{2R_f + dR + R_1} - V_0.$$

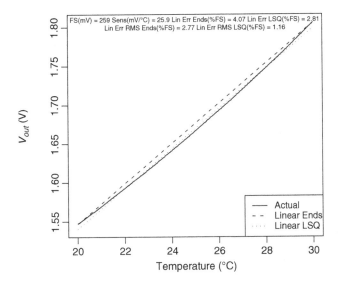

FIGURE 4.3   Transducer response using a linearized thermistor is still nonlinear.

For a given change in temperature dT, we can use $dR = SdT$

$$dV = V_s \frac{R_f}{2R_f + SdT + R_1} - V_0 \tag{4.7}$$

For a numeric example, suppose the thermistor is $10\,k\Omega$ nominal (at $25°C$) with slope $S = \left.\dfrac{dR}{dT}\right|_{T=T_0} = -0.46\dfrac{k\Omega}{°C}$. Let us calculate the response for $26°C$ (or $+1°C$ change with respect to nominal). The change in thermistor resistance is $dR = SdT = -0.46\dfrac{k\Omega}{°C} \times (26 - 25)°C = -0.46\,k\Omega$ and the thermistor now has a resistance of $10 - 0.46 = 9.54\,k\Omega$. For instance, for $V_s = 5\,V$, $R_1 = 10\,k\Omega$, $R_f = 10\,k\Omega$, the transducer nominal output voltage is $V_0 = 5\dfrac{10}{10 + 10 + 10} = 1.66\,V$, and thus the change of transducer output is $dV = 5\dfrac{10}{10 + 9.54 + 10}\,V - 4\,V = 1.70 - 1.66\,V = 40\,mV$.

It is important to point out that even though the thermistor response has been linearized, the transducer response is still nonlinear because the voltage divider equation is nonlinear (Figure 4.3).

BALANCED SOURCE VOLTAGE DIVIDER

We start with an extension of the simple voltage divider circuit as given in Figure 3.18 of Chapter 3 by powering it with a balanced source (Figure 4.4). Using the superposition principle, we can determine the output voltage as the sum of the effect of each source considered separately

$$V_{out} = V_s\left(\frac{R_f}{R_t + R_f} - \frac{R_t}{R_t + R_f}\right) = V_s\left(\frac{R_f - R_t}{R_t + R_f}\right) \tag{4.8}$$

Now, we select the fixed resistor R_f to have a value equal to R_0 which is the nominal value of a resistive sensor at the nominal value of the measurand. To be specific in this illustration, we will use a thermistor as example. The nominal value R_0 is the value of the sensor resistance R_t at the

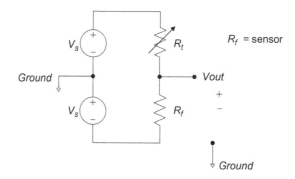

FIGURE 4.4　Balanced source voltage divider.

nominal operating value of temperature, say T_0. The transducer output at nominal temperature is zero $V_{out}(T_0) = V_0 = V_s\left(\dfrac{R_f - R_0}{R_0 + R_f}\right) = 0$. A small departure of temperature from the nominal, say a decrease $-dT$ would cause an increase $+dR$ in the thermistor resistance, and therefore the output voltage would change to

$$V_{out} = V_s\left(\frac{R_f - R_f - dR}{R_f + R_f + dR}\right) = V_s\left(\frac{-dR}{2R_f + dR}\right) = \frac{V_s}{2}\left(\frac{-dR}{R_f + \dfrac{dR}{2}}\right) \tag{4.9}$$

For instance, for $S = -0.46\dfrac{k\Omega}{°C}$ 1°C change $dR = -0.46$ kΩ. With $V_s=5$ V, $R_f=10$ kΩ, the output would be $V_{out} = \dfrac{5}{2}\left(\dfrac{0.46}{10 - \dfrac{0.46}{2}}\right) = 0.118$ V. When the resistance variation dR is very small compared to the nominal, such that $2R_f \gg dR$, we have

$$V_{out} \approx V_s\left(\frac{-dR}{2R_f}\right) = -\frac{V_s}{2R_f}dR \tag{4.10}$$

Denote by dV the departure of V_{out} with respect to zero and move dR to the left side $\dfrac{dV}{dR} \approx -\dfrac{V_s}{2R_f}$ or the slope of the line shown in Figure 4.5a. Note that when dR is negative, dV is positive and decreases, whereas when dR is positive, then dV is negative and decreases. Now, when we include the sensor response to temperature, we note that as temperature increases, dR becomes negative and then dV is positive and increases with temperature. However, as temperature decreases, dR is positive and then dV is negative and decreases with temperature.

The slope of the line shown in Figure 4.5b or sensitivity of the transducer is given by

$$\frac{dV}{dT} = \frac{dV}{dR}\frac{dR}{dT}\bigg|_{T=T_0} \approx -\frac{V_s}{2R_f} \times S \tag{4.11}$$

Sometimes this circuit is referred to as "half-bridge" but the use of this term for this circuit is not to be confused with a "half-bridge" defined as a two-sensor Wheatstone bridge as we will discuss later.

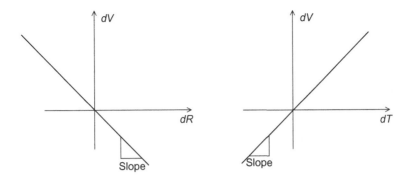

FIGURE 4.5 Circuit response dV/dR and transducer response to temperature dV/dT.

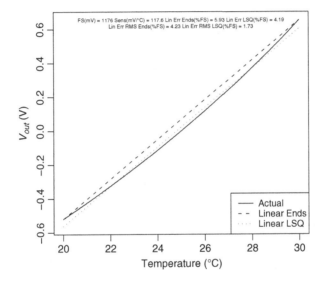

FIGURE 4.6 Output voltage of the balance source voltage divider with linearized thermistor.

As a numeric example, using $V_s = 5$ V, $R_f = 10$ kΩ, $S = -0.46$ kΩ/°C, we obtain a transducer sensitivity of $\dfrac{dV}{dT} \approx -(-0.46)\dfrac{5}{2 \times 10} = 115$ mV/°C.

This is of course only an approximation valid for small values of dR. In Lab 4, of the companin Lab Manual we write an R program to analyze the response with more detail using Equation (4.9) and the linearized thermistor to obtain Figure 4.6 assuming the same parameter values.

When using this voltage divider, the inverse calculation of temperature from output voltage changes in the following manner. First, obtain the thermistor resistance given the output voltage from Equation (4.8). Start by rearranging terms $R_t + R_f = \dfrac{V_s}{V_{out}}\left(R_f - R_t\right)$, then isolating terms in R_t

$\left(\dfrac{V_s}{V_{out}} + 1\right)R_t = R_f\left(\dfrac{V_s}{V_{out}} - 1\right)$, which leads to the thermistor resistance

$$R_t = R_f \frac{\left(\dfrac{V_s}{V_{out}} - 1\right)}{\left(\dfrac{V_s}{V_{out}} + 1\right)} \tag{4.12}$$

Equation (4.10) could be used to derive a simple approximation for the thermistor resistance

$$R_t = R_0 + dR \approx R_0 - \frac{2V_{out}R_f}{V_s}$$
(4.13)

To illustrate numerically, when $dR \ll R_f$, we would have $V_{out} \ll V_s$. Take for example $V_{out} = 0.1\ V_s$. Assume $R_f = R_0$ for simplicity. Then Equation (4.12) yields $R_t = 0.818 \times R_f$, whereas Equation (4.13) yields $R_t \approx 0.8 \times R_f$. As dR becomes much smaller, say $V_{out} = 0.01\ V_s$, Equation (4.12) yields $R_t = 0.98 \times R_f$ and Equation (4.13) yields the same value $R_t \approx 0.98 \times R_f$.

As derived in Chapter 3, once we have the thermistor resistance, we complete the inverse calculation of temperature with

$$T = \left(T_0^{-1} + B^{-1} \ln\left(\frac{R_t}{R_0} \right) \right)^{-1}$$
(4.14)

obtained from the B model and requiring thermistor parameters T_0, B, and R_0.

ONE-SENSOR CIRCUIT: QUARTER-BRIDGE

Now consider a Wheatstone bridge circuit (Figure 4.7). Using voltage division twice, we can determine the output voltage as the potential difference between the two midpoints

$$V_{out} = V_s \left(\frac{R_f}{R_t + R_f} - \frac{R_f}{R_f + R_f} \right) = V_s \frac{R_f - R_t}{2 \times (R_t + R_f)}$$
(4.15)

As we have usually done, we select the fixed resistor R_f to have a value equal to R_0, which is the nominal value of a resistive sensor at the nominal value of the measurand. For example, using a thermistor, the nominal value R_0 is the value of the sensor resistance R_t at the nominal operating value of temperature, say T_0. The output voltage at nominal temperature is zero as seen by $V_{out} = V_s \left(\frac{R_0 - R_0}{R_0 + R_0} \right) = 0$.

A small departure of temperature from the nominal, say $-dT$, would cause an increase $+dR$ in the thermistor resistance, and therefore the output voltage would change to

$$V_{out} = V_s \left(\frac{R_f}{R_f + R_f + dR} - \frac{R_f}{2R_f} \right) = V_s \left(\frac{2R_f - (2R_f + dR)}{2(2R_f + dR)} \right)$$

which reduces to

$$V_{out} = \frac{V_s}{4} \times \frac{-dR}{R_f + \frac{dR}{2}}$$
(4.16)

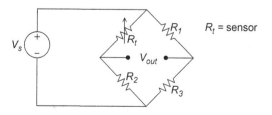

FIGURE 4.7 Wheatstone bridge circuit: quarter-bridge.

Note that this is half of the output for the balanced source divider given by Equation (4.9).

When the resistance variation dR is very small compared to the nominal, such that $2R_f \gg dR$, we have

$$V_{\text{out}} \approx -\frac{V_s}{4R_f} dR \qquad (4.17)$$

Denote by dV the departure of V_{out} with respect to zero and move dR to the left side $\dfrac{dV}{dR} \approx -\dfrac{V_s}{4R_f}$.

Naturally, the behavior of the sign of dV with respect to the sign of dR is the same as described above for the balanced source divider. Similarly, we assume that the sensor response to temperature dR/dT evaluated at the nominal value is $S = \dfrac{dR}{dT}\Big|_{T=T_0}$, therefore

$$\frac{dV}{dT} = \frac{dV}{dR}\frac{dR}{dT}\Big|_{T=T_0} \approx -\frac{V_s}{4R_f} \times S \qquad (4.18)$$

This circuit is often called a "quarter-bridge" because the Wheatstone bridge contains one sensor out of the possible four resistances. Note also that the sensitivity has the divisor four corresponding to a quarter of the total V_s/R_f, and that the linearity error in Equation (4.16) would be the same as of Equation (4.9).

As a numerical example, consider a thermistor with a nearly linear relationship of slope $S = -0.46$ kΩ/°C at a nominal $R_0 = 10$ kΩ at $T_0 = 25$°C. We build a quarter-bridge with a reference resistance of $R_f = 10$ kΩ and a voltage source of $V_s = 5$ V. Approximately, the sensitivity of the transducer is $\dfrac{dV}{dT} \approx \left(-\dfrac{V_s}{4R_f}\right)S = \left(-\dfrac{5}{4\times 10}\right)(-0.46)\dfrac{\text{Volt}}{°C} = 57.5\dfrac{\text{mV}}{°C}$, which is half of the sensitivity of the balanced source divider. This is of course only an approximation valid for small values of dR. In Lab 4, we use an R program to analyze the response in more detail using Equation (4.15) and the linearized thermistor for the same parameter values (Figure 4.8).

When using this voltage divider, the inverse calculation of temperature from output voltage changes in the following manner. First, obtain the thermistor resistance given the output voltage by

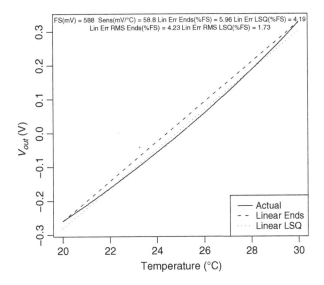

FIGURE 4.8   Output of quarter-bridge with linearized thermistor.

inverting Equation (4.15). Start by isolating terms in R_t as follows $2 \times (R_t + R_f) = \dfrac{V_s}{V_{out}}(R_f - R_t)$, then

moving terms $\left(\dfrac{V_s}{V_{out}} + 2\right) R_t = R_f \left(\dfrac{V_s}{V_{out}} - 2\right)$, which leads to the thermistor resistance calculation

$$R_t = R_f \frac{\left(\dfrac{V_s}{V_{out}} - 2\right)}{\left(\dfrac{V_s}{V_{out}} + 2\right)} \tag{4.19}$$

Equation (4.17) could be used to derive a simple approximation for the inverse equation

$$R_t = R_0 + dR \approx R_0 - \frac{4 V_{out} R_f}{V_s} \tag{4.20}$$

As described above for the balanced source divider, once we have the thermistor resistance, we complete the inverse calculation of temperature with Equation (4.14) obtained from the B model and requiring thermistor parameters T_0, B, and R_0.

TWO-SENSOR CIRCUIT: HALF-BRIDGE

We configure the bridge to have two thermistors, located in opposite arms in such a way that the change of voltage in the R_t resistor of each arm is of opposite sign (Figure 4.9). The output signal is now twice as much as in quarter-bridge circuit, as we will demonstrate shortly.

$$V_{out} = V_s \left(\frac{R_f}{R_t + R_f} - \frac{R_t}{R_t + R_f} \right)$$

Assuming that the two thermistors have identical responses, using the change dR from nominal in R_t

$$V_{out} = V_s \left(\frac{R_f}{2R_f + dR} - \frac{R_f + dR}{2R_f + dR} \right) = V_s \left(\frac{-dR}{(2R_f + dR)} \right) = \frac{V_s}{2} \left(\frac{-dR}{R_f + \dfrac{dR}{2}} \right) \tag{4.21}$$

This is the same as Equation (4.15) except that the response is twice as large, i.e., it has a divisor 2 instead of a 4. In other words, we have increased sensitivity by a factor of 2. The linearity error remains the same.

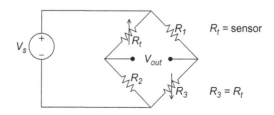

FIGURE 4.9 Wheatstone bridge circuit: half-bridge.

TWO SENSORS WITH OPPOSITE EFFECT: HALF-BRIDGE

A *strain gage* sensor is a resistor printed on a film. Because the resistor is very thin, it can stretch and shrink depending on whether we apply a tension or compression force along the longitudinal axis. Therefore, as we apply tension, the resistance increases because of elongation. However, as we apply compression, the resistance decreases because of shortening.

   We can locate two strain gages in the same arm of a bridge but having opposite responses, one in compression and the other in tension (see Figure 4.10).

$$V_{out} = V_s \left(\frac{R_{t2}}{R_{t1} + R_{t2}} - \frac{R_f}{R_f + R_f} \right)$$

Now when R_{t2} has an increase dR, the opposite occurs for R_{t1} experimenting a decrease $-dR$; therefore

$$V_{out} = V_s \left(\frac{R_f + dR}{R_f + dR + R_f - dR} - \frac{R_f}{2R_f} \right) = V_s \left(\frac{R_f + dR}{2R_f} - \frac{1}{2} \right) = \frac{V_s}{2} \left(\frac{dR}{R_f} \right) \qquad (4.22)$$

produces the beneficial effect of linear bridge response because the dR has disappeared from the denominator. The only nonlinearity error remaining is that of the sensor.

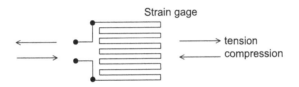

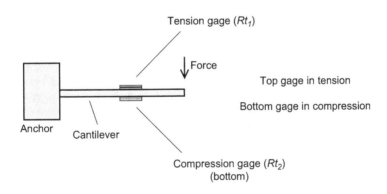

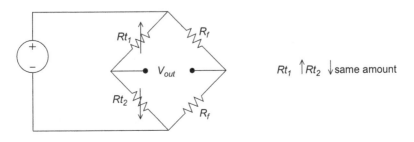

FIGURE 4.10   Strain gages and Wheatstone bridge circuit: half-bridge.

FOUR-SENSOR CIRCUIT: FULL-BRIDGE

Now we will use four strain gages, arranged as two pairs of sensors; each pair has one sensor in compression and the other one in tension, giving opposite responses (see Figure 4.11). Following a process like the one in the previous section, we can demonstrate that

$$V_{out} = V_s \left(\frac{R_f + dR}{R_f + dR + R_f - dR} - \frac{R_f - dR}{R_f + dR + R_f - dR} \right)$$

$$= V_s \left(\frac{R_f + dR}{2R_f} - \frac{R_f - dR}{2R_f} \right) = V_s \left(\frac{dR}{R_f} \right) \tag{4.23}$$

obtaining twice the sensitivity of the half-bridge, shown by Equation (4.22), and a linear circuit response. Again, the only nonlinearity error remaining is that of the sensor.

A summary of the bridge circuits we have discussed is shown in Figure 4.12. At the top-left panel (a), we have only one sensor and is described by Equation (4.16), $V_{out} = \frac{V_s}{4} \frac{-dR}{R_f + \frac{dR}{2}}$ with linearity error of 0.5%. Next, at the top-right (b), we add a second sensor in the opposite arm and make it vary in the same direction. In this case, Equation (4.21) predicts the output as $V_{out} = \frac{V_s}{2} \frac{-dR}{R_f + \frac{dR}{2}}$ which

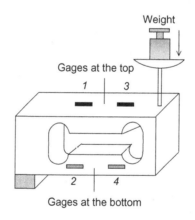

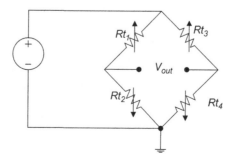

FIGURE 4.11   Strain gages form a load cell or Wheatstone bridge circuit: full-bridge.

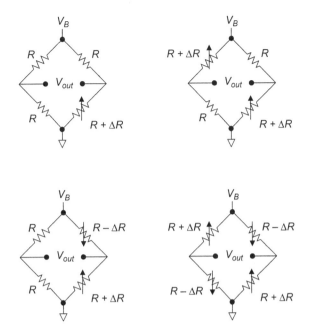

FIGURE 4.12 Summary of four bridge circuits: (a) only one sensor, (b) two sensors in opposite arms vary in the same direction, (c) two sensors in the same arm vary in opposite directions, and (d) four sensors arranged such that each pair in the same arm vary in opposite directions.

is twice as high as the one sensor bridge, and the linearity error is the same 0.5%. The bottom-left panel (c) circuit has two sensors in the same arm varying in opposite directions. In this case, we use Equation (4.22) $V_{\text{out}} = \dfrac{V_s}{2}\dfrac{dR}{R_f}$ thus achieving a null linearity error. Finally, at the bottom-right panel (d), we have four sensors arranged such that each pair in the same arm varies in opposite direction and described by Equation (4.23) $V_{\text{out}} = V_s\dfrac{dR}{R_f}$, which is twice as high in amplitude with zero linearity error.

Strain gage bridges have several applications in environmental monitoring. For example, they are used to estimate water depth from pressure, to make *lysimeters* allowing to measure soil water by weight, and to measure strain in a *dendrometer*, making it proportional to plant stem diameter increments.

ZERO ADJUST AND RANGE ADJUST

In Figure 4.13, we show a bridge circuit with a potentiometer R_z to adjust the zero when the resistances are not identical and an extra resistance R_r to adjust the range. Denoting by p the fraction of R_z in the lower branch, we have $R_r + R_z \times (1-p)$ in parallel with $R_1 = R_f$ and $R_r + R_z \times p$ in parallel with $R_3 = R_f$. In this case, the output changes to

$$V_{\text{out}} = V_s\left(\frac{R_f}{R_t + R_f} - \frac{R_f \| (R_r + R_z \times p)}{R_f \| (R_r + R_z \times p) + R_f \| (R_r + R_z \times (1-p))} \right)$$

Here we are using double vertical lines $\|$ to denote resistances in parallel. This equation can be analyzed for the effect of R_r and R_z, but for brevity we will not discuss it in this book.

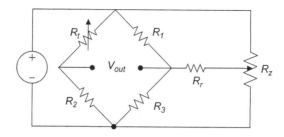

FIGURE 4.13   Wheatstone bridge with offset and range adjustments.

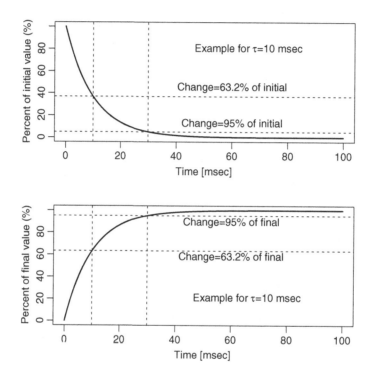

FIGURE 4.14   Transducer time constant.

SENSOR SPECIFICATIONS

As discussed in Chapter 3, we can distinguish three major types of specifications (Brown and Musil 2004). These are static, dynamic, and environmental specifications. We covered static specifications in Chapter 3. Dynamic specifications are those applying when the sensor is not at steady state conditions or the measurand is changing. It is more relevant when the principle of operation depends on achieving equilibrium, as happens in electrochemical sensors as we describe in the next section. We will discuss three dynamic specifications: response time, time constant, and damping ratio.

Response time is the time required to respond to 95% of imposed step change, whereas a similar measure, *time constant* denoted by Greek letter τ (tau), is the time required to respond to 63.2% (value derived from exponential rate) of the imposed step change. See Figure 4.14. *Damping ratio*, denoted by Greek letter ζ (zeta), is the ratio of actual damping to critical damping. It is a coefficient in a second-order differential equation that arises from the response of electrical circuits with inductance and capacitance or from mechanical systems with mass and spring

$$\frac{d^2 x}{dt^2} + 2\zeta\omega_n \frac{dx}{dt} + \omega_n^2 x = 0 \tag{4.24}$$

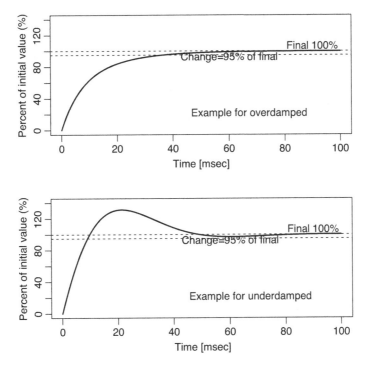

FIGURE 4.15 Transducer damping ratio.

where x is the displacement in mechanical systems or voltage in electrical systems, ω_n is the natural frequency of oscillation. The response can be underdamped for $\zeta < 1$, overdamped for $\zeta > 1$, or critically damped for $\zeta = 1$. See Figure 4.15.

Environmental specifications relate to performance under harsh conditions, such as:

- *Temperature*: often given as an operating range, say −10°C to +50°C, and sometimes given as sensitivity to extreme temperature.
- *Relative Humidity*: often given as an operating range, say 0%–90%, and sometimes given as tolerance to condensation.
- *Pressure*: effects of either high pressure (e.g., when submerged) or low pressure (e.g., at high elevation).
- *Vibration*: given by amplitude and frequency of vibrations that the transducer can tolerate.

ELECTROCHEMICAL SENSORS

Electrochemical sensors produce a voltage as a response to a concentration change in a chemical sample. Examples include *ion-selective electrodes* that generate a voltage based on a specific ion concentration difference between two phases separated by a membrane or interface selective to the ion under measurement (Pallás-Areny 2000; Vanýsek 2004).

According to the *Nernst* equation, we can calculate the voltage or emf E at which the electric field due to charge separation balances or equilibrates with the concentration gradient

$$E = \frac{RT}{zF} \ln\left(\frac{a_1}{a_2}\right) = 2.303 \frac{RT}{zF} \log\left(\frac{a_1}{a_2}\right) \tag{4.25}$$

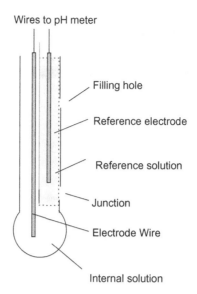

FIGURE 4.16   Glass electrode.

where R is the gas constant (8.31 J/mol K), T is the temperature in K, F is the Faraday constant (96,500 C), z is the valence of the ion of interest, and a_1, a_2 are the activities on each side of the interface. For liquid solutions, activity is concentration multiplied by the activity coefficient. In general, for low concentrations, the activity coefficient is near unity and therefore activity is approximately the same as concentration.

Glass electrodes are electrochemical sensors based on a solution-filled glass bulb. The glass allows an ion-exchange process with the solution, thus producing an electrical potential, which is a function of the ion concentration of the sample. The thickness of the glass is important because it affects electrical resistance. For example, to measure pH or the concentration of hydronium ions, a glass electrode is composed of alkali metal ions, which through an ion-exchange process with hydrogen ions in the solution produce an electrical potential proportional to the logarithm of the concentration of free hydrogen ions.

The typical glass probe has two electrodes in one device (Figure 4.16). One electrode is the sensor electrode, which has the ion-selective interface and contains a solution with a known concentration of the ion under measurement. The other electrode is a reference electrode, which has a porous interface, allowing free diffusion between the sample and its own solution, thus no concentration difference. Therefore, this electrode produces a constant potential reference. The difference of potential between the two electrodes is the voltage from the probe, which is then a function of concentration difference between the sensor electrode solution and the sample.

Using the Nernst Equation (4.25), the electric potential is

$$E = E_0 + \frac{RT}{zF}\ln(a) = E_0 + 2.303\frac{RT}{zF}\log(a) \tag{4.26}$$

where E_0 is a constant (the standard electrode potential) and a is the activity of the ion in the sample.

For example, for pH, the ion of interest is hydronium, with $z=1$, and $\log(a)=$pH

$$E = E_0 + \frac{2.303 \times RT}{F}\text{pH} = E_0 + k(T) \times \text{pH} \tag{4.27}$$

At 25°C, this voltage sensitivity k is 59.12 mV per pH unit, so the range is from about -7×60 mV $= -420$ mV to about $+7 \times 60$ mV $= +420$ mV. Because the probe has relatively high output impedance, it must connect to high input impedance circuits (~MΩ).

Glass electrode probes response time depends on wall thickness and geometrical shape of the bulb because this affects impedance. Spherical bulb probes respond to 95% of the value in one second, whereas flat bulbs may take longer (~5 seconds). Hemispherical bulbs are intermediate in response time.

Glass electrode probes must be stored moist and calibrated often. A probe will age, and its performance will degrade with time and therefore has a finite usable life. In addition to shape and wall thickness, response time of a pH probe depends on surface conditions; therefore, response time increases with probe age (Ross 2014).

Electrochemical sensors have applications in water quality, soil chemistry, and in general environmental monitoring involving chemical processes (Artiola 2004). It is often of interest to measure oxidation-reduction potential (ORP) of liquids. An ORP sensor produces a voltage proportional to the potential of the measured solution to act as oxidizing or reducing (loss or gain of electrons from other substances). ORP is related to pH and the ORP sensor is similar to the pH sensor but made to produce a voltage proportional to ORP instead of pH.

Electrochemical sensors can also be used for air quality measurements. The gas to be analyzed is dissolved in a solution at a known pH, which is passed through ion-selective electrode. The ion concentration proportional to the concentration of pollutant is absorbed and measured electronically.

EXAMPLE: DYNAMIC SPECIFICATIONS AND A POTENTIOMETER-BASED WIND DIRECTION

A wind vane converts wind direction to circular position, which moves the slider of a low-torque potentiometer (Figure 4.17). Therefore, resistance changes with wind direction. The potentiometer is powered by a voltage source, say 12 V, and the output voltage signal becomes a function of position since the potentiometer itself is a voltage divider. There is a threshold of wind speed to start movement. This threshold must be low, for example, 0.2 m/s. Dynamic considerations are important. The sensor must have fast dynamic response and an appropriate damping ratio (say 0.25–0.5) to minimize oscillation persistence but obtain fast response. In addition, it has to survive high wind speed (say 60–100 m/s).

The desired operating range is 0–360°, but typically, for a 360° range there is a dead zone of 10–20° or a north gap (or going from 0 to 360). Some designs allow eliminating the gap by using signal conditioning (topic covered later in this chapter) and data processing (Chapter 5). A heating element that is regulated using a temperature sensor helps to withstand icing conditions.

DIELECTRIC PROPERTIES

When we place an insulator or non-conductive material (glass, plastic, ceramic) between two plates of conductive material (metal), we make a capacitor. The rate of change of charge dq/dt on the plates increases with the rate of change of voltage dv/dt across the plates according to capacitance C, i.e.,

$$i = \frac{dq}{dt} = C\frac{dv}{dt} \tag{4.28}$$

Capacitance has units of Farad (F) and depends on the *dielectric constant* or *permittivity* ε of the insulator material and the geometrical arrangement according to

$$C = \varepsilon\frac{A}{d} \tag{4.29}$$

where A is the area of the plates and d is the separation between the plates. Permittivity is the ratio of electric flux density and the electric field strength E; it has units of F/m. It reflects the relation between the electric field E and charge separation or displacement in opposite directions within

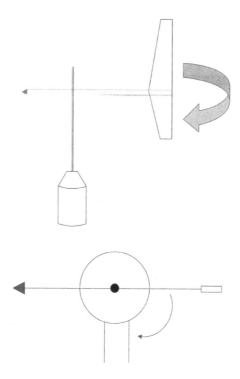

FIGURE 4.17   Wind vane and potentiometer to measure wind direction.

the material. Permittivity of vacuum ε_0 is 8.85 pF/m. A dielectric material has permittivity much higher than vacuum and thus its relative permittivity $\varepsilon_r = \dfrac{\varepsilon}{\varepsilon_0}$ is much larger than 1 (Pallás-Areny 2000). Note that we can write $C = \varepsilon_r C_0$ where C_0 is the capacitance obtained when there is a vacuum between the plates.

Many insulators show changes in permittivity (and a decrease in resistivity) as their water content increases. Likewise, permittivity changes with applied pressure and temperature. Therefore, besides the simple use as insulator, dielectric materials can be used as sensors. In this case, we seek to affect the dielectric property of the sensor by the measurand; for example, changes of dielectric properties of ceramic or plastic by moisture, pressure, and temperature. A common environmental application is in measuring soil moisture content, since dielectric of soil changes with water content. We will discuss this application with more detail in Chapter 13.

EXAMPLE: PIEZOELECTRIC SENSORS

The piezoelectric effect is reversible and consists of electric polarization of material under stress and its reverse, a strain due to the application of a voltage (Pallás-Areny 2000). These materials can be used to measure pressure by producing a voltage, which depends on pressure. They also offer an opportunity to generate a small amount of power for sensor networks (Chapter 8).

EXAMPLE: SOIL TENSIOMETER

In addition to measuring soil water content, it is important to measure soil matric potential Ψ_T, which is given in terms of negative pressure in kPa. One device to measure this variable is built with an access tube, a porous cup, and a pressure gage. Water moves in and out of the cup due to changes in matric potential, and because the tube is sealed, the pressure inside represents the matric potential (Yolcubal et al. 2004). In some tensiometers the pressure is sensed by piezoresistive sensors that are inserted in a Wheatstone bridge to convert the change of resistance into an output voltage V_{out} that depends on V_{in}, the bridge voltage source (Cobos 2007; UMS 2011). Figure 4.18 shows the sensor ceramic cup and three tensiometers installed at various soil depths.

SIGNAL CONDITIONING

By signal conditioning, we mean the process of improving the quality of the transducer's output signal. This includes linearization of the circuit response, amplification in cases when the signal may be too small, filtering when we need to reduce noise or unwanted signals that may mask the real signal, and to provide regulated voltage for the transducer. Signal conditioning achieves compatibility between the transducer and the subsequent stages of data acquisition. In Chapters 2 and 3, we have discussed converting the output of the transducer to a digital value that can be input to an MCU for processing and storage (Figure 4.19). We will further discuss A/D conversion in terms of signal conditioning. Besides linearization by circuit techniques covered in this chapter, we can correct nonlinearity of the transducer using software in the MCU.

OPERATIONAL AMPLIFIERS

This section provides a very basic and quick review of operational amplifiers (op-amps for short) and the circuits we can build using op-amps, such as non-inverting, inverting, and differential amplifiers (Sheingold 1980).

An op-amp has two inputs, inverting and non-inverting, denoted by – and + signs on circuit diagrams, one output, and two power connections, positive and negative, denoted often by +Vcc and

FIGURE 4.18   Soil tensiometers: (a): ceramic cup. (b): three tensiometers installed at various depths.

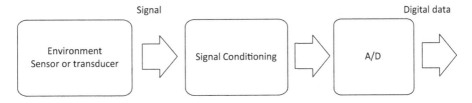

FIGURE 4.19   From transducer analog signal to a digital signal.

$-Vcc$, which are connected to the power rails. Whenever we do not need negative swing of the output, it suffices to have only positive power rail and ground.

We connect circuit elements to the input and output pins to configure a variety of amplifiers. Central to circuit configuration is the concept of feedback and a simple model of an op-amp; related to the op-amp's high input impedance and low output impedance. This simple model is that current drawn by the op-amp at the inputs (inverting and non-inverting) is negligible and that these two pins are virtually at the same voltage. We can use feedback by connecting the output via a circuit element to the inverting input. In op-amp circuit diagrams, many times we omit the required power supply connections for schematic clarity.

For example, consider the inverting amplifier (Figure 4.20). Because the two inputs are virtually at the same voltage, the inverting output is near ground. Because there is negligible current drawn at the inverting input, the current $\dfrac{V_{in} - 0}{R_1}$ through the input resistance R_1 is the same as the current $\dfrac{-V_{out} - 0}{R_2}$ through the feedback resistance R_2. Therefore, we find the gain $\dfrac{V_{out}}{V_{in}}$ as the ratio of feedback to input resistance

$$\frac{V_{out}}{V_{in}} = -\frac{R_2}{R_1} \tag{4.30}$$

Following similar reasoning, we can derive the gain of the non-inverting amplifier as

$$\frac{V_{out}}{V_{in}} = 1 + \frac{R_2}{R_1} \tag{4.31}$$

The differential amplifier (Figure 4.21) has two arms

$$V_{out} = \frac{R_2}{R_1}\left(1 + \frac{R_1}{R_2}\right)\left(\frac{R_4}{R_4 + R_3}\right)V_2 - \frac{R_2}{R_1}V_1 \tag{4.32}$$

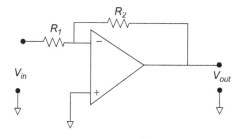

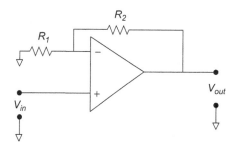

FIGURE 4.20   Inverting (a) and non-inverting (b) amplifier using an op-amp.

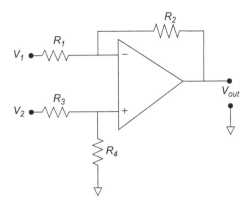

FIGURE 4.21   Differential amplifier using an op-amp.

As example, the LM741 is a widely used op-amp due to its low cost and versatile characteristics. It is composed of 18 transistors integrated together on a silicon chip inserted into an 8-pin package. The LM741 op-amp has limited output current and can yield medium gain (20–100) for many signals.

LINEARIZATION OF THE BRIDGE CIRCUIT OUTPUT

One use of op-amps is to linearize the output of a bridge. Bear in mind that using this approach, we only linearize the circuit not the sensor response, i.e., not the change in resistance vs. the measurand. Figure 4.22 shows possible method to linearize the response of a quarter-bridge and a half-bridge (Kester 2015; Pallás-Areny 2000). The top circuit is a differential amplifier with V_s input in both arms and the sensor in the feedback loop. It forces the difference at midpoint of the bridge to be zero.

$$V_{\text{out}} = \frac{R_f + dR}{R_f}\left(1 + \frac{R_f}{R_f + dR}\right)\left(\frac{R_f}{2R_f}\right)V_s - \frac{R_f + dR}{R_f}V_s \qquad (4.33)$$

Note that when $dR=0$, we get zero V_{out} response

$$V_{\text{out}} = \frac{R_f}{R_f}\left(1 + \frac{R_f}{R_f}\right)\left(\frac{R_f}{2R_f}\right)V_s - \frac{R_f}{R_f}V_s = V_s - V_s = 0 \qquad (4.34)$$

We can simplify Equation (4.33) to be

$$V_{\text{out}} = \frac{R_f + dR}{R_f}V_s\left(\left(1 + \frac{R_f}{R_f + dR}\right)\left(\frac{R_f}{2R_f}\right) - 1\right) = -\frac{V_s}{2}\frac{dR}{R_f} \qquad (4.35)$$

This is the same as $V_{\text{out}} = -\frac{V_s}{2} \times \frac{R_t - R_0}{R_f}$. Interestingly, the response is linear with respect to dR and the denominator has been reduced by half, i.e., from 4 to 2. In other words, we get the same magnitude of the response as a two-sensor bridge (half-bridge), but it has the advantage of being linear with respect to dR and of using only one sensor. Because the output is that of an op-amp, it has low impedance, and this is convenient for the subsequent processing of the signal.

In Lab 4, we use the R system to analyze the linearized quarter-bridge and obtain the output voltage plus sensitivity and linearity error. As an example, we use a thermistor with slope $S=-0.46$ kΩ/°C, nominal $R_0=10$ kΩ, $T_0=25$°C in a bridge with $R_f=10$ kΩ, and a voltage source of $V_s=5$ V. Figure 4.23

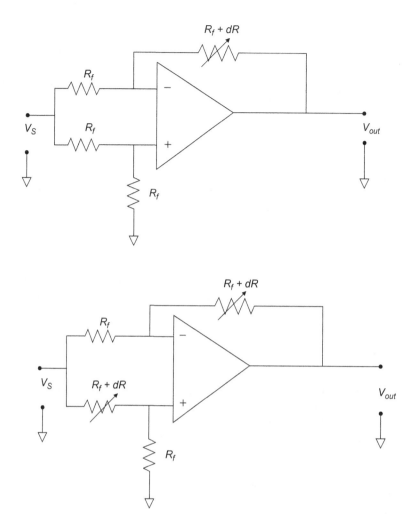

FIGURE 4.22   Using an op-amp to linearize a bridge output signal. Top: quarter-bridge, Bottom: half-bridge.

shows the output voltage with sensitivity 115.5 mV/°C and low linearity error. It is important to realize that the op-amp only linearizes the circuit and that therefore if we were to consider the full nonlinear response of the thermistor, the output would be nonlinear. To emphasize this point, we show Figure 4.24 produced with the B model for the thermistor, where we see higher linearity error.

To linearize a two-sensor bridge, we can use the same approach (Figure 4.22). In this case

$$V_{\text{out}} = \frac{R_f + dR}{R_f}\left(1 + \frac{R_f}{R_f + dR}\right)\left(\frac{R_f}{(R_f + dR) + R_f}\right)V_s - \frac{R_f + dR}{R_f}V_s$$

We can demonstrate that

$$V_{\text{out}} = -\frac{V_s}{R_f}dR$$

This has twice the sensitivity as the previous circuit since the denominator has gone from 2 to 1. In essence, we get the same magnitude of the response as a four-sensor bridge (full-bridge) while still preserving the advantage of being linear when using only two sensors. The op-amp will have

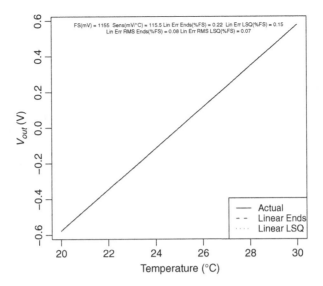

FIGURE 4.23   Output of the linearized quarter-bridge with linearized thermistor.

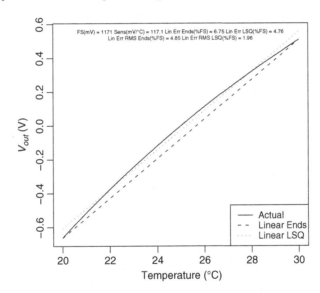

FIGURE 4.24   Op-amp linearized bridge output using the B parameter model for the thermistor.

to have both negative and positive power rails because the output must be able to swing negative. Because the output is that of an op-amp, it has low impedance. Do not forget that we have only linearized the circuit, and therefore the nonlinear response of dR vs. the measurand remains.

When using the linearized quarter-bridge, the inverse calculation of temperature from output voltage changes in the following manner. First, obtain the thermistor resistance given the output voltage, we write the thermistor resistance as $R_t = R_0 + dR$ and use Equation (4.35) to solve for dR. We obtain

$$R_t = R_0 + dR = R_0 - \frac{2V_{\text{out}}R_f}{V_s} \tag{4.36}$$

As described above for the balanced source divider, and the quarter-bridge, once we have the thermistor resistance, we complete the inverse calculation of temperature with Equation (4.14) obtained from the B model and requiring thermistor parameters T_0, B, and R_0.

COMMON-MODE REJECTION

Ideally, a differential amplifier would be able to eliminate unwanted signal in the bridge circuit because the signs of the signals are opposite for different arms of the bridge circuit. This property is the Common Mode Rejection (CMR); in other words, canceling out the unwanted signal in both arms of the bridge. The common mode signal is amplified but with much lower gain A_{cm} than the differential gain A_d.

$$V_{out} = A_d(V_+ - V_-) + \frac{1}{2}A_{cm}(V_+ - V_-) \tag{4.37}$$

The CMR is specified as the ratio of both gains expressed in dB, i.e., 20 times the log base 10 of the ratio.

$$CMR = 20\log_{10}\left(\frac{A_d}{|A_{cm}|}\right) \tag{4.38}$$

Thus, the higher this number the better is the amplifier in terms of minimizing the common-mode signal. For example, when the ratio of gains is 10, we obtain 20 dB, when the ratio is 100, we get double the dB, which is 40 dB. The CMR is an important specification that varies with gain of the amplifier (Morrison 1992).

INSTRUMENTATION AMPLIFIER

An instrumentation amplifier (in-amp for short) is similar to a differential amplifier but includes high input impedance in each arm, which is accomplished with op-amps (Kitchin and Counts 2006; Pallás-Areny 2000). One great advantage of the in-amp is that it keeps the bridge balanced. We connect the output of a quarter-bridge to an instrumentation amplifier (Figure 4.25).

Denoting by G the gain of the amplifier, the output is now

$$V_{out} = \frac{V_s}{4}\left(\frac{-dR}{R_f + \frac{dR}{2}}\right)G$$

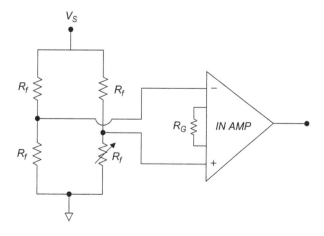

FIGURE 4.25   Amplifying the bridge output signal using an in-amp.

This gain G is set with a single resistor R_G. An in-amp also has excellent CMR. For example, some in-amps can have about 100-dB CMR for a gain programmed at $G=10$.

When using an amplifier of gain G to increase the magnitude of V_{out}, the thermistor resistance equation for the inverse calculation of temperature needs to be modified by the gain G. For example, for the quarter-bridge using the simplified equation (4.20)

$$R_t \approx R_0 - \frac{4V_{out}R_f}{V_s \times G} \tag{4.39}$$

Suppose that a quarter-bridge output has a range of –10 to 10 mV, or FS of 20 mV and it connects to an in-amp to amplify the output to a signal varying between –100 and 100 mV, or a resulting FS of 200 mV. Therefore, we need a gain of $G = 200/20 = 10$. Suppose we employ an in-amp with a gain formula of $G = \dfrac{49.4\ k\Omega}{R_G} + 1$. We would calculate the required R_G from this formula to obtain the desired G of 10.

SPECTRUM

Let us take the time to introduce the concept of signal *spectrum*, which we need to explain the effect of interference of unwanted signals with monitoring equipment. First, note that a signal $x(t)$ has a *time-domain* representation that is simply the amplitude that varies with time. The variation of this amplitude has time periodicities that are captured as a variety of frequencies expressed in Hertz (Hz) or s^{-1}. The spectrum of the signal is formed by the values of all frequencies present in the signal. It allows for a *frequency-domain* representation of the signal, which consists of amplitude and phase that varies with frequency.

Mathematically, this conversion from time-domain to frequency-domain is obtained by the *Fourier transform*. This general concept applies to a variety of processes, such as electromagnetic and sound waves. When applied to electromagnetic waves, the spectrum includes important regions such as ultraviolet, visible light, infrared, and radio frequency waves.

Please recall from Chapter 3 that pure alternating current (AC) is a sine wave of a given frequency. We can express it mathematically as $x(t) = A_x \cos(\omega t)$ where A_x is the amplitude and ω is the angular frequency and given by $\omega = 2\pi f$ in units of rad/s, where f is the frequency. For example, in the United States, AC power has $f = 60$ Hz and $\omega = 2\pi \times 60 = 377$ rad/s. When a signal has only one frequency, say 60 Hz, the spectrum is simply a spike at this frequency. A relatively more complicated spectrum occurs when including multiples of this fundamental frequency or *harmonics* that would show in the spectrum as multiple spikes. Simply put, the signal is the sum of sinusoids at this set of frequencies. This is the basis of *Fourier series*. Signals that are more complicated would include a continuous of frequencies and not simply spikes at given frequencies and can be represented by the Fourier transform.

NOISE

In general, noise is an unwanted signal. In monitoring equipment, this can occur from radio signals and power lines. It can be continuous such as a power hum, random such as spurious radio signals, or transient such as the operation of a power switch (Morrison 1992, 1986). Many sources of noise are random yet repetitive over a longer period.

One way to characterize noise is by the RMS or root mean square

$$X = \left(\frac{1}{T} \int_0^T x^2(t)\,dt \right)^{\frac{1}{2}} \tag{4.40}$$

Certain periodic waveforms have easy-to-recall RMS values. For example, a sinusoidal signal described by the waveform $x(t) = A_x \cos(2\pi ft)$ has the RMS value of $A_x/\sqrt{2}$ or 70% of the amplitude A_x. Here f is the frequency and is given in Hz. An example of a power line hum would be a sinusoidal at 60 Hz.

Let us consider a signal containing harmonics of the fundamental frequency. For example, if the fundamental is 60 Hz, and a signal has significant components at 180 and 300 Hz, we say that it has a third harmonic and fifth harmonic.

$$v(t) = \underbrace{v_0}_{\text{DC}} + \underbrace{v_1 \cos \omega t}_{\text{Fundamental}} + \overbrace{v_2 \cos 2\omega t}^{\text{Even}} + \overbrace{v_3 \cos 3\omega t}^{\text{Odd}} + \cdots \qquad (4.41)$$

When we have only odd harmonics

$$v(t) = v_1 \cos \omega t + v_3 \cos 3\omega t + v_5 \cos 5\omega t + \cdots \qquad (4.42)$$

The total RMS is the square root of the sum of all squares

$$I = \sqrt{I_1^2 + I_2^2 + I_3^2 + \cdots} \qquad (4.43)$$

Total harmonic distortion (THD) is with respect to the fundamental

$$\text{THD} = \frac{\sqrt{I_2^2 + I_3^2 + I_5^2 \cdots}}{I_1} \qquad (4.44)$$

that is the same as the square root of the sum of the squares of the ratios of the harmonics to the fundamental

$$\text{THD} = \sqrt{\left(\frac{I_2}{I_1}\right)^2 + \left(\frac{I_3}{I_1}\right)^2 + \left(\frac{I_5}{I_1}\right)^2 + \cdots} \qquad (4.45)$$

ELECTRIC FIELD AND ELECTROSTATIC SHIELDING

The electric and magnetic fields control electrical phenomena and therefore affect a monitoring system. Voltages have associated electrical fields and currents have associated magnetic fields (Morrison 1992, 1986). The electric field is the basis for understanding how noise affects analog circuits and thus becomes important for monitoring. Although magnetic field interference becomes more relevant at frequencies above 100 kHz, monitoring equipment can be affected by the magnetic field near a transformer.

Figure 4.26 (top) illustrates how the electric field of a conductor at potential V_1 has lines directed to ground (reference voltage, $V = 0$). The bottom part of the same figure shows that these field lines can be isolated and contained using conductor 2 as a shield. Most electric fields are contained within capacitors. Non-contained fields can cause interference and therefore the geometry setting is important to understand interference and its control (Morrison 1992). Fields that interfere with a circuit performance originate from power frequencies and their harmonics are considered near field; other random sources from transmitters are far fields. Frequency has an important implication for this distinction.

Circuits are enclosed in a metal conducting enclosure to terminate all electric fields on the enclosure, which is connected to ground to avoid coupling unwanted fields to the circuit. Whenever a wire leaves the enclosure, this wire may pick up interference. To avoid this effect, we surround the conductor by a shield, which is connected to the ground. This way the interference field flows into

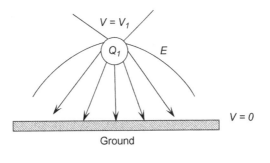

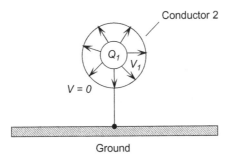

FIGURE 4.26 Top Electric field between conductor 1 and ground. Bottom Conductor 2 shields the electric field of conductor 1.

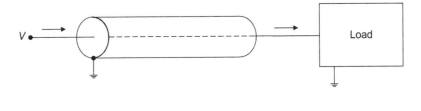

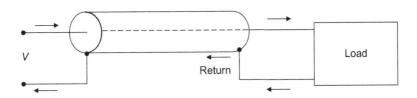

FIGURE 4.27 Grounded shield (Top) and coaxial cable (Bottom).

shield instead of the signal conductor. The enclosure is connected to the reference conductor only once to avoid current loops. The cable shield connected to ground at a single point is adequate for low-frequency signals (Figure 4.27).

However, a *coaxial cable* is needed for high-frequency signals. In order to confine the fields emanating from a pair of wires, one of the conductors wraps around the other as a shield or sheath, forming a coaxial line. A dielectric (non-conductor material) isolates the wires. The sheath can be a braid or a smooth tube. The current in the center conductor should return on the outer sheath for a truly coaxial operation (Figure 4.27 bottom). Otherwise, if current returns on another path, the fields leave the conductors. In other words, most of the current flows on the inside surface of the sheath where the field terminates. Therefore, both ends of the coaxial cable are connected, the use of more expensive coaxial cable is important for high-frequency signals. A coaxial cable grounded at both ends allows for other signal return paths (Figure 4.27 bottom). However, if the fields are confined to the cable, the current will not return via those paths.

ISOLATION

In signal conditioning terms, *isolation* consists of passing the electrical signal from the transducer to the next stage without an electrical connection. Isolation can be performed by using various media and devices, such as magnetic field (e.g., a transformer), and light (optical coupling). Advantages of isolation include breaking ground loops, blocking high-voltage surges, and rejecting high common-mode voltages.

COLD-JUNCTION COMPENSATION

One specific signal conditioning technique is cold-junction compensation, which is required when using thermocouples as sensors (Sheingold 1980). Recall that thermocouples sense temperature as the difference in voltage when joining two different metals (see Chapter 3). The same principle applies at the connection between the thermocouple and the next circuit in the system. Cold-junction compensation provides the temperature at this connection for appropriate correction. This is done for example as a step before using an in-amp (Figure 4.28) (Kitchin and Counts 2006).

A/D CONVERTER (ADC)

We have discussed ADCs in Chapters 2 and 3. In this chapter, we cover additional details. ADCs have various capabilities to accept analog inputs: *unipolar* or *bipolar*, *single-ended* or *differential*. Unipolar means that the voltage is always positive (say from 0 to 1 V), whereas for bipolar, the voltage can be positive and negative (say from −0.5 to 0.5 V). ADCs able to read bipolar signals are configured to encode signed numbers, e.g., use "two's complement" which means one bit is reserved for the sign. Thus, half the numbers are negative and the other half positive. For example, using 11 bits, you will have 10 bits and the sign bit. A single-ended input is a single input voltage referred to ground, whereas a differential input consists of two input voltages and the converted signal is the difference between the two voltages.

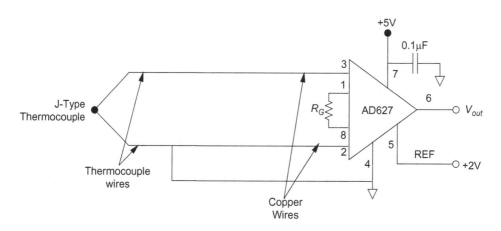

FIGURE 4.28 Cold-junction compensation using an in-amp.

Other ADC specifications are the sampling rate in *SPS* (samples per second), and *signal-to-noise ratio* (SNR) which we discuss next. For many environmental applications, a single conversion is sufficient, however some applications (e.g., reading output of a spectrometer) demand many conversions.

In some applications, the analog signal is sampled at a fixed rate, say $fs = 1/T$ where T is the sampling interval and each time converted to digital. In these cases, an ADC is also specified by bandwidth, which is established by the sampling rate. To reconstruct the original signal from its sampled version, the sampler should have a rate fast enough to capture the fast-varying components of the signal. Being more specific, the sampling criterion requires the sampler to operate at the *Nyquist rate*, which is twice or double the highest frequency component of the signal. For example, a 60-Hz sinusoid would demand the sampler to operate at 120-Hz sampling rate.

The SNR is the ratio of power of the signal to the power of the noise,

$$\text{SNR} = \frac{P_{\text{signal}}}{P_{\text{noise}}} \tag{4.46}$$

alternatively, assuming the same impedance for both, it is the ratio of the squares of the waveform RMS amplitude

$$\text{SNR} = \left(\frac{A_{\text{signal}}}{A_{\text{noise}}} \right)^2 \tag{4.47}$$

Expressing in dB

$$\text{SNR} = 20 \log_{10} \left(\frac{A_{\text{signal}}}{A_{\text{noise}}} \right) \text{dB} \tag{4.48}$$

For example, a 10-fold in amplitude from signal to noise means SNR=20dB, and 100-fold means 40dB.

Consider an ADC with n bits. At any step, the *quantization noise* is distributed within 1 least significant bit (LSB), or taking the midpoint of the step the error is distributed between −1/2 LSB and 1/2 LSB. The signal-to-quantization-noise ratio or SQNR is

$$\text{SQNR} = 20 \log_{10} \left(2^n \right) = n20 \log_{10}(2) \approx 6n \text{ dB} \tag{4.49}$$

We can think of this metric as the maximum SNR for an ADC. For example, 8 bits would have SQNR of 48dB, whereas $n = 16$ bits would have SQNR of 96dB.

It is important to match the voltage range of transducer vs. the voltage range of ADC and select an ADC based on resolution needed.

CURRENT LOOP: 4–20 MA

A low-voltage signal from a transducer is inadequate for connection to A/D processing circuitry when the transducer is located far away from the processor because of voltage drops in the wire. One manner to send the transducer low-voltage output signal to the processor is to convert transducer output voltage to current at a signal-conditioning module before transmission. At the receiving end, this current is detected by measuring voltage drop across a resistor of known value. This voltage becomes the input to the ADC, and the resistor may not be an actual resistor but the input resistance of the ADC. A typical range of current is 4–20mA; the non-zero 4-mA value for the minimum makes easy to detect an open circuit condition since this condition corresponds to 0mA (Pallás-Areny 2000).

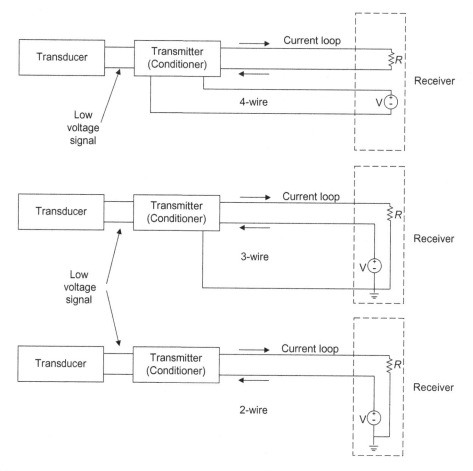

FIGURE 4.29   Current loop 4–20-mA configurations. Top: Four-wire. Middle: Three-wire. Bottom: Two-wire.

The advantage of current loop transmission of transducer output over voltage is that the signal is less sensitive to voltage drops and unwanted thermocouple effects in the lead wires. Consequently, the wires can be thinner. Using twisted pair wires reduces possible magnetic interference. Wire length effects are also minimized by using a receiver resistor of low value (e.g., 250 Ω).

Current loop configurations are four-, three-, and even two-wire (Figure 4.29). In the four-wire configuration, the power supply connects to the conditioner (transmitter) using two wires, and the conditioner to the readout resistor with two wires. In the three-wire configuration, a return wire is shared, whereas in the two-wire configuration, the supply is in series with the readout resistor and the conditioner (Pallás-Areny 2000). A series resistor is included to provide safety in the event of a short circuit. A typical supply voltage is 24 V and it needs to overcome drops in the conditioner (12 V), in the readout resistor (5 V), plus 2 V in the lines, and 5 V in the series safety resistor. The conditioner is not the source of the power for the current in the loop. Rather, it regulates the current according to the transducer signal (ACROMAG 2014).

The relationship between transducer output voltage and current is established by calibration at the conditioner representing a linear correspondence between the measurand units (engineering units) span and current span.

The AD693 (Analog Devices) is an example of Integrated Circuit (IC) for signal conditioning providing 4–20-mA transmitter capabilities. It accepts low-level inputs from a variety of transducers, such as Resistance Temperature Detectors (RTDs), bridges, and pressure sensors and generates a standard 4–20-mA, two-wire current loop. The IC has a voltage reference and auxiliary amplifier for transducer excitation.

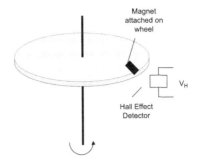

FIGURE 4.30 Translating rotation into pulses by a magnet and Hall effect sensor.

PULSE SENSORS

The output of a pulse sensor is a train of voltage pulses with varying frequency, which depends on the measurand. The transducer modulates the frequency of the pulse wave. The pulses consist of transition from low to high and high to low-voltage levels. A traditional cup anemometer is an example, as wind speed increases the cup anemometer rotates at higher spins per minute, and then we count these spins using a circuit or microprocessor. Another example is a water flow sensor that uses a propeller to measure how much liquid is moving through it. A circuit or processor counts the turns of the propeller to obtain a value proportional to water velocity.

These sensors use the concept of translating fluid movement or kinetic energy into mechanical rotation, and mechanical rotation into electrical signals. The conversion to electrical signal can be achieved with switch closures, or by a magnet and Hall effect sensors (Figure 4.30).

EXERCISES

Exercise 4.1

Consider a thermistor in a voltage divider circuit as in Figure 3.23 of Chapter 3 with $R_1 = R_f = 10$ kΩ, and the sensor is a thermistor with a nearly linear relationship of slope -100 Ω/°C in the range 20°C–30°C at a nominal $R_0 = 10$ kΩ at $T_0 = 298$ K nominal temperature. Calculate sensitivity for a source of $V_s = 12$ V.

Exercise 4.2

Consider a balanced source voltage divider circuit. The reference resistance is $R_f = 10$ kΩ, and the sensor is a thermistor with a nearly linear relationship of slope -100 Ω/°C in the range 20°C–30°C with nominal $R_0 = 10$ kΩ at $T_0 = 25$°C. Calculate approximate sensitivity for a voltage source of $V_s = 1.5$ V.

Exercise 4.3

Consider a quarter-bridge Wheatstone bridge circuit. The reference resistance is $R_f = 10$ kΩ and the sensor is a thermistor with a nearly linear relationship of slope -100 Ω/°C in the range 20°C–30°C with nominal $R_0 = 10$ kΩ at $T_0 = 25$°C. Calculate approximate sensitivity for a voltage source of $V_s = 1.5$ V.

Exercise 4.4

Suppose that a quarter-bridge output has a range of -10 to 10 mV. Determine transducer output full-scale and the in-amp gain needed to amplify the output signal to a signal varying between

−500 and 500 mV. Suppose you employ an in-amp with a gain formula of $G = \dfrac{49.4 \text{ k}\Omega}{R_G} + 1$. Calculate gain resistance R_G.

Exercise 4.5

Suppose a current signal has odd harmonics of 1, 3, and 9 with RMS values of 5, 4, and 3, respectively. What is the RMS total current? What is the THD?

Exercise 4.6

Consider an ADC of $n = 10$ bits and a full-scale analog signal of −0.5 to 0.5 V. Determine: (1) the least significant bit voltage in volts and as percent of full-scale, (2) the signed integers that can be represented, (3) the SQNR.

REFERENCES

Acevedo, M.F. 2024. *Real-Time Environmental Monitoring: Sensors and Systems,* Second Edition – Lab Manual. Boca Raton, FL: CRC Press, Taylor & Francis Group. 463 pp.

ACROMAG. 2014. *Whitepaper: Introduction to the Two-Wire Transmitter and the 4–20 MA Current Loop.* Accessed April 2015. http://www.acromag.com/sites/default/files/Acromag_Intro_TwoWire_Transmitters_4_20mA_Current_Loop_904A.pdf.

Artiola, J. F. 2004. "Environmental Chemical Properties and Processes." In *Environmental Monitoring and Characterization,* edited by J. F. Artiola, I. L. Pepper and M. L. Brusseau, 241–261. Burlington, MA: Elsevier Academic Press.

Brown, P., and S. A. Musil. 2004. "Automated Data Acquisition and Processing." In *Environmental Monitoring and Characterization,* edited by J.F. Artiola, I.L. Pepper and M.L. Brusseau, 49–67. Burlington: Academic Press.

Cobos, D. R. 2007. "Measuring UMS Tensiometers with Non-UMS Control and Data Acquisition Systems." In *Application Note,* edited by Decagon Devices. Pullman, WA: Decagon Devices.

Kester, W. 2015. *Practical Design Techniques for Sensor Signal Conditioning, Section 2 Bridge Circuits.* Accessed March 2015. http://www.analog.com/media/en/training-seminars/design-handbooks/49470200sscsect2.PDF.

Kitchin, C., and L. Counts. 2006. *A Designer's Guide to Instrumentation Amplifiers.* Norwood, MA: Analog Devices.

Morrison, R. 1986. *Grounding and Shielding Techniques in Instrumentation.* New York, NY: John Wiley & Sons.

Morrison, R. 1992. *Noise and Other Interfering Signals.* New York: John Wiley & Sons.

Pallás-Areny, R. 2000. *Amplifiers and Signal Conditioners.* Boca Raton, FL: CRC Press.

Ross, M. 2014. *pH Electrode Performance.* Accessed 2014. http://www.eutechinst.com/tips/ph/14.pdf.

Sheingold, D., ed. 1980. *Transducer Interfacing Handbook.* Norwood, MA: Analog Devices, Inc.

UMS. 2011. *T4/T4e Pressure Transducer Tensiometer.* Munich, Germany: UMS.

Vanýsek, P., 2004. *The Glass pH Electrode.* Accessed May 2023. https://www.electrochem.org/dl/interface/sum/sum04/IF6-04-Pages19-20.pdf.

Yolcubal, I., M. L. Brusseau, J. F. Artiola, P. Wierenga, and L. G. Wilson. 2004. "Environmental Physical Properties and Processes." In *Environmental Monitoring and Characterization,* edited by J. F. Artiola, I. L. Pepper and M. L. Brusseau, 207–239. Burlington: Elsevier Academic Press.

5 Dataloggers and Sensor Networks

INTRODUCTION

This chapter covers data acquisition systems (DAS) emphasizing dataloggers and sensor networks applicable to environmental monitoring. We discuss the real time clock (RTC) since it is an important element of dataloggers providing timestamps to the data records, and various serial communication interfaces with emphasis on connecting dataloggers and multiple intelligent sensors together as a sensor network. In this regard, we discuss RS-232, Universal Asynchronous Receiver Transmitter (UART), RS-485, Serial Peripheral Interface (SPI), Inter-Integrated Circuit (I2C), and Serial Digital Interface (SDI-12). We conclude with a detailed examination of wiring and programming for an example of a commercially available environmental datalogger. These circuits and topics are expanded by computer and hands-on exercises in Lab 5 of the companion Lab Manual (Acevedo 2024).

DAS

A DAS is an MCU-based system that can read sensors and transducers using signal conditioning and A/D converters (ADCs). In addition, it includes data storage and data retrieval, to be explained in this chapter, and optionally telemetry capabilities (Chapter 6). Therefore, a DAS is the first step in going from transducers to data processing components (Figure 5.1). Further data processing components include computers that will perform calculations on data for warnings and alerts, trends, archival, and other purposes.

An important example of a DAS is a *datalogger* or device that records data over time and can operate autonomously for prolonged periods. The terms datalogger and DAS are often used as synonyms. However, technically there are differences: a DAS is a more general concept that would encompass data logging. Therefore, a datalogger is a DAS, but a DAS is not always a datalogger. We characterize a datalogger as a stand-alone DAS with low-drift RTC and slow sampling rates (1–100 Hz). For the purposes of this book, we will focus on dataloggers.

DATALOGGERS

Most dataloggers employed in environmental monitoring include the capabilities of conditioning signals from transducers, convert analog responses to digital form, process the data to engineering units, a basic level of statistical summary, and store data in files that can be retrieved or transmitted when needed. In addition, many dataloggers include some level of actuator capabilities to control other devices that may be associated with the measurement system, such as switches and relays. With these added control capabilities, a datalogger is a stand-alone *measurement and control* system. An important component of a datalogger is a microprocessor that automates the major DAS functions. In Figure 5.2, thick arrows depict interaction functions of the datalogger with the power system, transducers, controlled devices, and ports to communicate with the datalogger to upload programs and to download data. Thin arrows depict the internal interactions with memory, clock, ports, signal conditioning, and ADC components.

Some dataloggers are designed to perform specific tasks; however, general-purpose dataloggers are programmable so that the users can adapt the system to their own needs. Therefore, the datalogger may have a programming language, which the user can use to write scripts or programs that will

DOI: 10.1201/9781003425496-5

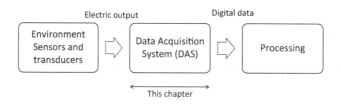

FIGURE 5.1 DAS is first step from transducers to data processing.

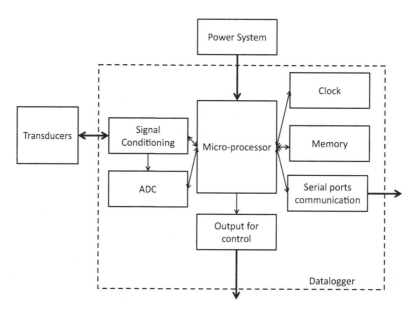

FIGURE 5.2 Block diagram of a typical datalogger.

perform the desired tasks. Many times, this language is high-level, meaning that there are functions available to facilitate writing the program.

APPLICATIONS IN ENVIRONMENTAL MONITORING

Dataloggers for environmental monitoring must operate under potentially rugged outdoor conditions. Once deployed, dataloggers should operate autonomously and should be able to remain unattended for the intended period of operation. This period can occasionally be short, but it is long in most cases.

Examples include weather stations (wind speed/direction, temperature, relative humidity (RH), solar radiation), hydrographic stations (water depth, water flow), water quality (pH, conductivity, dissolved oxygen), soil moisture (SM), air quality and emissions (ozone), offshore buoys for weather, tides, other environmental conditions, wildlife habitat conditions, groundwater levels, and many others.

ANALOG CHANNELS

Dataloggers have *analog channel* inputs to read sensors and transducers. Two major types are *single-ended* (SE) input and *differential* (DIFF) input. An SE input allows inputting the voltage of a transducer with respect to ground. A DIFF input is formed by using two SE inputs (High and Low) and measuring the voltage between these two points (Figure 5.3).

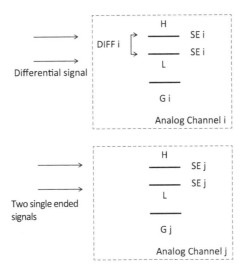

FIGURE 5.3 Typical analog input channel of a datalogger. A single voltage sensor output uses an SE terminal, whereas a DIFF signal uses both leads of the channel as a DIFF channel.

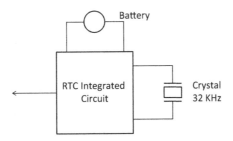

FIGURE 5.4 RTC.

RTC

Because dataloggers must provide a timestamp to each data record, they require an RTC. This electronic circuit keeps track of current time, and it is not just an element to synchronize circuits. An RTC uses a crystal oscillator of a given frequency, e.g., ~32 kHz. The RTC is powered from an alternate source such as a backup battery, so that it keeps running in the event of main power failure (Figure 5.4). For environmental applications, the RTC should be low power and have little drift. Some RTC chips include an embedded crystal.

An RTC-integrated circuit (IC) provides counters for tenths/hundredths of seconds, seconds, minutes, hours, days, date, month, year, and century. These counters are implemented as registers using BCD (binary-coded decimal) format.

Natural BCD consists of simply representing a decimal digit by four bits as follows.

```
Decimal
Digit          Binary
0              0 0 0 0
1              0 0 0 1
2              0 0 1 0
3              0 0 1 1
4              0 1 0 0
5              0 1 0 1
6              0 1 1 0
7              0 1 1 1
8              1 0 0 0
9              1 0 0 1
```

Note that taking advantage of all four bits we can represent a hexadecimal (hex) digit

```
Hex
Digit          Binary
0              0 0 0 0
1              0 0 0 1
2              0 0 1 0
3              0 0 1 1
4              0 1 0 0
5              0 1 0 1
6              0 1 1 0
7              0 1 1 1
8              1 0 0 0
9              1 0 0 1
A              1 0 1 0
B              1 0 1 1
C              1 1 0 0
D              1 1 0 1
E              1 1 1 0
F              1 1 1 1
```

Many systems have 8-bit bytes. Each byte is composed of two 4-bit *nibbles*, the low nibble (the least significant bits or rightmost bits) and the high nibble (the most significant bits or leftmost bits). BCD can be in *uncompressed* or *packed* formats. The uncompressed format uses a byte for each digit, and the digit is coded by the low nibble of the byte, whereas in the packed format, there are two digits per byte, one per nibble. Numbers exceeding the range of an even number of bytes use a contiguous byte.

For example, a four-digit number 3241 is coded by two bytes 32–41

```
   3    2      4    1
0011 0010   0100 0001
```

```
   0    5      3    2      4    1
0000 0101   0011 0010   100 0001
```

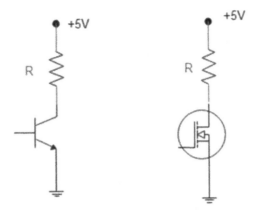

FIGURE 5.5   Pull-up resistors. Left: BJT circuit. Right: MOSFET circuit.

To represent a signed number, we use the low nibble from a byte; the sign of a number with several digits will be in the low nibble of the rightmost or least significant byte. The convention is to use hex digits A through F for the sign, A, C, and E are positive+sign, whereas B and D are negative- sign, and F is unsigned. C and D are common for+and – respectively. For example, the five-digit number +32416 is coded by three bytes as 32–41–6C where the last byte contains the sign. Number -32416 is 32–41–6D. BCD registers in an RTC will count from zero to nine and return to zero on the count of nine.

PULL-UP RESISTORS

When using a Bipolar Junction Transistor (BJT), "open-collector" means that the collector is floating. A common BJT transistor is of NPN type, constructed by a P-type semiconductor element, the base in between two N-type semiconductor elements, the emitter and the collector. For example, an NPN type transistor circuit would be able to sink current to ground from a+5-V source upon applying voltage at the base (Figure 5.5). When using a Metal–Oxide–Semiconductor Field-Effect Transistor (MOSFET), "open-drain" means the drain is floating. A field-effect transistor has three elements, the source, gate, and drain; applying voltage to the gate controls the conductivity between source and drain. For example, a MOSFET transistor circuit would be able to sink current to ground from a +5-V source upon applying a voltage at the base (Figure 5.5). The pull-up resistor is connected from the collector or the drain to a high potential or a source. The collector will acquire this potential until the transistor is turned on. Similar concepts apply to "pull-down" resistors, but these are connected to the low voltage or ground, and enabling a low state when the transistor is turned on.

SERIAL COMMUNICATION, DATALOGGERS, AND SENSOR NETWORKS

Typically, dataloggers store collected data in files, which are retrieved locally or remotely. Users copy data files to other devices but not move them; this way, files are preserved and available for retrieval by multiple users. Useful access to data files is via serial port connections, such as RS-232 and USB, or by Ethernet connections using appropriate protocols, such as TCP/IP (Transmission Control Protocol/Internet Protocol). We have discussed these in Chapter 2.

Some sensors and transducers have some digital processing capabilities of their own, e.g., intelligent sensors, and are designed to communicate serially with a datalogger. These sensors may have a MCU to specify calibration, correction, and other functions. Communication may be implemented in a variety of interface standards, such as RS-232, UART, RS-485, SPI, I2C, and SDI-12. We will discuss these serial communication methods in the next six sections.

RS-232

We discussed RS-232 in Chapter 2 focusing on connections of peripherals to PCs and communication between two PCs. A reduced RS-232 specification uses fewer lines (Figure 5.6). For example, a five-wire connection will consist of only TXD, RXD, RTS, CTS, and GND. A three-wire connection consists only of TXD, RXD, and GND, without the hardware control exerted by RTS, and CTS. This reduced connection can be used to establish a two-way connection between a sensor and a datalogger. Minimally, for one-way applications, such as a sensor sending a signal, a two-wire TXD and GND connection will suffice.

RS-485

RS-485 is a standard defined by the Electronic Industries Association and widely used communication interface in data acquisition, as well as control applications, particularly to interconnect multiple devices. It is a standard interface for the physical layer or first OSI model layer (Chapter 2). RS-485 works in differential mode; it compares the voltage difference between two lines A and B, instead of the voltage level of each line. The lines have opposite polarity, are balanced, and carried on twisted pair wires, increasing noise immunity; thus, the interface cable does not need to be shielded. With RS-485, up to 32 devices can be connected in a network, communicate over long distances (up to 1200 m) and at fast data rates; up to 35 Mbps for short distances and 100 kbps for long distances.

RS-485 allows configuring a network with a one-line multi-drop topology (Figure 5.7). At the OSI network layer or layer 3, most communication *protocols* implemented on RS-485 networks consist of one *master* node and multiple *slave* nodes (e.g., *Modbus*). This terminology implies that only the master initiates transmission and slave devices process requests and responses. However,

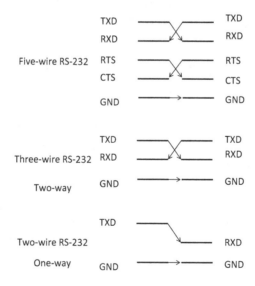

FIGURE 5.6   Reduced RS-232 connections: five-wire, three-wire, and two-wire.

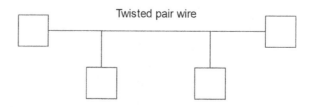

FIGURE 5.7   RS-485 network. Multi-drop topology.

RS-485 can be used for protocols allowing every node to start transmission. In this case, there may be data collision and thus requires error detection.

Electrically, each device has an RS-485 driver that can be switched to transmission mode or to receiving mode. Under a one-master multiple-slaves protocol, the master can switch on the driver on its own, whereas a slave node is in receiving mode except when allowed by the master to switch its driver on to transmission mode. Low-cost RS-485 *transceivers* suffice for small networks over short distances.

A device, such as an MCU or an SBC with UART (e.g., Arduino, Raspberry Pi), can connect to an RS-485 network using a transceiver (e.g., MAX485), with four TTL lines RO, $\overline{RE}$, DE, DI that connect to the device and the two A, B lines for the RS-485 bus. The TTL lines stand for receiver output (RO), receiver enable (RE, which is marked as with a bar or inverted because it is active when low), driver enable (DE), and data input (DI). The Tx line of the UART connects to DI and the UART Rx connects to RO. As noted above, the $\overline{RE}$ line is low when active, while DE is high when active and thus can be connected to 0 V and 5 V, respectively, if the device will be always receiving or always transmitting. Otherwise, the DE and $\overline{RE}$ lines can be connected to an I/O pin of the device to enable it for transmitting (when high=5 V) or receiving (when low=0 V). Additionally, one pull-up resistor is used for the A line and one pull-down resistor for the B line connected at only one end of the RS-485 bus; the resistance values for pull-up and pull-down resistors depend on the length of the RS-485 wires.

As mentioned above, an example of protocol for RS-485 networks is *Modbus*, which is a serial communication protocol based on one master or host and multiple slaves. Recently the terminology changed to *client* devices and *server* devices. A client initiates communication, whereas a server processes requests from clients and sends responses.

As an example of use in environmental monitoring, consider the sensor JXBS-3001-EC that combines the measurement of SM, electrical conductivity (EC), and temperature and interfaces with other sensors and dataloggers using RS-485 (Figure 5.8). We can form a network with these sensors using RS-485 and connect for example with an SBC or MCU using a driver (e.g., MAX485) as indicated above. We will discuss this example in Chapter 6 in the context of network protocols.

SPI

Serial Peripheral Interface (SPI) is synchronous, i.e., uses a clock to control serial communication, and it is based on a master-slave protocol. The master is the controlling device (e.g., MCU) and a slave (e.g., sensor) communicates as instructed by the master. Data are transferred without interruption with any number of bits in a continuous stream. There is no acknowledgment and no error checking (Circuit Basics 2022b).

There are four SPI lines: MOSI (Master Output/Slave Input) used by the master to send data to slave, MISO (Master Input/Slave Output) used by a slave to send data to the master, SCLK (Clock) that is the clock signal, and SS/CS (Slave Select/Chip Select) used by the master to select the slave which will receive data.

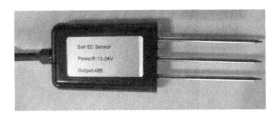

FIGURE 5.8 Example of sensor providing output based on RS-485.

Two devices, one master and one slave, would be connected directly MOSI to MOSI, MISO to MISO, SCLK to SCLK, and SS/CS to SS/CS. More than one slave can be connected using two separate CS lines as illustrated in Figure 5.9 for two slaves. However, it is possible to connect them in daisy chain where the slave 1 MISO connects to slave 2 MOSI, and slave 2 MISO connects back to the master MISO.

I2C

Inter Integrated Circuit (I2C) is a serial communication interface that combines features of SPI and UARTs. It can connect multiple slaves to a single master (like SPI) and only uses two wires to transmit data between devices (like UART) (Figure 5.10). However, I2C permits to have multiple masters connected to single or multiple slaves (Circuit Basics 2022a). For example, this allows more than one MCU logging data to a single memory card.

I2C is synchronous, meaning that bits at the output are synchronized to bits sampling by a clock line which is controlled by the master. The two lines are SDA (Serial Data) used for the master and slave to send and receive data and SCL (Serial Clock) that carries the clock signal. There is not a

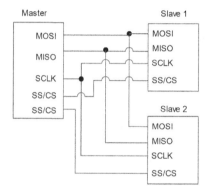

FIGURE 5.9   SPI example. One master and two slaves selected by different CS lines.

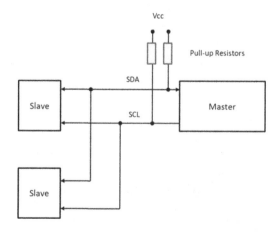

FIGURE 5.10   I2C example. One master and two slaves.

line dedicated to select a slave, and therefore device selection is achieved by embedding an address frame in the message, just after the start bit. The slave will acknowledge that its address match using an ACK (acknowledge) bit (Figure 5.11). Pull-up resistors on the SDA and SCL lines are often needed when connecting devices to an SBC or an MCU.

As an example of use in environmental monitoring, consider a set of embedded devices for water quality measurements (Atlas Scientific 2022). These devices provide UART or I2C interface between an MCU or SBC and a probe, for example pH, EC, temperature, and dissolved oxygen. Setting the devices in I2C mode and using pull-up resistors allows connecting one or more devices to the I2C bus. For example, when measuring two types of brackish water, we can use a set of EC and temperature devices for each water type and connect all four devices on the I2C bus to an SBC, such as a Raspberry pi. Once we provide a unique address to each device, we can write code (e.g., in Python) that polls readings from each device.

SDI-12

SDI is an asynchronous standard that specifies how a device (e.g., datalogger) communicates with intelligent sensors. The SDI specification includes electrical interface characteristics (such as number of conductors, voltage levels, and line impedance), communication protocol, and timing requirements.

A detector with a built-in MCU is an example of sensor producing output data in SDI-12 format (SDI-12 Support Group 2013). SDI-12 specifies three wires, one for power, one for ground, and one for serial data. SDI-12 stands for SDI at 1200 baud. SDI-12 allows low-cost and low-power operation to connect multiple sensors (Figure 5.12).

As an example of use in environmental monitoring, consider a TEROS 12 sensor that combines the measurement of SM, EC, and temperature and interfaces with other sensors and dataloggers using SDI-12 (METER 2022). We can wire multiple TEROS 12 on the SDI-12 bus and connect to an SBC, such as a Raspberry Pi, using an adapter SDI-12 to UART or USB (Liu Dr Electronic Solutions LLC 2022); for example, an SDI-12 to USB adapter to connect to the USB port of the SBC. Once we provide a unique

Message								
Start	Address	Read/ Write	ACK/ NACK	Data Frame	ACK/ NACK	Data Frame	ACK/ NACK	Stop
1 bit	7-10 bits	1 bit	1 bit	8 bits	1 bit	8 bits	1 bit	1 bit

FIGURE 5.11   I2C message.

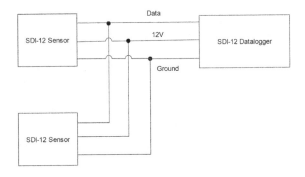

FIGURE 5.12   Basic SDI-12 wiring for two sensors. The same scheme applies for more than two sensors.

address to each device, we can write code (e.g., in Python) that polls readings from each device. From a practical point of view, it is not convenient to have many devices on the same adapter because a faulty device will bring down the entire SDI-12 bus, and it would be difficult to identify the faulty sensor.

MCUs AS DAS

MCUs are useful for monitoring because they are suited to interact with sensors (i.e., have on-chip ADC), communicate with a computer (via serial ports), and using *interrupt* signals can respond to events from the system under measurement and control. An interrupt signal suspends the current process and goes to a service routine. This is particularly useful to wake an MCU from a low-power sleep state to perform a certain task, such as to make a measurement and store data, or when connecting a computer to retrieve data. This feature allows reducing power consumption.

Figure 5.13 summarizes a typical design for an MCU-based DAS. Transducers are connected using signal conditioning as in conventional microprocessor-based DAS. Computers connect via serial communication channels. Here we assume that the MCU has on-chip RTC. However, some popular MCUs have only a timer on-board, which resets itself upon loss of power. In this case, we need to use an RTC IC or add-on to perform clock functions. The MCU requires a crystal oscillator not shown in the diagram.

Figure 5.14 illustrates a typical interrupt-driven data-measuring scheme. After initialization, the MCU waits for an interrupt signal. The interrupt can come from the timer, when it is time to take a measurement or from a computer (such as an SBC or PC) that may be connected to the MCU. Upon a timer interrupt, the MCU will run a service to acquire a measurement and store it in memory. Upon a computer interrupt, the MCU will run the corresponding service that may consist of querying the stored data, sensor calibration, or a reset to the system.

CONDITIONS AND ENCLOSURES

To implement a practical and reliable DAS for environmental monitoring, it is important to consider harsh or challenging environmental conditions for the electronic equipment. In terrestrial environments, the most challenging are moist, humid, hot, dusty, and corrosive conditions. Consequently, equipment and components follow accepted standards and specifications. An

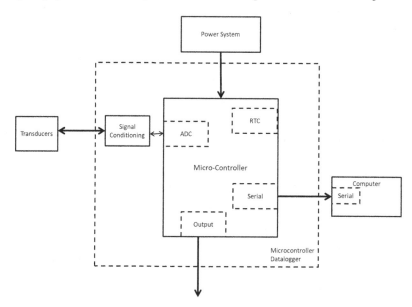

FIGURE 5.13   MCU-based DAS.

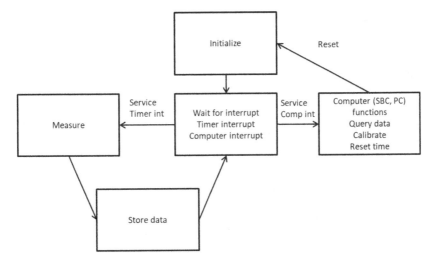

FIGURE 5.14 Basic interrupt-driven mode of MCU-based DAS.

FIGURE 5.15 Examples of submersed sensor plus datalogger.

important consideration is equipment protection and enclosure specifications to protect against harsh environmental conditions.

Standards are established for example by the National Electrical Manufacturers Association (NEMA) and the International Electrotechnical Commission (IEC). For instance, NEMA ICS6 enclosure standards, and IEC standard 60529 or IP Codes (ANSI/IEC 60529-2004). In particular, NEMA4X or IEC IP65/66 enclosures offer substantive protection, including corrosion-resistant properties. Other effects of environmental conditions on the system should be considered. For example, the design of mounting structure against hazardous wind gusts.

Excessive moisture may lead to electronic equipment failure due to several mechanisms, commonly corrosion Water condensation with dissolved ions is more conductive and creates potential for short circuits. To protect against high humidity conditions inside the enclosure, it is customary to use a desiccant, i.e., a hygroscopic material that maintains dryness in its surroundings by absorbing water (Multisorb 2023). A common desiccant is silica, which is inert and non-toxic.

Particularly challenging is to deploy a datalogger under submersed conditions for prolonged time. Some of the required waterproof enclosures are made of aluminum or stainless steel. For example, some dataloggers that are built to measure water level are submersed together with the sensor inside the same enclosure (Figure 5.15).

A DATALOGGER EXAMPLE: THE CR1000

After discussing datalogger features in general, it is useful to present an example, such as the Campbell Scientific's CR1000 datalogger. For this purpose, we will use the operation manual (Campbell Scientific Inc. 2013) and a set of weather and SM transducers. The manual contains detailed technical specifications and for easy reference, a brief overview of some specifications for the CR1000 is given in Table 5.1. The CR1000 is a programmable datalogger with non-volatile memory that can be used for example in weather measurement applications. The supply voltage is 12 V DC nominal (it can vary between 9.6 and 16 V DC), and the power consumption is ~28 mA when active at its fast scan rate (100 Hz) or ~17 mA (at slower rate 1 Hz) and can go down to 0.6 mA in sleep low power mode. The nine-pin DCE port acts as DTE when using a null-modem cable (recall RS-232 concepts from Chapter 2).

Table 5.2 shows an example of wiring a set of sensors to measure weather and SM, it shows typical wires for sensors and datalogger channels and covers a variety of wiring types for the sake of illustration. An RH sensor is powered by the switched 12-V supply because it requires more power

TABLE 5.1
CR1000 Specifications (Brief)

Specification	CR1000
Channels	8 DIFF, 16 SE
Communication port	1 RS-232 DCE port (9-pin D-sub)
Comm. ports for sensors	4 RS-232 ports (2-pin each Tx/Rx)
Data Storage	4 MBytes
Input voltage range	±5.0 V
Scan rate	1 Hz, 100 Hz
A/D Conversion	13-bit
Programming	CRBasic

TABLE 5.2
Example of Wiring Weather Sensors to a Datalogger

Sensor	Sensor Wire	Datalogger (e.g., CR1000) Channel
RH	Power (e.g., 7–28 V)	Switched 12 V
	RH signal	SE (e.g., SE4)
	Ground	Switched power GND (e.g., G)
Air temperature	Temp signal	SE (e.g., SE3)
	Ground	Signal ground
Wind gage	Wind speed signal	Pulse input (e.g., P1)
	Wind speed ground	Signal ground
Wind vane	Wind direction signal	SE (e.g., SE1)
	Wind direction excitation	Switched voltage excitation (e.g., VX1)
	Wind direction ground	Signal ground
Barometric pressure	Signal	SE (e.g., SE2)
	Ground	Signal ground
Solar radiation	Signal	DIFF (e.g., DIFF3-H)
	Reference	DIFF (e.g., DIFF3-L)
	Ground	Channel ground
Rain gage	Signal	Pulse input (e.g., P2)
	Signal return	Signal ground

at 7–28 V DC, which would exceed the 5-V supply; the ground is wired to the switched supply ground, to return higher current. The signal is connected to an SE channel. In some cases, a return for the signal is needed and would be wired to a signal ground of the datalogger; this case is not shown in the table. An air temperature sensor is wired to an SE terminal and a signal ground. Often you will find combined RH and temperature sensors, which can be powered from the same supply. The same SE configuration is shown in the table for barometric pressure.

Also shown in Table 5.2 are a wind gage for wind speed and a wind vane for wind direction, which are often combined into one device. Wind speed is wired to a pulse input since the sensor produces pulses as the rotating magnet triggers pulses by a Hall effect sensor. In many cases, sensor pulse output requires a pull-up resistor. The rain gage is also connected to a pulsed input since it is based on closures of a switch as the tipping bucket movement is detected. Wind direction is given by a potentiometer, which is connected as a voltage divider, or as we have discussed in Chapter 4, sometimes referred to as half-bridge. Thus, the signal is wired to an SE terminal and powered by the switched voltage excitation, e.g., VX, meant for bridge measurements in the CR1000. The solar radiation sensor is connected to a DIFF channel instead of SE since it produces a DIFF signal.

Soil sensors wiring is given in Table 5.3, where the moisture sensor is a voltage divider (again often named half-bridge) powered by the excitation or bridge measurement supply and output connected to an SE terminal, whereas the tensiometer is connected to a DIFF channel since the output is one from a full bridge. It could be powered from the excitation supply, but here we are using 5 V since it requires more current and using G as return.

As an example of programming dataloggers, we describe some fundamentals of programming the CR1000. This datalogger's programs start with declarations of variables and units. Instruction Public is used for variables and instruction Units for units. For example, to declare the battery voltage and an SM variable

```
'Declare variables and units
Public Batt_Volt
Public SM

Units Batt_Volt=Volts
Unit SM=mV
```

TABLE 5.3
Example of Wiring Soil Sensors to a Datalogger

Sensor	Sensor Wire	Datalogger (e.g., CR1000) Channel
SM sensor	Analog out	SE (e.g., SE 7)
	Excitation	Switched voltage excitation (e.g., VX2)
	Ground	Signal ground
Soil tensiometer	Supply positive	Power 5 V or
	Supply negative	Switched power GND (e.g., G)
	Signal positive	DIFF (e.g., DIFF8-H)
	Signal negative	DIFF (e.g., DIFF8-L)

A line preceded by single quote symbol is a remark. Then a program proceeds to define Data Tables

```
'Define data tables
DataTable(GBC,True, -1)
        DataInterval(0,5,Sec,10)
        Sample(1,Batt_volt,FP2)
        Sample(1, SM, FP2)
EndTable
```

Then the main program

```
'Main program
BeginProg
        Scan(1,Sec,1,0)
                'default battery measurement instruction
                Battery(Batt_Volt)
                'Half Bridge measurement
                BrHalf(SM,1,mV2500,1,1,3,2500, False,10000, 250,2500,0)
                'Call datatable to store measurements
                CallTable(GBC)
        NextScan
EndProg
```

In the main program, we used a BrHalf instruction to read SM. The sensor is, for example an EC-5 by Decagon Devices (Figure 5.16) with three leads connected to a channel using one of the SE inputs. An EC-5 sensor has a two-prong design and when supplied an excitation voltage (white wire) produces a signal (red wire) in mV.

Once the data table is stored in memory and we retrieve as a text file, we would have results like this

TIMESTAMP	Batt_Volt	SM
TS	Volts	mV
8/12/2010 11:40	13.03	37.09
8/12/2010 11:45	13.03	37.07

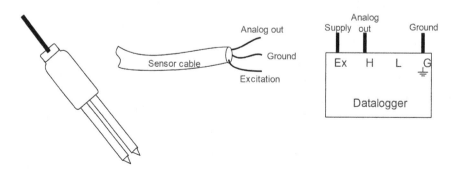

FIGURE 5.16   SM sensor and connection to a datalogger.

The EC-5 is meant to measure volumetric water content (VWC) S_w in m^3/m^3 or percentage. Therefore, we need to convert the mV from the EC-5 to S_w. The calibration equation for mineral soil

$$S_w = (11.9 \times 10^{-4}) \times mV - 0.401 \qquad (5.1)$$

where S_w is the VWC (m^3/m^3) and the mV is the output SM of the sensor. The water content is expressed as percent multiplying S_w by 100. We can include this calculation in the datalogger program. First, change Units in the unit's declaration

```
Units SM=Perc
```

in addition, add the calculation to the main program after the BrHalf instruction

```
'        BrHalf(SM,1,mV2500,1,1,3,2500, False,10000, 250,2500,0)
         'Convert from mV to volumetric content in percent
         SM=(0.00119*SM-0.401)*100
```

This is a good opportunity to introduce the specifications of an SM sensor using the EC-5 probe as an example. Accuracy is $0.03\,m^3/m^3$ up to a soil conductivity of 8 dS/m and its resolution is 0.25%. Range is from zero to saturation. Power requirements are 2.5 V DC – 3.6 V DC at 10 mA. Operating Environment is from –40 to +60°C. Sensor dimensions are 8.9 cm × 1.8 cm × 0.7 cm.

There are many types of data inputs and corresponding measurement instructions of the CR-1000. In the following, we will briefly describe the syntax of some commonly used functions.

VoltSe to measure SE voltages

```
VoltSe(Dest, Reps, Range, SEChan, MeasOfs, SettlingTime, Integ,
Mult, Offset)
e.g.: VoltSe(RH, 1, mV2500, 1, False, 0, 250, 0.1, 0)
```

VoltDiff to measure DIFF voltages

```
VoltDiff(Dest, Reps, Range, DiffChan, RevDiff, SettlingTime, Integ,
Mult, Offset)
e.g.: VoltDiff(P_mb, 1, mV250, 6, true, 0, _60Hz, 1.84, 600)
```

BrHalf to measure voltage divider (or also called half-bridge)

```
BrHalf(Dest, Reps, Range, SEChan, Vx/ExChan, MeasPEx, ExmV, RevEx,
SettlingTime, Integ, Mult, Offset)
e.g.: BRHalf(SM1_1, 3, mV2500, 9, VX1, 3, 2500, False, 10000, 250,
2500, 0)
```

BrFull to measure full-bridge sensors

```
BrFull(Dest, Reps, Range, DiffChan, Vx/ExChan, MeasPEx, ExmV, RevEx,
RefDiff, SettlingTime, Integ, Mult, Offset)
e.g.: BRFull(TM, 2, mV2500, 7, VX1, 3, 2500, False, 10000, 250,
2500, 0)
```

PulseCount to measure sensors producing a pulse output

```
PulseCount(Dest, Reps, PChan, PConfig, POption, Mult, Offset)
e.g.: PulseCount(RAIN, 1, 2, 2, 0, 0.01, 0)
```

The following is an example of a complete program. First the variables declaration

```
'CR1000 data from Discovery Park Weather Station
'Created by Short Cut (2.5)

'Declare Variables and Units
Public Batt_Volt
Public AirTC
Public RH
Public SlrW
Public SlrkJ
Public Rain_in
Public WS_mph
Public WindDir
Public BP_mmHg
Public OutString As String * 200
```

Then dimensions of arrays

```
Dim i, Record(15), curTime(9), outRecord(15) As String
```

Followed by units

```
Units Batt_Volt=Volts
Units AirTC=Deg C
Units RH=%
Units SlrW=W/m²
Units SlrkJ=kJ/m²
Units Rain_in=inch
Units WS_mph=miles/hour
Units WindDir=Degrees
Units BP_mmHg=mmHg
```

and the definition of data tables

```
'Define Data Tables
DataTable(DP_WS,True,-1)
      DataInterval(0,5,Min,10)
      Sample(1,Batt_Volt,FP2)
      Sample(1,AirTC,FP2)
      Sample(1,RH,FP2)
      Average(1,SlrW,FP2,False)
      Maximum(1,SlrW,FP2,False,False)
      Minimum(1,SlrW,FP2,False,False)
      StdDev(1,SlrW,FP2,False)
      Totalize(1,SlrkJ,IEEE4,False)
      Totalize(1,Rain_in,FP2,False)
      Average(1,WS_mph,FP2,False)
      Maximum(1,WS_mph,FP2,False,False)
      Minimum(1,WS_mph,FP2,False,False)
      StdDev(1,WS_mph,FP2,False)
      Sample(1,WindDir,FP2)
      Average(1,BP_mmHg,FP2,False)
EndTable
```

This follows bythe main program that contains loops

```
'Main Program
BeginProg
      Scan(10,Sec,1,0)
            'Default Datalogger Battery Voltage Batt_Volt:
            Battery(Batt_Volt)
            'HMP50 Temperature & Relative Humidity Sensor AirTC and RH:
            VoltSE(AirTC,1,mV2500,1,0,0,_60Hz,0.1,-40.0)
            VoltSE(RH,1,mV2500,2,0,0,_60Hz,0.1,0)
            If (RH>100) And (RH<108) Then RH=100
            'LI200X Pyranometer measurements SlrkJ and SlrW:
            VoltDiff(SlrW,1,mV7_5,2,True,0,_60Hz,1,0)
            If SlrW<0 Then SlrW=0
            SlrkJ=SlrW*2.0
            SlrW=SlrW*200.0
            'TE525/TE525WS Rain Gauge measurement Rain_in:
            PulseCount(Rain_in,1,1,2,0,0.01,0)
            '03001 Wind Speed & Direction Sensor WS_mph and WindDir:
            PulseCount(WS_mph,1,2,1,1,1.677,0.4)
            If WS_mph<0.41 Then WS_mph=0
            BrHalf(WindDir,1,mV2500,5,1,1,2500,True,0,_60Hz,355,0)
            If WindDir>=360 Then WindDir=0
            'CS105 Barometric Pressure Sensor measurement BP_mmHg:
            PortSet(1,1)
            VoltSE(BP_mmHg,1,mV2500,6,1,0,_60Hz,0.184,619.725)
            BP_mmHg=BP_mmHg*0.75006
            'Call Data Tables and Store Data
            CallTable(DP_WS)
```

```
              If TimeIntoInterval(3,5,Min) Then
                OutString = ""
                RealTime(curTime)
                For i=1 To 6
                    OutString = OutString & curTime(i) & ","
        Next i

                GetRecord(Record(),DP_WS,1)
                For i=1 To 15
                    If Record(i) < 0 Then
                        outRecord(i) = CHR(45) & FormatFloat(ABS(Record(i)),"%.3
                        g")
                    Else
                        outRecord(i) = FormatFloat(ABS(Record(i)),"%.3g")
                    EndIf
                    OutString = OutString & outRecord(i) & ","
        Next i

                OutString = OutString & CHR(13) & CHR(10) & ""
                SerialOpen(COMRS232,115200,0,0,10000)
                SerialOut(COMRS232,OutString,"",0,100)
              EndIf
        NextScan
EndProg
```

This program generates the following example of data file, which is shown with wrapped lines of text

```
TOA5    CR1000_DPWS   CR1000 15720  CR1000.Std.15 CPU:Wthr_DPWS_
ShortCut.CR1 60724  DP_WS
TIMESTAMP       RECORD Batt_Volt      AirTC  RH     SlrW_Avg       SlrW_Max
SlrW_Min        SlrW_Std         SlrkJ_Tot       Rain_in_Tot  WS_mph_Avg
WS_mph_Max      WS_mph_Min       WS_mph_Std      WindDir        BP_mmHg_Avg
TS      RN      Volts  Deg C  %     W/m?   W/m?   W/m?   W/m?   kJ/m?
inch    miles/hour      miles/hour      miles/hour      miles/hour     Degrees
mmHg
                Smp    Smp    Smp    Avg    Max    Min    Std    Tot
Tot     Avg     Max    Min    Std    Smp    Avg
8/12/2010 11:40         85136  13.03  37.09  33.95  803    810    797.5
3.466   240.9406        0      2.473  4.593  0      1.376  138.6  744.5
8/12/2010 11:45         85137  13.03  37.07  34.64  812    815    808
2.149   243.4854        0      3.614  5.599  0.568  1.258  157.1  744.5
8/12/2010 11:50         85138  13.03  36.28  37.03  818    823    811
2.848   245.5307        0      4.783  8.28   3.251  1.204  203.7  744.5
8/12/2010 11:55         85139  13.04  37.63  32.59  826    831    821
2.574   247.7273        0      3.036  7.611  0      1.648  120.9  744.5
8/12/2010 12:00         85140  13.03  37.26  33.14  837    843    831
3.495   251.1464        0      5.04   8.62   2.748  1.416  241.6  744.4
8/12/2010 12:05         85141  13.03  37.33  33.58  845    848    842
1.392   253.5751        0      4.721  7.947  2.245  1.463  137    744.4
8/12/2010 12:10         85142  13.03  37.68  33.29  856    861    848
4.192   256.8434        0      2.092  5.599  0      1.685  80.6   744.4
8/12/2010 12:15         85143  13.03  37.77  33.31  865    867    862
1.356   259.4824        0      4.391  7.779  0.568  2.087  127.1  744.3
8/12/2010 12:20         85144  13.03  38.51  32.46  870    873    867
2.29    261.0869        0      1.327  3.586  0      1.149  193.5  744.3
```

Datalogger manufacturers offer support software that helps users in a variety of tasks, from wiring, programming, and viewing stored data. For example, Campbell Scientific offers its Logger Net software that can help configure a connection to download data, build a program using templates by a method called Shortcut, edit programs in an editor window, and viewing stored data.

REMOTE TELEMETRY UNIT OR REMOTE TERMINAL UNIT (RTU) AND SUPERVISORY CONTROL AND DATA ACQUISITION (SCADA)

The terms RTU and SCADA are common in industrial applications, such as the energy industry, and large facilities for environmental applications, such as municipal water treatment plants. An RTU is an MCU or SBC device that monitors and controls devices (e.g., sensors) in the field. Therefore, an RTU is similar to a datalogger but often with more control capabilities. Multiple RTUs in the field are connected to an SCADA. An SCADA typically includes a Human Machine Interface (HMI), which an operator uses to monitor and control the processes under the SCADA scope. Large SCADA systems encompass many RTUs distributed over a large area. Then, the HMI acts as central location to monitor and control a large area.

EXERCISES

Exercise 5.1

Consider the seconds register at address 00 hour of an RTC chip 1307

	Bit7	Bit6	Bit5	Bit4	Bit3	Bit2	Bit1	Bit0
A		1	0	1	0	1	0	0
B		0	0	1	0	1	0	1
C		1	0	0	0	0	0	1

What are the number of seconds for cases A, B, and C? Hint: use the BCD code for the high nibble (bits 4–6) and low nibble (bits 0–3); for example, A is 54 seconds.

Exercise 5.2

Draw a diagram to connect one master and three slaves using SPI. Label the devices as well as the lines and indicate the direction of each line using arrowheads.

Exercise 5.3

Draw a diagram to connect one master and three slaves using I2C. Label the devices as well as the lines and indicate the direction of each line using arrowheads. Include pull-up resistors.

Exercise 5.4

Develop a CR1000 program to monitor the voltage from the battery and SM from two sensors. Assume that each sensor is a voltage divider (half bridge) with an output voltage in mV. Write the header and a couple of lines that exemplify the data collected by the program.

REFERENCES

Acevedo, M.F. 2024. *Real-Time Environmental Monitoring: Sensors and Systems*, Second Edition – Lab Manual. Boca Raton, FL: CRC Press, Taylor & Francis Group. 463 pp.

Atlas Scientific. 2022. *Environmental Robotics, Embedded Solutions*. accessed September 2022. https://atlas-scientific.com/embedded-solutions/.

Atlas Scientific. 2023. *EZO-EC Embedded Conductivity Circuit*. Atlas Scientific,, Accessed May 2023. https://files.atlas-scientific.com/EC_EZO_Datasheet.pdf.

Campbell Scientific Inc. 2013. *CR1000 Measurement and Control System, Operator's Manual. Revision 5/13*. Logan, UT: Campbell Scientific Inc.

Multisorb. 2023. *Dessicants*. Accessed June 2023. https://www.multisorb.com/desiccants/.

Circuit Basics. 2022a. *Basics of the I2C Communication Protocol*. accessed February 2022. https://www.circuitbasics.com/basics-of-the-i2c-communication-protocol/.

Circuit Basics. 2022b. *Basics of the SPI Communication Protocol*. accessed February 2022. https://www.circuitbasics.com/basics-of-the-spi-communication-protocol.

LiuDr Electronic Solutions LLC. 2022. *Your Online Electronics Store with Products Designed by Dr. Liu*. accessed September 2022. https://liudr.square.site/.

METER. 2022. *TEROS12 Advanced Soil Moisture Sensor + Temperature and EC*. accessed September 2022. https://www.metergroup.com/en/meter-environment/products/teros-12-soil-moisture-sensor.

SDI-12 Support Group. 2013. *SDI-12 A Serial-Digital Interface Standard for Microprocessor-Based Sensors*. Utah: SDI–12.

6 Wireless Technologies
Telemetry and Wireless Sensor Networks

INTRODUCTION

In this chapter, we study radio communications for a variety of wireless applications to environmental monitoring, particularly on telemetry from a remote station and wireless sensor network (WSN) systems. We focus on the physical principles and available technology. For this purpose, we cover topics in transmission, reception, and antennas. Telemetry relates to collecting data from a remote station. A WSN is a network of wireless-enabled MCU-based single-board computers (SBCs) interacting with sensors. WSNs have environmental monitoring applications allowing data collection with finer spatial and temporal resolution. These concepts and systems are expanded by computer and hands-on exercises in Lab 6 of the companion Lab Manual (Acevedo 2024).

RADIO WAVE CONCEPTS

ELECTROMAGNETIC WAVES

Electromagnetic (EM) waves are fluctuations of the electric and magnetic fields that propagate in space, mostly in the air that surrounds us for many environmental monitoring applications. To characterize a wave, we use the frequency f of the fluctuation (in Hz or s^{-1}), which when multiplied by the wavelength λ (length of the wave, in m) equals to the speed of propagation, which is the speed of light c (3×10^8 m/s). The relationship is

$$c = f\lambda \qquad (6.1)$$

Note that the units (m/s) will be the same in the right-hand side as in the left-hand side. In other words, the frequency would be the number of cycles that go through a fixed point in space every second; wavelength is the distance between a point of the wave and a similar point in the next cycle, say between peaks (Figure 6.1).

As an example, a 3-kHz wave would have a wavelength of

$$\lambda = \frac{c}{f} = \frac{3 \times 10^8}{3 \times 10^3} \frac{\text{m/s}}{\text{1/s}} = 1 \times 10^5 \text{ m} = 100 \text{ km}$$

whereas a 300-GHz wave would have a wavelength of

$$\lambda = \frac{c}{f} = \frac{3 \times 10^8}{300 \times 10^9} \frac{\text{m/s}}{\text{1/s}} = 1 \times 10^{-3} \text{ m} = 1 \text{ mm}$$

One more characteristic of the wave is the amplitude or height of the fluctuation or height of the peak, or half the peak-to-peak height. Wavelength (or alternatively frequency) determines how the

DOI: 10.1201/9781003425496-6

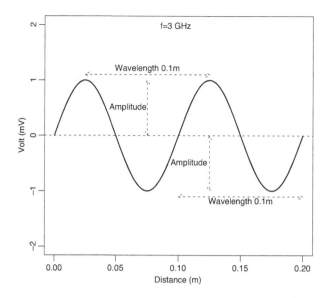

FIGURE 6.1   Wavelength and amplitude.

wave interacts with objects in the propagation path and size of antennas to transmit and receive the waves.

RADIO WAVES

In the broadest sense, radio waves are EM waves in the frequency range of 3 kHz to 300 GHz, corresponding to wavelength ranging from 1 mm to 100 km as calculated in the examples above. However, we restrict the term *radio* to frequencies up to 1 GHz or longer than 30 cm (but shorter than 100 km) and use the term *microwave* for waves in between 1 and 300 GHz, or between 1-mm and 30-cm long. Figure 6.2 illustrates where these frequencies (and its wavelengths) are in the EM wave spectrum and its relationship with the immediately higher frequencies (infrared, visible, and ultraviolet).

The International Telecommunication Union of the United Nations establishes names for various radio wave ranges. For example, HF (high frequency), VHF (very high frequency), and UHF (ultra-high frequency) ranges. HF radio or shortwave is from 3 MHz to 30 MHz, including Citizens Band at about 27 MHz. Next, VHF radio is in between 30 MHz and 300 MHz or 1 m and 10 m. Within this range, we find FM radio around 100 MHz or about a 3-m long wave. UHF is in between 300 MHz and 3 GHz or 10 cm and 1 m, including 900 MHz used for phones and WSN and Internet of Things (IoT) devices to be described later in this chapter.

PROPAGATION

Propagating radio waves consist of electric- and magnetic-field components oscillating perpendicular to each other as well as perpendicular to the direction of propagation. As we setup a communication link, we need to understand what happens to the waves as they travel from source to destination. For example, several processes can affect the waves: reflection, diffraction, and refraction (Figure 6.3). All these processes depend on the characteristics of the terrain, land use (rural vs. urban), topography (hilly vs. flat), and land cover (forest vs. grassland).

Reflection: We are familiar with visible EM waves (light) reflecting from shiny surfaces, such as mirrors. Radio waves reflect the same way (Figure 6.3, top). Wetland surfaces are good reflectors of

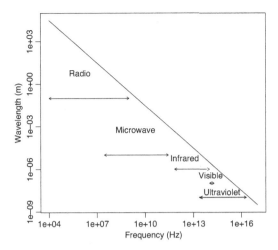

FIGURE 6.2 EM frequency spectrum from radio to ultraviolet.

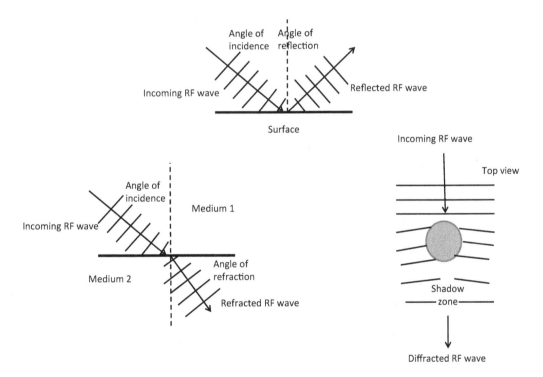

FIGURE 6.3 Reflection (top), diffraction (left), and refraction (right).

radio waves, such as a lake, the sea, or wetlands. In urban areas, metallic surfaces of buildings are also good reflectors.

Refraction: Likewise, we are familiar with how light refracts as we go from air to water, for example the apparent bend of a partly submerged stick. Radio waves refract as well and tend to bend or change direction when going to media with different refractive index (Figure 6.3, left).

Diffraction: This is a less intuitive process but can visualize it when placing an obstacle in the path of waves traveling in water. The waves seem to go around the object; in essence, primary waves generate secondary waves that fill the void of fluctuations behind the obstacle (Figure 6.3, right).

Huygen's principle explains this process by stating that each point of a primary wave is a source of a secondary wave.

Absorption: Waves are dampened as they go through a medium or material because the material absorbs radiation. Absorption depends on frequency of the waves and the material itself. For example, in the microwave range waves are absorbed by metal (electrons move freely and thus absorb energy) and water (molecules are excited and absorb energy).

PROPAGATION MODELS

We will discuss two major models of propagation, the *free-space* model, and the *two-ray* model. For both models, we calculate received power from the transmission.

FREE-SPACE PROPAGATION MODEL

This model assumes that there are no obstacles along the propagation path of the radio waves, and thus the waves do not undergo any changes due for instance to reflection and diffraction. Hence, there is only one path of propagation, referred to as the *Line of Sight* (*LoS*) path.

Received power P_r is given by an equation proposed in the 1940s by Friis

$$P_r = \frac{P_t G_t G_r \lambda^2}{(4\pi)^2 d^2} \tag{6.2}$$

where P_t=transmitted power of the signal (W), G_t=transmitting antenna gain, G_r=receiving antenna gain, λ=wavelength of the signal, and d=distance between transmitter and receiver. This equation is derived from the assumption that the received signal is calculated on a sphere of radius d centered on the transmitter.

Note that received power varies inversely with the square of distance, $P(d) \propto \frac{1}{d^2}$ Doubling the transmission distance reduces the power by one quarter of the original value $P(2d) \propto \frac{1}{(2d)^2} = \frac{1}{4} P(d)$.

Equivalently, converting to dB, $10 \times \log(P(2d)) = 10 \times \log\left(\frac{P(d)}{4}\right) = 10 \times \left(\log(P(d)) - \log(4)\right)$ or

$P_{dB}(2d) = 10 \times \log(P(d)) - 10 \times 0.602 = P_{dB}(d) - 6.02$ dB which states that doubling the transmission distance reduces the power by 6 dB

$$P_{dB}(2d) \approx P_{dB}(d) - 6 \text{ dB} \tag{6.3}$$

We use the term *path loss* (*PL*) to denote the reduction in signal strength of a radio signal as it propagates through the medium. It relates to the increasing spread of the wave with distance, implying that signal strength reduces as it covers longer distance. In the free-space propagation model, PL is *Free Space Loss* (FSL) and derives from the Friis equation by inverting

$$\frac{P_t}{P_r} = \frac{(4\pi)^2 d^2}{G_t G_r \lambda^2} \tag{6.4}$$

This free-space model is a good predictor of PL for short transmission distance.

Expressing this equation in dB $PL_{dB} = 10 \times \log\left(\dfrac{P_t}{P_r(d)}\right) = 10 \times \log\left(\left(\dfrac{4\pi d}{\lambda}\right)^2\right) - 10 \times \log(G_t G_r)$

or $PL_{dB} = 20 \times \log\left(\dfrac{4\pi d}{\lambda}\right) - 10 \times \log(G_t G_r) = 20 \times \log\left(\dfrac{4\pi f d}{c}\right) - 10 \times \log(G_t G_r)$.

Assume f_{MHz} is frequency in MHz and d_{km} is distance in km

$$PL_{dB} = 20 \times \log\left(\dfrac{4\pi d_{km} f_{MHz} \times 10^3 \times 10^6}{c}\right) - 10 \times \log(G_t G_r) \text{ and then } PL_{dB} = 20 \times \log\left(\dfrac{4\pi}{3}10\right)$$

$+20 \times \log(d_{km} \times f_{MHz}) - 10 \times \log(G_t G_r)$. Thus, we obtain

$$PL_{dB} = 20 \times \log(f_{MHz}) + 20 \times \log(d_{km}) + 32.45 - 10 \times \log(G_t G_r) \tag{6.5}$$

Each ten-fold increase in gain 10^n reduces the loss by 10 dB since $10 \times \log(G_t G_r) = 10 \times \log(10^n) = 10n$. For unity gain $G_t G_r = 1$, we have the popular formula

$$FSL_{dB} = 20 \times \log(f_{MHz}) + 20 \times \log(d_{km}) + 32.45 \tag{6.6}$$

which states that loss increases with frequency and distance. For example, at 900 MHz and for 1 km, $FSL_{dB} = 20 \times \log(900) + 20 \times \log(1) + 32.45$ dB = 91 dB. A gain $G_t G_r = 100$ reduces the loss by 20 dB, and we now have $PL_{dB} = 91 - 20$ dB = 71 dB.

PL is proportional to d and therefore for a fixed frequency, it increases by 6 dB every doubling of distance since $PL_{dB}(2d) = PL_{dB}(d) + 20 \times \log(2) = PL_{dB}(d) + 6$ dB. For example, at 900 MHz, for 2 km, and unity gain yields $FSL_{dB}(2) \approx 91 + 6$ dB = 97 dB.

Subtracting the loss from transmitted power, received power in dB is simply

$$P_{r_{dB}} = P_{t_{dB}} - PL_{dB} \tag{6.7}$$

It is convenient to express power in dBm, which is a ratio of power to a reference value (1 mW). For example, a transmitter with power 1 W represents 30 dBm.

$$P_{t_{dBm}} = 10 \times \log\left(\dfrac{1000 \text{ mW}}{1 \text{ mW}}\right) = 30 \text{ dBm} \tag{6.8}$$

Therefore, the reference 1 mW is $10 \times \log(1) = 0$ dBm. A negative value represents a fraction of 1 mW; for example, 0.1 mW, is $10 \times \log(10^{-1}) = -10$ dBm.

Since the loss is a unit less ratio, received power in dBm is transmitted power in dBm minus the loss in dB

$$P_{r_{dBm}} = P_{t_{dBm}} - PL_{dB} \tag{6.9}$$

For example, for $P_t = 1$ W at 900 MHz, over 1 km, and gain $G_t G_r = 100$, we have $P_{r_{dBm}} = 30 - 71$ dBm = -41 dBm. This is a fraction of 1 mW.

Weak signals have high negative values; for example, 10^{-8} mW is $10 \times \log(10^{-8}) = -80$ dBm. Typically, we want the received signal to be better than -60 dBm to have good throughput and reliable connections.

TWO-RAY PROPAGATION MODEL

In addition to one LoS path between transmitter and receiver, there exists a ground reflection path (Figure 6.4) and the received power from each are combined to obtain total received power. For long distances, the two-ray model is a better predictor than the free-space model. This calculation is more complicated; however, for long distances, received power is approximately

$$P_r(d) \approx \frac{P_t G_t G_r h_t^2 h_r^2}{d^4} \tag{6.10}$$

where h_t and h_r are transmitter and receiver antenna heights. The approximation yields accurate results when the distance d is much larger than the *crossover distance* (d_c) defined by

$$d_c = \frac{4\pi h_t h_r}{\lambda} \tag{6.11}$$

In other words, Equation (6.10) can be used when $d \gg d_c$.

Note that received power vary with the fourth power of distance $P_r \propto \dfrac{1}{d^4}$, therefore doubling the transmission distance reduces the power by 1/16 of the original value $P(2d) \propto \dfrac{1}{(2d)^4} = \dfrac{1}{16}P(d)$. Equivalently, doubling the transmission distance reduces the power by 12 dB, $P_{dB}(2d) = P_{dB}(d) - 12.04$ dB.

PL in dB units predicted by Equation (6.10) is $PL_{dB} = 10 \times \log\left(\dfrac{P_t}{P_r(d)}\right) = 10 \times \log(d^4) - 10 \times \log\left(h_t^2 h_r^2\right) - 10 \times \log(G_t G_r)$, which reduces to

$$PL_{dB} = 40 \times \log(d) - 20 \times \log(h_t h_r) - 10 \times \log(G_t G_r) \tag{6.12}$$

Using distance in km,

$$PL_{dB} = 40 \times \log(d_{km}) + 120 - 20 \times \log(h_t h_r) - 10 \times \log(G_t G_r) \tag{6.13}$$

For example, for $h_t = h_r = 10$ m, $d = 100$ km, and unity gain $PL_{dB} = 40 \times \log(100) + 120 - 20 \times \log(10 \times 10) = 80 + 120 - 40 = 160$ dB. A gain $G_t G_r = 100$ reduces the loss by 20 dB, and we now have $PL_{dB} = 160 - 20$ dB $= 140$ dB. Subtracting the loss from transmitted power, received power in dBm is given by Equation (6.9). In this example, when $P_t = 1$ W, we have $P_{r\,dBm} = 30 - 140$ dBm $= -110$ dBm, which is a small fraction of 1 mW.

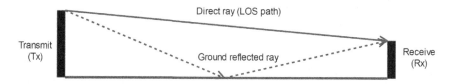

FIGURE 6.4 Two-ray propagation model.

FRESNEL ZONES

Two waves are *in phase* if they are in the same position in space; say, the peaks are at the same point. Conversely, they are *out of phase* if there is a difference in their position. This difference is the *phase shift* (Figure 6.5). When the waves are in phase, the phase shift is zero. The phase shift is a fraction of the wavelength since the difference is the same in all cycles. For example, a quarter of a wavelength $\lambda/4$ phase shift. This can be expressed as degrees from 0 to 360, or in radians 0 to π or a full cycle. Consequently, 360-degree phase difference is the same as 0-degree phase shift.

We will now look more carefully at what happens when a propagating signal experiences interference from obstacles causing reflection, refraction, and diffraction. These processes may cause a phase shift, which may enhance or reduce the signal. For example, the signal is enhanced if the reflected wave is in phase with the direct path signal; however, the direct path signal is reduced if it is out of phase with the reflected signal.

Fresnel zones provide a way to predict the enhancement or reduction by phase differences. Fresnel zones can be odd first, third, and so on, or even second, fourth, and so on. Odd-numbered zones cause out-of-phase interference, while even-numbered zones cause in-phase interference. In the first zone, phase shift is $0-\pi$, in the second zone $\pi-2\pi$, and then repeats.

Wave propagation is conceptualized as ellipsoid layers with extremes at the transmitter and receiver stations and longitudinal axis between these two points (Figure 6.6). The Fresnel zones are ellipsoidal layers with the inner one being the first zone and progressing outward. The cross-section of the ellipsoids at all values of distance are concentric circles such that the increasing radius corresponds to increase zone number. The first Fresnel zone is the innermost circle, and the others are annulus or donut-shaped going outward. The zones are used to examine propagation patterns, particularly disrupted vs. clear (Figure 6.7). To avoid the effect of interference and hence treat the path as free space, at least 60% of the first Fresnel zone should clear the obstacles.

The radius at a point such that d_1=distance of the point from one end (m), $d_2=d-d_1$=distance from the other end (m) is calculated using

$$r_n = F_n = \sqrt{\frac{n\lambda d_1 d_2}{d_1 + d_2}} \tag{6.14}$$

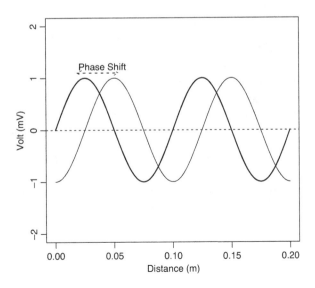

FIGURE 6.5 Phase shift.

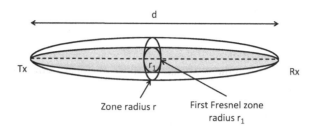

FIGURE 6.6   Fresnel zones.

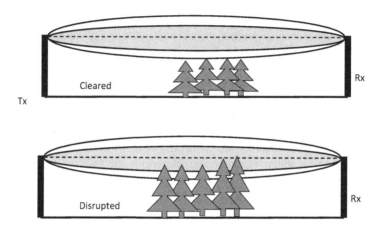

FIGURE 6.7   Disrupted vs. clear Fresnel zone.

where $r_n = F_n = n$th Fresnel zone radius (m). For the first zone, $n = 1$, we obtain

$$r_1 = F_1 = \sqrt{\frac{\lambda d_1 d_2}{d_1 + d_2}} \tag{6.15}$$

ANTENNAS AND CABLES

An antenna is a device that converts the RF signal propagating on the RF cable or transmission line to a wave propagating in space. For example, we can make a simple antenna by bending the open ends of the RF cable in opposite directions such that waves are in phase (Figure 6.8). The length of the bent is a quarter of the wavelength to produce a half-wave dipole antenna (WNDW 2013).

We now define several specifications that are very important when designing or selecting antennas (WNDW 2013):

Input Impedance: For efficient transfer of the energy from the radio to the antenna, all impedances of the radio, cable, and input impedance of the antenna must match. The impedance is typically 50 Ω. A metric that specifies impedance mismatch is *return loss (RL)*, which is the ratio of input power to the antenna P_i from the RF cable to reflected power from the antenna P_{ref} and given in dB

$$RL(dB) = 10 \times \log\left(\frac{P_i}{P_{ref}}\right) \tag{6.16}$$

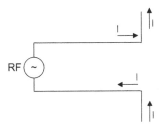

FIGURE 6.8   Dipole antenna.

Another way of describing impedance mismatch is by SWR (standing wave ratio) that takes the value 1 for perfect match and a larger value for mismatch.

Bandwidth: It is the range of frequency in Hz over which the antenna meets certain specifications, e.g., gain or RL. An antenna performs well over this range of frequency designed to correspond to the radio transmitter and receiver.

Directivity: A directional antenna focuses the energy on a given direction (anisotropic radiation), whereas an omni-directional antenna radiates in all directions with the same energy (isotropic radiation). Directional antennas need an alignment of position and orientation such that they face each other lying in LoS path to obtain optimum received power.

Gain: It is a relative measure with respect to an isotropic antenna (given in dBi) or to a half-wave dipole antenna (given in dBd). We can convert from dBd to dBi by adding the gain of a dipole antenna of 2.15 dBi.

The transmitter and the receiver are connected to their respective antennas by an RF transmission line or RF cable. For high frequencies, this cable is important to maintain the integrity of the signal. The simplest cable is bifilar or twin wires. As we discussed previously (Chapter 4), a better approach to the twin wire setup is to use a wire with a shield, and even better a coaxial cable for HF and higher. The coaxial cable inner conductor carries the RF signal while the outer conductor is a shield to contain the electric field and to prevent interference. Any RF cable attenuates the signal, and it is specified in dB per meter; implying that the transmitter should be close to the antenna.

FADE MARGIN

This margin is an allowance in a radio link gain so that it can accommodate expected fading. In other words, it is the amount that we could reduce the received power without degrading performance below a specified level. Even after optimizing a radio link in terms of LoS, antenna, and alignment, there are adverse conditions that are inevitable, such as extreme weather and new obstructions, such as leafing of trees in the growing season. We account for these uncertain conditions having ample fade margin. A fade margin of 20 dB is typically sufficient, it signifies that the signal is 100 times stronger than the minimum required.

POLARIZATION

The direction of the electric field vector determines polarization. Linear vertical polarization occurs when the electric field vector is vertical (pointing only upward or downward), whereas linear horizontal polarization corresponds to the electrical field vector aligning horizontally or pointing only sideways. This linear polarization scheme is generated by a dipole antenna placed either vertically or horizontally. When directions are combined, by having two perpendicular dipole antennas, the electric field vector can rotate in a circle or ellipse on a plane perpendicular to the direction of propagation (WNDW 2013).

Polarization helps to implement two links sharing the direct path between two points. One link will use one polarization direction, the other link the vertical polarization direction. Their corresponding antennas would have to match and reject each other polarization direction.

RADIO LINKS: COMMUNICATION CHANNEL

MODULATION: DIGITAL SIGNALS

In communication systems, radio waves carry a signal that contains information that a sender wants to transfer to a receiver. For this purpose, the radio link is a communication *channel* established at the frequency of the *carrier* wave. Modulation is the process of imposing a signal on the RF wave carrier by modifying its amplitude, frequency, or phase. For example, amplitude modulation (AM) consists of making the carrier amplitude vary with the signal, phase-shift keying (PSK) modifies the phase of the carrier. Demodulation is the process of extracting the signal from the modulated carrier. In general, signal suffers from attenuation, delay, and noise contamination when traveling through the channel.

Analog signals can directly modulate the carrier wave and then be transmitted. However, they are vulnerable to interference when traveling through the channel. A more robust method is to convert the signal into *symbols* so to reduce the effect of noise contamination. Many modern communication systems are digital where the symbols are based on binary 0s and 1s. As we know from Chapters 2 - 5, we can convert an analog signal to digital by sampling and using an ADC.

Some modulation methods are good for maximizing noise rejection, others for maximizing *capacity* (amount of information transmitted in bits/s), and yet others for maximizing *spectral efficiency* (number of bits/s per Hz of bandwidth). We often have a trade-off among these properties. Examples of modulation methods are Binary PSK (BPSK), Quadrature PSK (QPSK), and Quaternary AM (QAM). In BPSK, we use two values of phase separated by 180° and each phase represents a bit (0 and 1); we can transmit only one bit per symbol (low capacity), but it is very resistant to noise. In QPSK, we use four values of phase, coding 2-bit symbols (00, 01, 10, 11); we can transmit two bits per symbol thereby doubling the capacity, and it is still robust to noise. In QAM, we can transmit several bits in a symbol (more capacity) but is more vulnerable to noise. QAM can be implemented in several levels, such as 16-QAM (4 bits per symbol), 64-QAM (6 bits per symbol), and 256-QAM (8 bits per symbol).

CHANNEL PERFORMANCE

Several parameters can be used to characterize a channel, such as Data rate, Bit Error Rate (BER), throughput, and spectral width.

Data rate: number of bits per sec (bps) sent over the link. Desired rates are in the millions of bps (Mbps). *BER*: fraction of received bits that are in error. It should be low (e.g., 10^{-9}). As a signal loses power or encounters interference, the BER will increase.

Throughput: is a key measure of link performance calculated as the average rate of successful message delivery over a communication link. It is the ratio of total file size in bits delivered over the time taken for transferring (in sec). Therefore, throughput has same units as data rate (bps).

Spectral width: we can reduce spectral width in order to decrease interference from overlapping channels and increase power spectra density; however, this results in lower throughput.

MULTIPLEXING

Multiplexing allows sharing a link by many users. This is also called *Multiple Access* (MA), and it is accomplished in a variety of ways. For example, assigning a different frequency to each user as in FDMA (frequency division MA), or assigning different time slots to different users as in TDMA (time division MA), or assigning different codes to each user as in CDMA (Code division MA), or using different fading from different antennas as in SDMA (space division MA). A special case of sharing the link is making it *duplex*, providing for two simultaneous directions: downlink and uplink. This is done by dividing time (TDD, time division duplexing) or frequency (FDD, TDD) between the down and the up directions.

SPREAD SPECTRUM

Spread spectrum is a method by which a signal's bandwidth is "spread" over the spectrum to obtain a signal with larger bandwidth. This technique allows MA while reducing interference. Two major techniques are *direct-sequence* spread spectrum (DSSS) and *frequency hopping* spread spectrum (FHSS).

WI-FI

Recall that we discussed Wi-Fi in Chapter 2 and its role in networking. By calculating PL using Equation (6.6), we can compare the effect of transmission distance between 2.4 and 5 GHz and note that 2.4 GHz allows for longer distances than 5 GHz. When using 2.4 GHz, the bandwidth is about 100 MHz, which divided into 20 MHz channels would make most of them overlapping (except three). Therefore, there is possible channel interference when using 2.4 GHz (Figure 6.9). Contrastingly, when using 5 GHz, the bandwidth is larger allowing for higher data rates and more non-overlapping channels (reducing interference) of 40 MHzbandwidth.

Standard 802.11b calls for a DSSS technique called complementary coded keying; the bit stream is coded and then modulated using QPSK. The 802.11a and g systems use 64-channel orthogonal frequency division multiplexing, which divides the available band into channels. The transmitter

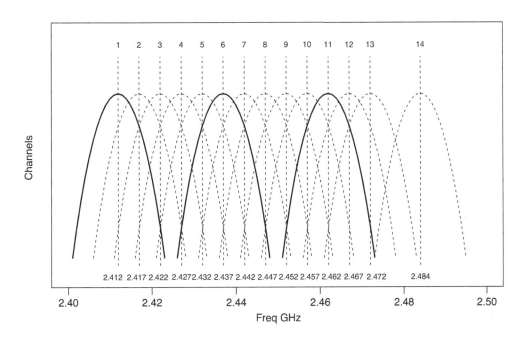

FIGURE 6.9   Wi-Fi 2.4 GHz. Channel bandwidth is 22 MHz, channel separation is 5 MHz (except between channels 13 and 14). Channels 1, 6, and 11 are non-overlapping.

encodes the bit streams on the 64 carriers using BPSK, QPSK, or one of two levels of QAM (16- or 64-QAM).

The classic NanoStation and NanoStation Loco (Ubiquiti Networks) are Wi-Fi products that operate in the 2.4-GHz frequency range and supporting the standards IEEE 802.11 b/g. Updates of these classic products are now UISP NanoStation AC 5 GHz and UISP NanoStation 5 AC Loco. A NanoStation allows to establish a wireless link between any two remotely located stations. It has a built-in directional antenna of 10-dBi gain, with 3-dB horizontal beam of width 60° and vertical beam width of 30°. One can also connect an external antenna for longer transmission range. The Nanostation2 uses vertical and horizontal polarization and in addition uses an adaptive antenna polarity technology, which can improve the link quality by switching antenna polarities statically or dynamically.

The maximum transmit power of 26 dBm allows it to cover the range of up to 15 km. The power supply required is 12 V, 1 A with 4-W maximum power consumption. The unit can be powered via the same Ethernet cable through which it is connected to other devices like a PC, router, or SBC. The data rate can go up to 54 Mbps and the throughput up to 25 Mbps. Maximum receiver sensitivity is −97 dBm (Ubiquiti 2009). The Nanostation2 has three rate modes: Quarter (5 MHz), Half (10 MHz), and Full (20 MHz). Rx Sensitivity is inversely proportional to data rate.

As a manner of illustrating the above concepts, we summarize a Wi-Fi radio link (Gurung 2009) between a weather station at the University of North Texas (UNT), Discovery Park (DP) campus (ground elevation 217 m, latitude, longitude 33.26°, −97.15°), and the EESAT building (roof elevation 242 m, latitude, longitude 33.21°, −97.15°) at UNT main campus. For this link, there is clear LoS for a $d = 4.74$ km. Azimuth angles to direct antennas are 1.4° and 181.4° at the EESAT and DP stations, respectively.

First, to decide on a propagation model to estimate received power, calculate crossover distance, using $f = 2.422$ GHz and antenna heights of 9 m (height of antenna mounting pole at DP) and 25 m (height at the roof of the EESAT building)

$$d_c = \frac{4\pi h_t h_r}{\lambda} = \frac{4\pi h_t h_r f}{c} = \frac{4\pi(25)(9)(2.422 \times 10^9)}{3.0 \times 10^8} = 22.83 \text{ km} \tag{6.17}$$

resulting in $d < d_c$, therefore select the free-space model and calculate FSL using Equation (6.6) $FSL(\text{dB}) = 20 \times \log(2.422 \times 10^3) + 20 \times \log(4.74) + 32.45 = 113.65$ dB. Using antenna gains $G_t = 10$ dBi and $G_r = 10$ dBi represents a gain $G_t G_r = 100$. The PL in dB is $PL_{dB} = 113.65 - 20$ dB ≈ -93 dB. With transmitted power $P_t = 26$ dBm, and using Equation (6.9), we get $P_{r \, \text{dBm}} = 26 - 93$ dBm $= -67$ dBm. Recall that the negative sign means that the received power is less than 1 mW.

After performing a field survey using a Wi-Fi Spectrum Analyzer tool, Channel 3 was selected being the least crowded at both sites. Radio-link modeling included a Digital Elevation Model (DEM) to consider topography of the land. The simulation results are received power −69.5 dBm, PL 114.5 dB, fade margin 22.5 dB, and worst Fresnel 1.2F1. Using a graph of receiver sensitivity as a function of data rate, 13.5 Mbps was selected which allows for ample fade margin. Spectral width was selected as 5 MHz, polarization was set as vertical (two-way), and IEEE mode selected was 802.11g to avoid unnecessary interference from 802.11b networks.

With these parameters, low-cost Nanostation N2 devices were installed at both locations using 25-m Cat5 Ethernet cable at the EESAT and 12-m Cat5 Ethernet cable at DP. After alignment of antennas, received power was −66 dBm and fade margin was 22.5 dBm, considered ample enough to counteract hostile environmental conditions. Throughput tests were conducted using IPERF (network performance measurement tool written in C++) and the Nanostation built-in tool. Tests included two transmit (Tx) power levels (26 and 11 dBm) and two data rates (54 and 13.5 Mbps), for a total of four experimental conditions transmitting a 12-Mb file in 60 s. There was no significant change in throughput (~0.4 Mbps) when the data rate is changed while keeping Tx power fixed at 11 dBm.

However, throughput doubles to ~ 3.75 Mpbs when the data rate is changed, while transmit power is 26 dBm. As we see from this example, Wi-Fi represents a practical technology to implement relatively long-distance links between a remote monitoring station and a base station. Later in the chapter we will discuss other radio options for long distance based on frequencies in the 900 MHz range.

CELLULAR PHONE NETWORK AND SATELLITE LINKS

A convenient method to transmit data from a remote station to a base station is to connect the monitoring station to the cellular phone network. This of course assumes that there is coverage by a cellular phone network provider and transmission is subject to fees charged by the provider. For practical purposes, the station will act as a regular mobile user of the network. There is a variety of cellular modem equipment that includes router capabilities to establish a LAN as part of the remote station; for example the Peplink industrial rugged series (Peplink 2023).

Satellite communication links allow for wireless access to remote areas not covered by the cellular networks. Some systems allow for Internet connectivity; for example Starlink (2023) and others aim to collect information from remote environmental monitoring stations and sensors; for example Argos (2023). Argos locates the stations on Earth by using the Doppler effect. This system is key to large programs over large areas such as Tropical Ocean-Global Atmosphere program, Tagging of Pacific Pelagics, World Ocean Circulation Experiment, and others. Argos transmitters can track long-distance movement of oceanic animals such as mammal and turtles. We will discuss wildlife tracking in Chapter 14.

WIRELESS SENSOR NETWORKS (WSNs)

During the 1990s, given the advancement in wireless technology and MCU-based SBCs, researchers started to develop networks of wireless-enabled MCU-based SBCs interacting with sensors. These networks became known as WSNs. In a decade, several technologies became available; e.g., U.C. Berkeley iMotes and Crosbow Mica2/Z, and potential environmental monitoring applications were identified (Culler et al. 2004; Mainwaring et al. 2002) allowing data collection with finer spatial and temporal resolution. Since then, WSN technology has evolved considerably as we will describe in this chapter.

Typically, we have a collection of *sensor nodes* communicating by radio among them and with a *base station node*. Each sensor node routes its logged data to the base station node, which assembles and stores the entire dataset for transmission to a PC or an SBC (Figure 6.10). You can think of this network as a spatially distributed datalogger. Figure 6.11 illustrates examples of WSN topology. Besides *point-to-point*, the simplest arrangement is a *star* for which all nodes connect directly to the base station node. In *string multi-hop*, nodes relay data from node to node to the base station node. A *cluster* arrangement consists of nodes connected in star to *cluster heads*, which in turn connect as a star to the base station node. All nodes are connected to each other in a *mesh* arrangement, which offers the most complete connectivity. As WSN are added to remote environmental monitoring systems, we face the challenge of how to power the nodes of the WSN. We cover this topic in Chapter 7, which is devoted to electric power.

WSN STANDARDS AND TECHNOLOGIES

As mentioned above, WSN standards and technologies have been evolving substantially, we will briefly describe some of the most relevant ones (see summary in Table 6.1) (Rawat et al. 2013). *IEEE 802.15.4* covers layers 1–2 and is generic with multiple applications. *ZigBee* covers layers 3–7 and is applied for example to control, monitoring, and home automation. *Bluetooth Low Energy* (BLE) has its own stack protocol that maps to Open Systems Interconnection (OSI) layers, it is

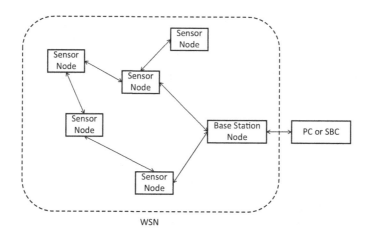

FIGURE 6.10   WSN.

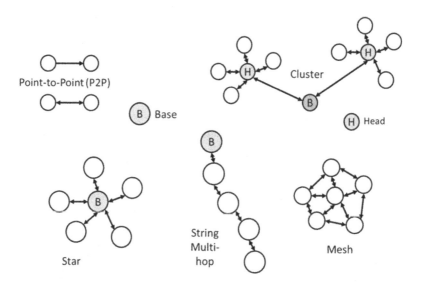

FIGURE 6.11   WSN topology examples.

widespread and applies to phones and entertainment. *Z-Wave* covers layers 1–2 and maps to layers 3, 4, and 7 with applications to home automation. *ANT* covers layers 1–4 and maps to higher layers; it is like BLE, but it is sensor-oriented and popular in fitness and health. *Wavenis* covers layers 1–2, by Open Standard Alliance, used for long distance (1–4 km), low power, and relatively low data rate (100 kbps). *EnOcean* covers layers 1–3 and maps to higher layers, has relatively long distance range (300 m), relatively low data rates (125 Kbps), and specifies battery-less ultra-low power; equivalent to ISO/IEC14543-3-10, with applications to energy harvesting. *Wireless-HART* (Highway Addressable Remote Transducer Protocol) covers layers 1–2 and applies to relatively long distances. *Ultra Wide Band – Impulse Radio (IR-UWB)* for short distances and fast data rate applications. *IEEE 802.15.6* applied to wireless body area networks specifies high data rates (up to 10 Mbps), low power, and short range (5 m).

TABLE 6.1
WSN Specifications

	Frequency	Data rate	Range	Nodes	Topology	Power
IEEE 802.15.4 + ZigBee	0.868, 0.915, 2.4	0.25	100	65536	S, PP, M	Low
Bluetooth Low Energy (BLE)	2.4	1	10	8	PP	Ultra Low
Z-Wave	0.868, 0.908, 0.916	0.040	30	232	M	Low
Ant	2.4	1	30	65533	S, PP, M	Ultra Low
Wavenis	0.433, 0.868, 0.915	0.1	4000	100000	PP	Ultra Low
EnOcean	0.315, 0.868	0.125	300		S, PP, M	Ultra-Low
Wireless Hart	2.4	0.250	250	65536	M	Very low
IR-UWB	3.1, 10.6	10	5	1000	S, M	Ultra low
IEEE 802.15.6	0.4, 2.4, 3.1, 10.6	10	5			Low

IoT

As discussed in Chapter 2, distributed sensor networks, wire and wireless, connected to the Internet and combined with advances in communication, control, and data analysis, has led to the concept of the *IoT* or a large network of interconnected devices and physical objects ("things"). In the next section, we will go in detail on various devices that can be used for WSN and IoT.

WSN Nodes

In environmental monitoring applications, each sensor node performs three main tasks, data acquisition, data processing, and data reporting through wireless communications (Figure 6.12). To accomplish the data-acquisition task, the node collects sensor readings periodically. Then, in the data-processing task, sensor nodes may calibrate, aggregate, summarize, and compress the data. Lastly, during the data-reporting task, data are transmitted wirelessly to a base station node. Depending on the network topology, the nodes may transmit directly to the base station or through *multi-hop* (to be described later in the chapter). A variety of software services implement timing, communication, and networking protocols (Yang et al. 2008, 2010). When a node includes an RTC, sensor data are time-stamped upon sampling, acting as a datalogger. A network with multiple data-logging nodes would require nodes to be synchronized.

A node must include components to perform the above three functions: sensor, processing, and wireless components or units (Figure 6.13). We can use the knowledge gained from earlier chapters and the first part of this chapter, to understand how these three components work. The sensor module includes transducers and signal conditioning, the processing unit includes MCU with ADC to read analog transducers, and the wireless unit is a radio transceiver with internal or external antenna.

Initially, sensor nodes were based on two or more interconnected boards, such as Crossbow motes, which combined an IRIS processor/radio and an MDA300 sensor module. Nowadays, it is common to mount the radio unit on the same board, so that all components are integrated on a single board, as exemplified by the Moteino with RFM69 radio (Figure 6.14) and the NodeMCU with ESP8266 unit for processing and Wi-Fi. We will describe these and other available devices in the next section.

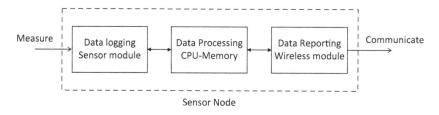

FIGURE 6.12 Functions of a sensor node.

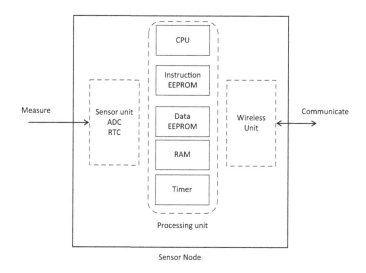

FIGURE 6.13 Components of a sensor node.

EXAMPLES OF DEVICES FOR WSN NODES AND IOT

MOTEINO

Moteino devices are small-footprint (e.g., 3 cm×2 cm) Arduino clones developed to consume low power and can be provided with radio capabilities that could be used to configure a WSN (LowPowerLab 2020). A Moteino device integrates datalogging, processing, and radio (Figure 6.14). On-board radio capabilities provided by an RFM69CW transceiver and wire antenna in 433-MHz, 868-MHz, and 915-MHz range. Compared to 2.4 GHz, these RFs allow longer transmission distances and can provide less attenuation in the field. We will work with Moteino devices following lab guide 6 of the companion Lab Manual.

ESP8266 AND ESP32

ESP8266 modules are low-cost and small-footprint devices based on the ESP8266 SoC that include Wi-Fi, TCP/IP, and MCU capability (Espressif 2022). The simplest ESP chips produced in the mid-2010s, the ESP-01, became popular to provide Wi-Fi and TCP/IP to other MCU boards, such as the Arduino. Development boards based on ESP8266 modules have a micro-USB connector that allows connection to a UART via USB, as well as access to GPIO pins. More recently, the family of ESP32 devices includes several series, ESP32, ESP32-C, ESP32-S2, and ESP32-S3. These devices include Bluetooth, more GPIO pins, ADC, and DAC. We will work with ESP8266 devices in Lab 6 of the companion Lab Manual.

FIGURE 6.14 Moteino by LowPowerLab.

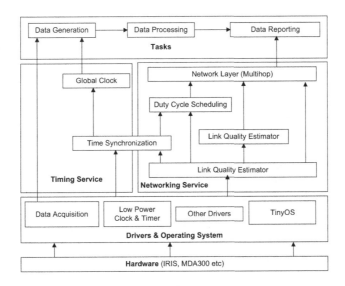

FIGURE 6.15 Functional diagram of a sensor node as implemented in the IRIS/MDA mote. From Yang, J. et al. (2010).

XBEE

Other radios that can be used for WSN include Xbee that is a brand name of Digi International and includes a series of radio modules; many based on the IEEE 802.15.4 standard.

LORA

LoRa is a radio modulation technique that encodes information using a chirped multi-symbol format; however, the term LoRa also refers to the devices and gateways that support the LoRa modulation, and to the LoRa communication network for IoT applications (Link Labs 2018). As a modulation technique, it allows cost-effective communication of low data-rate devices over longer distances. LoRa consists of a fractional-N–synthesized chirp generator which includes a fractional-N synthesizer and a digital ramp synthesizer.

NETWORK PROTOCOLS

MEDIA ACCESS CONTROL (MAC)

Recall from Chapter 2 that the OSI model has seven logical layers, where a layer serves the layer above it while being served by the layer below it. The MAC protocol is part or a sub-layer of layer 2 or data link layer. The MAC sub-layer provides addressing and access of multiple nodes to a shared medium. A medium is a general term that refers to a communication link, for example Ethernet. The MAC is an interface between layer 1 (Physical layer) and the logical link control sub-layer of layer 2 and emulates a full-duplex channel in a multi-point network.

MAC sub-layer protocols are categorized into two large groups, schedule-based and contention-based methods (Stallings 2004; Ye et al. 2002). In schedule-based protocols, wireless devices occupy different channels that are physically or logically independent. To achieve channel separation, we use time, frequency, and code division for MA methods (i.e., TDMA, FDMA, and CDMA, as described earlier in this chapter). In contrast, in contention-based protocols such as the carrier-sense MA (CSMA) method, wireless devices compete for a single shared channel. In CSMA, a node would check for existing traffic before transmitting on the channel.

MULTI-HOP WIRELESS COMMUNICATION

A WSN consists of nodes that communicate with each other and hop messages to the base station, which passes the messages to a PC or SBC, or "client" (Crossbow 2006b). By hopping data in this manner, we improve radio communications coverage and reliability while reducing power. Two nodes do not need to be in direct radio link; instead, nodes in between route messages. Similarly, messages can be re-routed when conditions impede communications between two nodes. Instead of limiting communication between the base station and one network node (the last link in the net), multi-hop allows several nodes receive and transmit information. Each hop is short distance.

LoRa-BASED PROTOCOLS

A MAC layer can be used to adopt the LoRa technology as the physical layer. For example, LoRaWAN is a low-power and wide-area MAC-based network protocol facilitating connection of WSN nodes and IoT devices, using LoRa as the physical layer, to the Internet (LoRa Alliance 2022). It is based on a star-of-stars topology in which gateways of each star relay messages between its end-devices and a server. The gateways are connected to the network server via IP and act as a bridge, converting RF packets to IP packets and vice versa. LoRaWAN is best suited for public networks because all devices are tuned to the same channel.

NETWORK PROTOCOL FOR ENVIRONMENTAL MONITORING

A suitable protocol for environmental monitoring applications would typically allow distributed sensor nodes to form a spanning-tree structure, rooted in a single data-collection base station or sink node, so that majority of the data traffic is from sensor node to the base station node.

Many existing protocols are intended for peer-to-peer applications where data are routed between any pair of nodes in the network. However, this type of protocol usually introduces a large amount of communication overhead and cannot exploit the unique tree structure to minimize energy consumption. Wireless communication and networking are among the most energy-consuming operations that a node performs. Therefore, the design of a protocol has an impact on energy efficiency.

A contention-based protocol is autonomous but energy-inefficient due to high collision rate in the shared channel and idle listening. A schedule-based protocol may eliminate these issues and achieve energy efficiency, but it may suffer interference from other types of devices operating in the same frequency band, especially in the unlicensed Industrial, Scientific, and Medical (ISM) bands (e.g., Wi-Fi) that sensor networks often employ.

In multi-hop WSN, though most of the traffic is many-to-one, nodes broadcast some of the control and signaling packets to neighboring nodes in order to establish multi-hop routes. Consequently, most of the MAC protocols designed for sensor networks adopt CSMA as the baseline mechanism but then implement time-slot scheduling algorithms to coordinate duty cycling.

Simple duty cycling scheme, where a mote keeps the radio on for a fraction of the time, and synchronous sleep scheduling makes the systems not scalable to capture spatial variation characteristics in a large area. Therefore, a hybrid MAC layer networking protocol was developed for a soil moisture environmental monitoring application. This hybrid protocol integrates CSMA and duty cycle scheduling to coordinate when sensor nodes go to sleep and achieve energy efficiency. Like many low-power MAC protocols available in the literature, this protocol divides the entire time axis into super frames, each of which is divided into time slots, as illustrated in Figure 6.16. At the beginning of a super frame, all nodes broadcast and receive packets competing for TDMA slots. A parent node assigns a TDMA slot to its child upon request. Then, sensor nodes turn off their radios and remain asleep except in their own active TDMA slots. This hybrid protocol strives to retain the flexibility of contention-based protocols while improving energy efficiency in multi-hop networks.

MQTT Protocol

MQTT is a lightweight messaging protocol useful to interconnect remote devices with reduced code footprint and network bandwidth (MQTT 2022). Its name originated from the acronym MQ Telemetry Transport, but the shorter name MQTT is most commonly used. The protocol defines how devices can exchange data over TCP/IP using the *publish/subscribe* pattern, in which the publisher and the subscriber do not communicate directly but via a *broker* that handles the connection, processing incoming messages and distributing them to the subscribers (HiveMQ 2022; Steve's Internet Guide 2022a).

The publisher and subscriber do not exchange IP addresses, do not need to run at the same time, and do not have to interrupt operation during publishing and subscribing. Instead, the broker filters incoming messages using the *topic* (subject) string embedded in the message; the subscribers will sign up for the topics of interest, and the broker will provide the subscribed topics. From a client/server perspective, both the publisher and subscriber are MQTT clients that connect to the MQTT broker. A client may range from a lightweight WSN node to an SBC running more complicated programs. Client libraries are available for many languages, including Arduino, Python, and PHP (MQTT.org 2022).

MQTT, at OSI layers 5 and 7, is based on TCP/IP, and thus clients and the broker must have TCP/IP (OSI layers 3 and 4). A connection is initiated by a client sending a message to the broker that responds with an acknowledgment message (CONNACK), remaining open until the client sends a

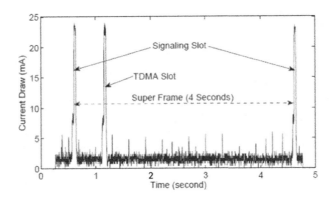

FIGURE 6.16 Time slot structure of a superframe in a hybrid MAC protocol. From Yang, J. et al. (2010).

disconnect message. On a LAN, clients on the same network as the broker, IP exchange is trivial, but in many IoT applications the devices are on separate networks. For example, when the broker is on an Internet public address, MQTT client's private IP is translated by a router to a public address, allowing Internet connection to the broker's public IP address.

A message is composed of a topic, used by the broker to filter messages to the subscriber, and a *payload*, which has the actual data to be exchanged. For example, for a message from a temperature sensor the topic could be "TempC" and the payload "26.7". Topics are structured by level separated by forward slash, e.g., in a WSN a message from sensor node 1 which measures temperature (among other variables), the topic could be "Node1/TempC" and the payload "26.7". A subscriber client connected to the broker would subscribe to this topic and thus the broker would deliver the message with topic "Node1/TempC" and payload 26.7. This client then would have computer code to use this information for whatever application is needed. Structuring topics as well as message payload are then very important in MQTT design (AWS 2022; HiveMQQT 2022; Steve's Internet Guide 2022b).

In lab 6 session of the Lab manual companion of this textbook, we will practice MQTT messaging using an example of WSN nodes implemented in ESP8266 acting as client publishers, one broker implemented in a Raspberry Pi, and one subscriber client running on the same Raspberry Pi.

WSN AND ENVIRONMENTAL MONITORING: PRACTICAL CONSIDERATIONS AND EXAMPLES

Radio Propagation and WSN

There are practical issues when deploying a WSN in the field, including receiver signal strength (RSS) and its relationship with battery power and distance between nodes. All these factors influence the number of nodes used and the configuration of the WSN such that we can establish reliable communication between neighboring nodes. Radio propagation is affected by many factors, particularly in field conditions with changing vegetation foliage and weather conditions.

In many WSN outdoor deployment for environmental monitoring, the antennas are not high above the ground (e.g., 1 m), and the distance between the sensor nodes is relatively small (up to 50 m). These settings are not common in most existing radio wave propagation models such as the ones we discussed earlier in this chapter. Some specific analysis can help have an insight into basic propagation mechanisms over a reflecting surface as well as the influence of foliage (Thelen et al. 2009).

Recall that crossover distance (d_c) determines what propagation model to apply. Assume a scenario of $h_t = h_r = 1$ m = height of the transmitting and receiving antennas and a frequency of 2.4 GHz. We obtain $d_c \sim 100$ m. Therefore, we would have to keep a distance less than 100 m to apply the free-space model, otherwise we should use the two-ray model. Note that if the antennas were to be lower say 0.5 m, the crossover distance shortens to 25 m; this would limit the application of free-space model vs. the two-ray model.

Let us then consider a two-ray model for WSN applications. Recall that the two-ray model Eqution (6.10). We can convert to a propagation loss L equation from the ratio $\dfrac{P_r}{P_t}$ in dB assuming isotropic antennas $L(dB) = -20 \times \log(h_t) - 20 \times \log(h_r) + 40 \times \log(d_m)$. Note that antennas lower than 1 m would contribute to the loss, whereas antennas higher than 1 m will decrease the loss. We can estimate that 1-m antenna heights would make the loss solely a function of distance. Also, note that the above equation implies a 12-dB reduction for a two-fold increase in distance.

Consider, for example 1-m antennas and 10-m distance. We would get a loss of 40 dBm; doubling this distance to 20 m would give 12 dB more for a loss of 52 dBm and doubling the distance gain (40 m) would add 12 dB more or a total of 64 dBm. For a perspective, this is about half the loss you would get for a 3-km Wi-Fi link using a free-space model. One more thing to think about is that in the two-ray model, we assume perfect reflection from the ground (or a coefficient equal to −1). However, reflection from the ground varies with soil wetness, near-ground vegetation cover, and of course the carrier frequency. In a vegetated environment, loss would increase with increasing frequency, height of the canopy with respect to antenna height, and moisture content of foliage (Thelen et al. 2009).

RADIO PROPAGATION EXPERIMENTS FOR WSN

The considerations of the previous section give us a framework to consider propagation issues when deploying a WSN in the field; but, because of the high variability of field conditions with topography, changing vegetation foliage, and weather conditions, we require experimentation to plan a reliable configuration of the network, including network topology.

Experiments include several variables. RSS is a measure of the received RF signal. Sensitivity is defined as the lowest RSS that allows obtaining complete data from a neighboring node. Packet receiver rate (PRR) is an indicator of the percentage data received from a neighboring node and is related to the RSS and the distance between sensor nodes. PRR decreases when transmission is weak, and data could be lost, and energy consumption will be higher due to re-transmission. A design goal is to find a configuration so that data retrieval is close to 100%.

For experimentation, we can use on-board elements of sensor nodes such as temperature, relative humidity (RH), RSS, PRR, and battery voltage. For example, research with the Chipcon CC1000 radio, part of the Mica2Dot sensor node indicated that radio waves are limited to 10m when potato crops are in flower, and radio waves propagate better at times of the day of high RH and rain (Thelen et al. 2009).

Experiments conducted at the UNT DP campus using crossbow motes have addressed issues of RSS, PRR, battery power, for varying distance between nodes, network topology, and in contrasting vegetation conditions. Results indicated that to maintain a given level of PRR (95%), the maximum distance decreases for increasing vegetation density; 70m in open field with short grass vegetation, 50m in tall grass with scattered trees, and 30m in forested areas (Chegwidden and Wood 2009). Air temperature and humidity affect the RSS, attaining better RSS as air temperature decreases and humidity increases. Experimental topology setup included parallel transects (two rows of nodes, covering 150 m×10m plot, each node was approximately 10m from its nearest neighbor nodes, making the furthest node 150-m away) or more interconnected topology.

Experiments at UNT have also addressed practical issues related to battery discharge according to ambient temperature and its relation to the use of enclosures and battery type. Temperature extremes (high and low) cause battery discharge; this effect has implications in enclosure design and is reduced using alkaline batteries (Arnold and De Lemos 2009).

EXAMPLE: WSN FOR SOIL MOISTURE IN A HARDWOOD BOTTOMLAND FOREST

During the early 2000s, we developed a WSN for distributed soil moisture monitoring in a hardwood bottomland forest, located in the Greenbelt Corridor (North Central Texas). The WSN was composed of IRIS/MDA motes and interfaces to an SBC (see Chapter 2) that serves as a Remote Field Gateway (RFG) Server, providing control and management of the WSN.

This early implementation of WSN for environmental monitoring employed the IRIS XM2110 that is a processor and wireless module based on the ATmega1281 MCU using a TinyOS operating system and a reduced set of C functions (Crossbow 2006a). It was low power, with a small footprint (6 cm×3 cm) and has a connector for an antenna. The radio uses 2.4GHz with power of 3 dBm. Operating distance range varies and can be more limited depending on field conditions (e.g., about 50m in a forest environment under dry conditions). The MDA300 interfaces with the IRIS via a 51-pin connector and has on-board air temperature and RH sensors and a wiring panel to connect external sensors. Figure 6.15 is a functional diagram of a sensor node implemented by an IRIS/MDA combination (Crossbow 2006b).

The SBC also coordinates the WSN with a traditional datalogger-based wired sensor system (Figure 6.17). We implemented the RFG server using an ultra-low-power SBC TS-7260 from Technology Systems, Inc. (Technologic 2014) which is Linux-based and supports remote

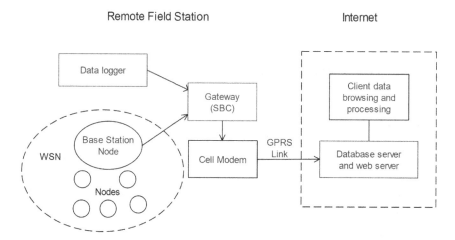

FIGURE 6.17   Integration of WSN with an SBC and dataloggers.

manipulation of the devices in the field. Sensor data collected from the distributed monitoring stations are stored in a local database at the SBC, which transmits data to a central data collection (CDC) server and stored in a centralized database (we will study database concepts in Chapter 10).

To minimize energy consumption, the environmental monitoring system undergoes a duty cycle between the active and sleep modes as explained earlier in this chapter. The SBC has RS232 serial ports, USB, and Ethernet ports facilitating its connection with the datalogger and the WSN gateway, modem, and other devices For several years, the long-haul wireless communication from the field to the CDC server was implemented using a GPRS (General Packet Radio Service) cellular modem. To be energy-efficient, the wireless modem is turned off during the system's sleep period. A photovoltaic panel and a rechargeable battery (we will cover power concepts in Chapter 7) power the SBC, datalogger, and modem.

The RFG server wakes up periodically to carry out data collection services. Upon boot up, the RFG server executes a series of scripts to initiate various services, including an event logging daemon, an FTP server, and an SSH terminal. Then, the SBC executes several data collection processes to start to poll data from the WSN node and dataloggers through RS232 ports and insert into a local MySQL database (Chapter 10).

Motes are deployed in weatherproof boxes, equipped with external antennas, and the boxes are installed 1 m above the ground on top of metal poles to avoid flooding water (Figure 6.18). Prior to the deployment, we conducted a site survey to measure the one-hop radio communication range between motes in the forest environment. As discussed earlier in this chapter, radio propagation characteristics vary significantly with vegetation characteristics and density. In summer surveys, at maximum Tx power, IRIS motes were able to transmit 50 m on average with 80% packet reception rate. Thus, we deployed motes with a maximum one-hop distance of about 30 m.

The WSN deployment consisted of about 30 motes equipped with soil moisture sensors, and about 20 radio-only motes that serve as relay points (Figure 6.19). Each one of the 30 sensor nodes collects data every 10 minutes from soil moisture sensors (connected by wire to the motes); simultaneously, the motes read on-board temperature and RH sensors. The network topology supports long-term hydrologic monitoring and modeling in the floodplain area alongside the Trinity River, providing an opportunity to collect a duplicated set of soil moisture variation along a cross-sectional transect from the river levee (higher elevation and sandy soil) to the central weather station (lower elevation and clay soil). Characterizing soil moisture variation with respect to elevation and soil type is vital to understanding vegetation distribution along the floodplain as well as responses to flooding (Figure 6.20).

FIGURE 6.18   Motes in the forest. Enclosure and deployment using an external antenna and AA batteries.

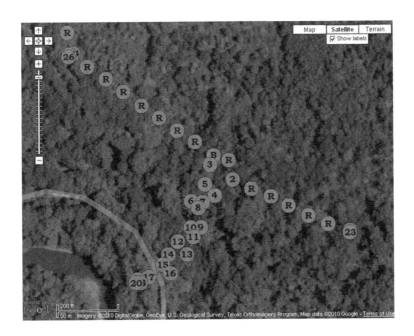

FIGURE 6.19   Topology of WSN to monitor soil moisture in a bottomland hardwood forest. Greenbelt corridor (North central Texas).

EXAMPLE: WSN FOR SOIL MOISTURE USING MOTEINO

Iyiola (2017) implemented a WSN based on Moteino devices with RFM69 radios operating at 915 MHz, using a star topology, with 1 gateway node and 9 nodes measuring air temperature and soil moisture, at distances ranging 10–30 m from the gateway. A node includes two types of sensors: the BME280, via a weather shield by Low Power Lab (2016b), combining air temperature, RH, and

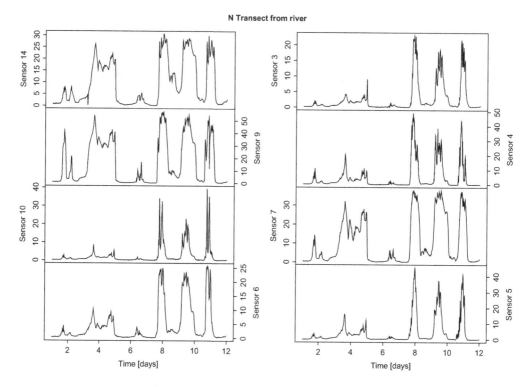

FIGURE 6.20   Example of soil moisture data collected by WSN in a bottomland hardwood forest. Greenbelt corridor (North central Texas).

barometric pressure sensor, and the ECHO-5, which measures soil moisture (Meter 2022). A shield also by Low Power Lab (2016a) powers the sensor module by a Lithium Polymer battery and a solar panel for battery charging. See more details on the topic of WSN power and battery charging in Chapter 7.

Humidity and pressure in the BME280 can be independently enabled and interfaces with the Moteino through the I2C or SPI interfaces. It has a current consumption of about 3.6 μA when taking readings and only 0.1 μA in sleep mode (Low Power Lab 2016b). Programming each sensor node with an automatic transmission control feature helped to dial down transmit power which was scheduled every 10 minutes. Preliminary tests indicated that RSS for a node decreased from −40 to −90 dBm as distance was varied from proximity to 200-m away from the gateway. The Moteino serving as gateway node communicates via serial using USB with a Raspberry Pi 3 that collects all data and runs a web server.

EXAMPLE: SOIL MONITORING USING ESP8266 AND MQTT

As an example of soil conditions monitoring, senior design students at UNT implemented a WSN for measuring soil moisture, electrical conductivity (EC), and temperature using NodeMCUs as WSN nodes that act as MQTT publishers and a Raspberry Pi 3 acting as an MQTT broker (Librado et al. 2021). Each WSN node is formed by a NodeMCU based on ESP8266 that monitors RS-485–based JXBS-3001-EC sensors (presented in Chapter 5) that read soil moisture, EC, and temperature. For that purpose, the ESP8266 was connected to the RS-485 bus using a MAX3485 driver. At each node, the ESP8266 captures data from the sensor and acts as an

MQTT client, in this case a publisher, transmitting data to the Raspberry Pi, which acts the broker. The database program InfluxDB (Influxdata 2022) installed in the Raspberry Pi was used to store and organize the data, which was accessed by a visualization provider (Grafana Labs 2022) acting as MQTT subscriber; in this manner, multiple devices can connect and access the data published by the nodes.

EXERCISES

Exercise 6.1

Calculate PL for a Wi-Fi 5-GHz low-band link between two sites separated by 3 km. The antennas are at 20 m above the ground for both locations. Then calculate received power in dBm assuming gains of 5 dBi for each antenna and 30-dBm transmitter power. We found a patch of 20-m tall trees at 0.3 km from one location. Determine if we clear the first Fresnel zone.

Exercise 6.2

Geographical coordinates of a monitoring station near Denton, Texas are 33°15′30.93″ N and 97° 2′20.15″ W. We want to establish a 2.4-GHz Wi-Fi link with another location with the same latitude but at 97°3′48.83″ W. Use Google Earth to calculate distance between these locations and whether land cover would allow LoS without elevating the antennas. Then, assuming antennas can be sufficiently elevated for LoS and clearance of first Fresnel zone, calculate PL.

Exercise 6.3

Consider a 101 × 101-cell segment of a DEM with 30-m cell size. We want to establish a 900-MHz radio link between stations located as follows. Station 1 is at the center of the cell in the SW corner; station 2 is at the center of the cell in the NE corner. The elevation values at these cells are 100 m for both. The cell at the center of the DEM (at row 50, column 50) is 110-m elevation. Calculate tower heights to obtain first Fresnel zone clearance. Calculate PL.

Exercise 6.4

Considering a WSN following the hybrid MAC protocol of Yang et al. (2010) which implies a super frame of 5 seconds consisting of a signaling slot of 0.1 second and a TDMA slot of 0.1 second. Suppose a node draws 20 mA when the radio is activated and 0.5 mA when the node is sleeping. We use two AA batteries of 1.6 V in series and 2000 mAh. There is a superframe every 30 minutes or two superframes per hour. Calculate total current drawn by a non-parent node (only one TDMA slot) in 24 hours. Calculate how many days it would take to discharge the batteries.

Exercise 6.5

Assume four nodes of a WSN at the corners of a rectangle of size 10 m×40 m oriented such that the longest side is East to West. Antennas are 1 m above the ground. The land is flat and cover is very short grass.

  a. How many pairs of nodes would you have? List the pairs.
  b. How many unique values of distance would you have among all node pairs? List and calculate these distances.
  c. Calculate all expected propagation losses assuming that the frequency is 900 MHz.
  d. Which node would receive the weakest signal from the NW corner?

Exercise 6.6

You are planning to deploy a soil moisture WSN in a forest over flat land using 36 nodes arranged such to cover 300 m × 300 m of land uniformly.

    a. What frequency would you prefer to use for this WSN deployment, 433 or 900 MHz, 2.4 or 5 GHz? Explain by discussing advantages and disadvantages among all these RF options for this application.

    b. Select antenna height for all nodes, calculate minimum crossover distance, and expected worst case of power loss between two nodes.

REFERENCES

Acevedo, M.F. 2024. *Real-Time Environmental Monitoring: Sensors and Systems*, Second Edition – Lab Manual. Boca Raton, FL: CRC Press, Taylor & Francis Group. 463 pp.

Argos. 2023. *ARGOS.* Accessed May 2023. https://www.argos-system.org/.

Arnold, D. and L. De Lemos. 2009. *The Effect of Extreme Temperature on the Lifetime of Battery Charge. RET Summer Internship Program 2009* Denton, TX: University of North Texas.

AWS. 2022. *MQTT Design Best Practices.* accessed December 2022. https://docs.aws.amazon.com/whitepapers/latest/designing-mqtt-topics-aws-iot-core/mqtt-design-best-practices.html.

Chegwidden, D. and S. Wood. 2009. *Wireless Sensor Networking, RET Summer Internship Program 2009.* Denton, TX: University of North Texas.

Crossbow. 2006a. *IRIS Datasheet.* San Jose, CA: Crossbow.

Crossbow. 2006b. *MoteWorks Getting Started Guide.* San Jose, CA: Crossbow Technology Inc.

Culler, D., D. Estrin, and M. Srivastava. 2004. Overview of sensor networks. *IEEE Computer* 37 (8):41–49.

Espressif. 2022. *ESP8266 Series of Modules.* accessed February 2022 https://www.espressif.com/en/products/modules/esp8266.

Grafana Labs. 2022. *Grafana: The Open Observability Platform.* accessed September 2022. https://grafana.com/.

Gurung, S. 2009. *Integrating Environmental Data Acquisition and Low Cost Wi-Fi Data Communication.* MS Thesis, Electrical Engineering. Denton, TX: University of North Texas. 258 pp.

Hive, M. Q. 2022. *MQTT Essentials. The Ultimate Guide to MQTT for Beginners and Experts.* accessed September 2022. https://www.hivemq.com/mqtt-essentials/.

Hive, M. Q. 2022. *MQTT Topics, Wildcards, & Best Practices: MQTT Essentials.* accessed December 2022. https://www.hivemq.com/blog/mqtt-essentials-part-5-mqtt-topics-best-practices/.

Influxdata. 2022. *InfluxDB v1.7.* accessed September 2022. https://docs.influxdata.com/influxdb/v1.7/introduction/getting-started/.

Iyiola, S. 2017. *Moteino-Based Wireless Data Transfer for Environmental Monitoring.* MS Thesis, Electrical Engineering. Denton, TX: University of North Texas. 69 pp.

Librado, A., N. Farrar, and Z. Sihalla. 2021. *Soil Quality Network (SQN) and Data Logging Node, Senior Design Project.* Denton, TX: University of North Texas.

Link Labs. 2018. *LoRa: A Breakdown of What It Is & How It Works.* accessed September 2022. https://www.link-labs.com/blog/what-is-lora.

LoRa Alliance. 2022. *What is LoRaWAN® Specification.* accessed September 2022. https://lora-alliance.org/about-lorawan/.

Low Power Lab. 2016a. *PowerShield.* https://lowpowerlab.com/guide/powershield/features/.

Low Power Lab. 2016b. *WeatherShield R2 released.* accessed November 2016. https://lowpowerlab.com/2016/09/09/weathershield-r2-released/.

LowPowerLab. 2020. *LowPowerLab.* accessed March 2020. https://lowpowerlab.com.

Mainwaring, A., J. Polastre, R. Szewczyk, D. Culler, and J. Anderson. 2002. *Wireless Sensor Networks for Habitat Monitoring.* Atlanta, GA: WSNA.

Meter. 2022. *ECH20 EC-5 Soil Moisture Sensor.* accessed April 2022. https://www.metergroup.com/environment/products/ec-5-soil-moisture-sensor/.

MQTT. 2022. *MQTT: The Standard for IoT Messaging.* accessed September 2022. https://mqtt.org/.

MQTT.org. 2022. *Libraries.* accessed September 2022. https://github.com/mqtt/mqtt.org/wiki/libraries.

Peplink. 2023. *MAX BR Series.* accessed May 2023. https://www.peplink.com/products/max-br-series/.

Rawat, P., K. Singh, H. Chaouchi, and J.-M. Bonnin. 2013. Wireless sensor networks: A survey on recent developments and potential synergies. *The Journal of Supercomputing* 68:1–50.

Stallings, W. 2004. *Wireless Communications & Networks*. Upper Saddle River, NJ: Prentice Hall.

Starlink. 2023. *Starlink Roam*. accessed May 2023. https://www.starlink.com/roam

Steve's Internet Guide. 2022a. *MQTT Publish and Subscribe Beginners Guide*. accessed December 2022. http://www.steves-internet-guide.com/mqtt-publish-subscribe/.

Steve's Internet Guide. 2022b. *MQTT Topic and Payload Design Notes*. accessed December 2022. http://www.steves-internet-guide.com/mqtt-topic-payload-design-notes/.

Technologic. 2014. *TS-7260 High Performance at Ultra-Low Power*. accessed April 2015. https://www.embeddedarm.com/products/board-detail.php?product=TS-7260.

Thelen, J., D. Goense, and K. Langendoen. 2009. *Radio Wave Propagation in Potato Fields*. accessed 2015. http://www.st.ewi.tudelft.nl/~koen/papers/winmee.pdf.

Ubiquiti. 2009. *Ubiquity Networks. NanoStation2 Datasheet*. accessed 2014. http://www.ubnt.com/downloads/ns2_datasheet.pdf.

WNDW. 2013. *Wireless Networking in the Developing World*. accessed 2023. https://wndw.net/download/WNDW_Standard.pdf

Yang, J, C. Zhang, X. Li, Y. Huang, S. Fu, and M. Acevedo. 2010. Integration of wireless sensor networks in environmental monitoring cyber infrastructure. *Wireless Networks* 16 (4):1091–1108.

Yang, J., C. Zhang, X. Li, Y. Huang, S. Fu, and M. Acevedo. 2008. "An Environmental Monitoring System with Integrated Wired and Wireless Sensors." In: *Wireless Algorithms, Systems, and Applications. WASA 2008. Lecture Notes in Computer Science*, edited by Y. Li, D.T. Huynh, S.K. Das, D.Z. Du, vol. 5258, pp. 224–236. Berlin, Heidelberg: Springer. https://doi.org/10.1007/978-3-540-88582-5_23

Ye, W., J. Heidemann, and D. Estrin. 2002. *An Energy-Efficient MAC Protocol for Wireless Sensor Networks*. New York, NY: IEEE Computer and Communications Societies.

7 Environmental Monitoring and Electric Power

INTRODUCTION

This chapter is devoted to the relationship of environmental monitoring and electric power systems, covering three major topics: understanding how to power a remote environmental station when its location precludes the use of the power grid, how to power Wireless Sensor Network (WSN) nodes, and applying environmental monitoring for the design and operation of renewable power systems. In terms of powering off-grid remote monitoring stations, we focus on photovoltaic (PV) solar panels since they offer a practical solution to this application; therefore, this chapter covers PV cells and panels. For WSN nodes in open areas, solar panels can be employed to recharge batteries, but in shaded areas, other energy harvesting techniques are required. Alternatives include harvesting or scavenging ambient energy, such as vibration, heat, and RF waves. Finally, this chapter covers the role of environmental monitoring in providing data useful to the design and operation of renewable power, for example solar, wind, and hydropower. A major contribution of monitoring is understanding the resource used by these forms of renewable power, e.g., solar radiation, wind speed, and water flow.

PV PANELS

The PV effect is the phenomenon of converting energy of photons of light to electrical current. First reported in 1839 by Becquerel, using silver-chloride or silver-bromide–coated electrodes, the PV effect was used in cells made from Selenium as early as 1880 but the conversion had very low efficiency (~1%) and the process was not well understood. Later in that decade, the *photoelectric* effect, or the production of current due light incident on metal, was discovered, and the intriguing differences in the effect due to wavelength of light were later explained by Einstein, for which he received the Nobel Prize.

In the 1950s, research demonstrated how to make PV cells from silicon crystals, and demand for space exploration applications in the 1960s advanced the technology considerably. Since the 1970s, there has been an increased use in residential, commercial, and utility-scale applications. Most remote off-grid monitoring stations use PV technology to provide power to its electronics system.

PV CELLS

A PV cell is made from a p-n junction using semiconductor material, mostly crystalline Silicon (Si) but also Germanium (Ge). These elements are combined with Boron (B), Phosphorous (P), Gallium (Ga), and Arsenic (As), among others. We can better understand the PV effect by using the energy band model of semiconductors. This model assumes that electrons can move from the last filled energy band to the conduction band and that jump requires an amount of energy at least equal to the "band-gap energy" or E_g which varies from material to material. For example, in silicon, $E_g = 1.12 \, \text{eV}$. One eV is an amount of energy equivalent to $1.6 \times 10^{-19} \, \text{J}$.

Light provides the energy to move electrons to the conduction band or in other words to jump the E_g gap. Of course, for this to happen, light must be of a frequency such that a photon must have energy of at least E_g to move the electrons. The photon energy E in J is directly related to the frequency ν of the electromagnetic wave by the Planck's constant $h = 6.6 \times 10^{-34}$ J-sec. This is to say $E = h\nu$. As we know frequency is the inverse of wavelength using the speed of light $c = 3 \times 10^8 \, \text{m/s}$

DOI: 10.1201/9781003425496-7

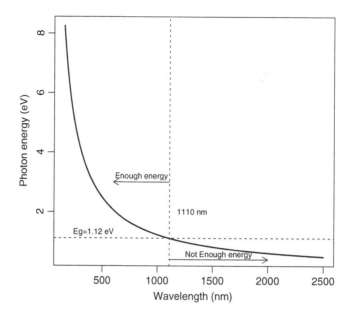

FIGURE 7.1   Required photon energy.

$v = \dfrac{c}{\lambda}$ Therefore, we can also write the energy of a photon as $E = \dfrac{hc}{\lambda}$. Using this relationship, we

can see that to overcome the bandgap, the photon energy should exceed Eg $E = \dfrac{hc}{\lambda} \geq E_g$.

For example, in Silicon

$$\lambda \leq \frac{hc}{Eg} = \frac{\overbrace{6.6 \times 10^{-34}}^{\text{J-s}} \times \overbrace{3 \times 10^{8}}^{\text{m/s}}}{\underbrace{1.12}_{\text{eV}} \times \underbrace{1.6 \times 10^{-19}}_{\text{J/ev}}} = 1.11 \times 10^{-6} \, \text{m} = 1110 \, \text{nm}$$

which means that light should have a wavelength shorter than 1110 nm or 1.11 μm to overcome the band-gap energy. However, a shorter wavelength with photon energy above this level is wasted (Figure 7.1) because the required level is equal to the usable (Masters 2013b). The percentage of the energy available from 1100 to 150 nm that is wasted represents about 40%. This critical wavelength 1.11 μm has a frequency of 270 THz or 0.27 PHz, which corresponds to waves just above the red part of the visible spectrum. As a reference to what we studied in radio waves, this frequency is 100,000 larger than Wi-Fi 2.4 GHz and which has a wavelength of 60 mm.

Solar radiation reaching Earth is distributed by wavelength according to Planck's law of black-body radiation increasing for short wavelengths (150–500 nm), reaching a peak at ~500 nm, and then decreasing as wavelength increases (Figure 7.2). The higher values of this density function correspond to the visible part of the spectrum. As we can observe from this figure, the right tail of this density (above 1100 nm) has energy lower than that required to excite the electrons in Silicon; the area under the curve for this tail amounts to about 20% of the total area. The energy corresponding to wavelengths shorter than 1100 nm represents then the remainder 80% and can excite the electrons, but as we know from the previous section only 60% is usable, thus $0.6 \times 80\% = 48\%$ is usable. Consequently, PV theoretical maximum efficiency for Silicon is slightly less than 50%. One manner to increase efficiency is to increase band-gap energy to maximize power, ideally in the 1.2–1.8 eV by adding elements to Silicon, thus making cells such as GaAs and CdTe (Masters 2013b).

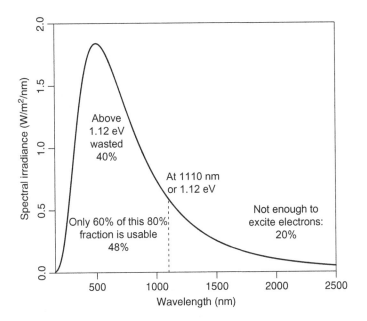

FIGURE 7.2 Solar radiation spectrum with required and usable photon energy.

However, the efficiency is lower because of other factors: PV cells get hot and radiate energy following the blackbody radiation law (7% loss) and there is a hole-saturation effect causing hole-electron pairs to recombine (~10% loss). Therefore, the theoretical efficiency is reduced to ~33%. The Shockley-Queisser limit states that the maximum efficiency is 33.7%. Other factors further reduce practical efficiency: radiation reaching the Earth surface is affected by atmospheric processes, including air mass path length, scattering (diffuse radiation), and reflection.

PV-CELL MODEL AND I-V CHARACTERISTICS

A p-n junction is formed by adjoining semiconductor material that has been differentially doped and forms a diode. The basic diode model relates current I and voltage V by an exponential function that depends on temperature T (in K), as follows:

$$I = I_0 \left[\exp\left(\frac{qV}{kT} \right) - 1 \right] \tag{7.1}$$

where the factor $\frac{q}{k}$ is 11.6×10^3, and I_0 is the current at $V = 0$. At reference ambient T of 25°C or 298K, we have $\frac{qV}{kT} = 38.9 \times V$. For example, for $I_0 = 1$ nA, we would have a I-V relationship of $I = \left[\exp(38.9 \times V) - 1 \right]$ nA. For small voltages, there is a small reverse current, whereas for larger voltages, there is a large increase of forward current as we approach the 0.5 V value.

A solar cell model consists of a diode and a current source that produces current when illuminated by radiation (Figure 7.3). When in short circuit, or zero load resistance, the cell would produce current I_{sc}, or short-circuit current. Its value at full sun conditions or fully illuminated becomes a specification of a solar cell; however, when the cell is in the dark, the current is just the inverse of the current of the diode I_d. When the cell is illuminated, the current is the I_{sc} minus the diode current I_d, that is to say $I = I_{sc} - I_d$. This is the simplest cell model; its I-V relation (Figure 7.4).

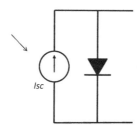

FIGURE 7.3   Simplest cell model: a current source and a diode.

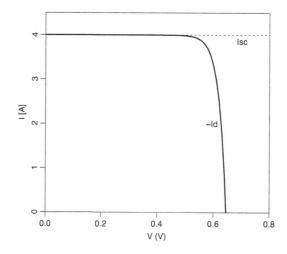

FIGURE 7.4   Simplest cell model: I-V characteristic.

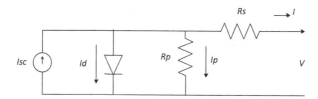

FIGURE 7.5   A more realistic model.

A more realistic model (Figure 7.5) includes a resistance R_s in series to account for lead and contact resistance and a parallel resistance R_p to account for a leak current path when the cell is shaded. Using Kirchhoff's Current Law and Ohm's law, we derive the equation for this model

$$I = I_{sc} - I_d - I_p = I_{sc} - I_0 \left[\exp(38.9 \times (V + I \times R_s)) - 1 \right] - \frac{V + I \times R_s}{R_p}$$

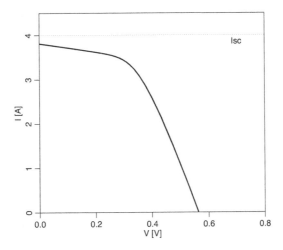

FIGURE 7.6 I-V of a more realistic model.

or more practical, separating V and I,

$$I = I_{sc} - I_0 \left[\exp(38.9 \times V_d) - 1 \right] - \frac{V_d}{R_p} \tag{7.2}$$

$$V = V_d - I \times R_s \tag{7.3}$$

To calculate current, we apply Equation (7.2) given V_d, then to calculate voltage, we use equation (7.3). These equations are implemented in a function of the R package renpow (Acevedo 2018), which is covered in lab 7 of the companion Lab Manual (Acevedo 2023). We can see the I-V characteristics in Figure 7.6.

When we look at power as the product of current and voltage, we see that power increases as voltage increases but then quickly decreases due to the decrease in current (Figure 7.7). Since I_{sc} is a function of illumination, as light decreases, the I_{sc} goes down and therefore the cell produces less current for the same voltage (Figure 7.8), until the current is zero at open-circuit voltage V_{oc} of about 0.6 V.

We can see that power also goes down as the current diminishes for the same voltage (Figure 7.9); in this figure, we have annotated the values of maximum power for each light level, and we can see how this maximum power decreases and occurs for a higher voltage as light decreases. For full illumination, the maximum power point (MPP) of ~1.2 W in this example occurs at about 4.5 V. We conclude that to optimize power output from the cell, we must modify the operating voltage to track the MPP as light changes.

FROM CELL TO MODULE

One PV cell outputs V_{oc} ~0.6 V, which is impractical to power most devices, therefore, to obtain larger voltages, cells are wired in series to obtain a module. For instance, 18 cells would output about V_{oc} ~0.6 V × 18 = 10.8 V, which would be considered a nominal 6-V module, with an operating voltage for MPP at full illumination of ~0.45 V × 18 = 8.6 V. Note that series wiring increases the

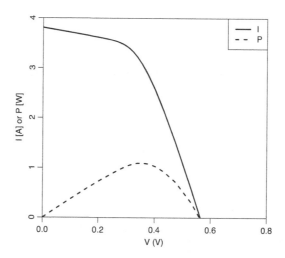

FIGURE 7.7 Current and power.

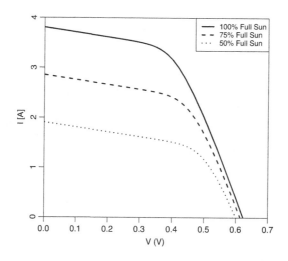

FIGURE 7.8 I-V as a function of light.

voltage but not the current, which must remain the same. A 6-V module would be useful to power small monitoring devices, such as a 3.3-V–based WSN node, such as the NodeMCU and Moteino devices, we studied in Chapter 6 as well as Lab 6 of the companion lab guides.

When needing higher voltage, for example 12 V to power more complex dataloggers with many sensors, 36 cells would output $V_{oc} \sim 0.6 \times 36 = 18$ V, which is referred to as a 12-V module, with an operating voltage of $\sim 0.45 \text{ V} \times 36 = 16.2$ V for MPP. In this case, multiplying the curve of each cell by 36, we obtain Figure 7.10 for the I-V curves and Figure 7.11 for power. Arrays, employed for higher power needs, are made wiring modules in series or parallel; when in parallel, we increase the

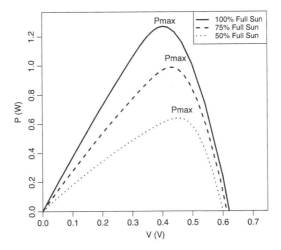

FIGURE 7.9 Power as a function of light.

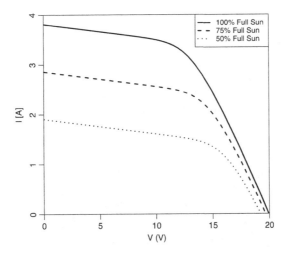

FIGURE 7.10 I-V for a module made from 32 cells.

current, I-V curves are added by current axis. Similarly, higher voltage modules, e.g., nominal 24-V modules, are made wiring more cells, but these are less typical in small off-grid monitoring stations.

A shaded cell in a module affects the entire module because the cells are in series. Assume we have n cells, and that some are shaded; we can calculate the decrease in voltage due to shad-

ing using the cell model (Figure 7.5). When we shade one cell, $V_{sh} = \dfrac{n-1}{n}V - I(R_p + R_s)$ and

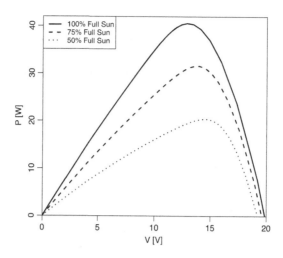

FIGURE 7.11   Power for a module made from 32 cells.

therefore the reduced voltage is $\Delta V = V - V_{sh} = V\left(1 - \dfrac{n-1}{n}\right) + I(R_p + R_s)$, or rearranging $\Delta V = \dfrac{V}{n} + I(R_p + R_s) \sim \dfrac{V}{n} + IR_p$. To avoid this issue, a bypass diode is placed in parallel with the cell and which shunts the shaded cell and decreases the voltage drop across R_p. A bypass diode can also be connected across a module. In addition, when using a parallel combination of strings, we can add a blocking diode to each string to avoid sending current to a malfunctioning string.

LOAD AND POWER

A resistive load is represented as a straight line on the I-V plane and when superimposed to the I-V curve for the PV module, it will determine operating points at each sun condition. For example, as shown in Figure 7.12, we obtain the plot for a 10-Ω load as a straight line with slope 1/10 and then find voltages for each intersection as shown in the figure. These are 5.6, 11.3, 14.3, and 14.8 V, respectively. Use $V^2/10$ to obtain the power for each sun condition, 3.1, 12.8, 20.4, and 21.9 W. These are not optimal power values, as we can see on the power vs. voltage graph (Figure 7.13); indeed, the voltage at which the MPP is established is 12.6, 13.8, 12.7, and 11.6 V, which corresponds to power 4.6, 14.0, 21.2, and 25.8 W, respectively. Similarly, we can look at the I-V lines for a battery charging and discharging and use these, for example to determine operating point by superimposing on I-V curves of PV module at various levels of sunlight.

CHARGING A BATTERY FROM A SOLAR PANEL

USING A VOLTAGE REGULATOR

In its simplest form, a battery charger circuit (Figure 7.14) consists of a voltage regulator to drop down the solar panel output voltage V_p to a voltage V_b required to charge the battery, and a diode (D1) to float the battery (i.e., stop the charging current) when the battery is full or $V_b > V_r$. The drop $V_p - V_b$ must exceed a minimum value, e.g., 2.5 V to charge the battery. For example, a 12-V battery

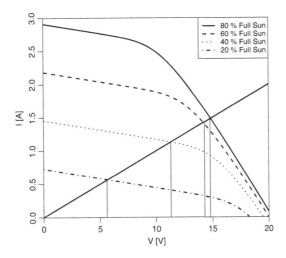

FIGURE 7.12 Resistive load and intersection points.

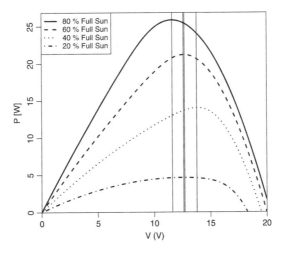

FIGURE 7.13 Power curves and MPPs.

requiring $V_b = 13.4\,\text{V}$ to charge means the panel must supply $V_b = 15.9\,\text{V}$. The charging voltage is fed back to the regulator to adjust it using a voltage divider. Diode D1 is a Schottky diode with a high reverse breakdown voltage exceeding by far the battery voltage. This type of diode is made of an n-type semiconductor and metal and has a low forward voltage. When $V_b < V_r$, the regulator will provide charging current and load current, whereas when $V_b > V_r$, the battery will float and will be the sole provider of current to the load.

USING A BUCK CONVERTER

Other common implementations of a solar charge circuit consist of a step-down or *buck* DC–DC converter dropping down the panel voltage using a pulse-width–modulated (PWM) signal activating a MOSFET switch, and an inductor to provide current to the battery. The diode is reversed when the switch is on, and the panel provides current to the inductor and the battery; conversely,

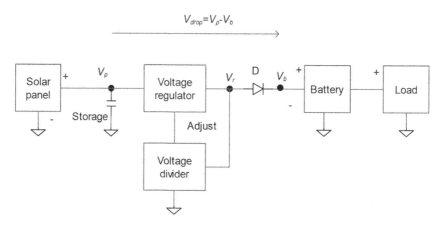

FIGURE 7.14   A very simple solar battery charger.

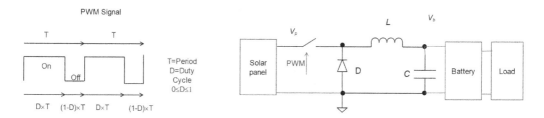

FIGURE 7.15   PWM switch and buck converter.

the diode is forward biased when the switch is off, and the inductor continues to provide current to the battery returning through the diode. The current through the inductor recovers during the next period when the switch is on. An additional capacitor will work together with the inductor as an LC filter to reduce the switching harmonics. The output voltage of the switch PWM is proportional to the input voltage $V_b = V_p \times D$ where D is the duty cycle, or fraction of time when the switch is on (Figure 7.15). For example, if we want the PWM to change from $V_p = 10\,\text{V}$ to $V_b = 8\,\text{V}$, it adjusts D to 0.8 such that $10\ \text{V} \times 0.8 = 8\ \text{V}$.

USING A BUCK-BOOST CONVERTER: MPP TRACKING (MPPT)

An interesting implementation of a solar panel battery charger is a buck-boost converter depicted in Figure 7.16 which is also based on a PWM switch. The goal of this circuit is to keep current in the inductor and voltage in the capacitor nearly constant. Under these conditions, the energy balance between that stored in the inductor and delivered to the battery is nearly equal and

thus, the output voltage delivered to the battery is related to the input voltage from the panel by $V_b = V_p \times \left(\dfrac{D}{1-D} \right)$ where the factor is depending on the duty cycle D of the pulses controlling the switch. This is an interesting result because it means we can buck the voltage when $D < 0.5$ or boost it when $D > 0.5$. For example, to obtain $V_b = 8\,\text{V}$ from $V_p = 10\,\text{V}$, we would need $\left(\dfrac{D}{1-D} \right) = 8/10 = 0.8$

and solving for D, we get $D = \dfrac{0.8}{1.8} = 0.44$. In contrast, if we want the converter to change from

$V_p = 11\,V$ to $V_b = 15\,V$, we should adjust D such that $\left(\dfrac{D}{1-D}\right) = 15/11 = 1.36$ and solving for D, we

get $D = \dfrac{1.36}{2.36} = 0.57$.

As already mentioned, the peak power in the P vs. V curve will vary as sunlight varies (Figure 7.13). To adjust the MPP, we can use a device to track this MPP, which means we need to increase or decrease the voltage depending on the changing light condition. This device then is called an MPP tracker or MPPT. The device is in essence a buck-boost converter as described above plus an algorithm to adjust the duty cycle as needed to track the MPP.

TILTING THE PANEL

Solar radiation received at the Earth's surface varies with latitude and other factors. A general pattern is that solar radiation decreases with increasing latitude. Recall that latitude is the angle α (in degrees) that a hypothetical line going through the center of the Earth makes with respect to the equatorial plane (Figure 7.17). Declination δ is the angle of the sun with respect to a plane parallel to the equator and placed at the latitude α. The maximum of this angle for each day is a function of the day number n in the year according to

$$\delta(n) = 23.45 \times \sin\left[\frac{2\pi}{n_y}(n - n_e)\right] \tag{7.4}$$

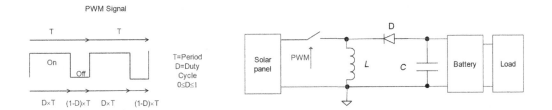

FIGURE 7.16 PWM switch and buck-boost converter.

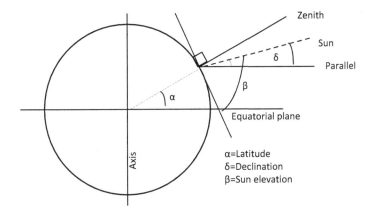

FIGURE 7.17 Latitude, declination, and sun angle definitions.

where n_y is the number of days in the year taking values of 365 for non-leap years and 366 for leap years; n_e is the day number for the vernal equinox in the northern hemisphere March 21; that is $n_e = 80$ for non-leap years or $n_e = 81$ for leap years.

Sun elevation angle β is the angle made by a line from the observer to the sun above the horizon, which in this figure is the angle that the sun makes with respect to the tangential plane normal to latitude. This angle varies during the day reaching a maximum, which is a function of the day number in the year. As we can see from Figure 7.17, declination, maximum sun elevation, and latitude for a given day number in the year are related by

$$\beta(n) = \delta(n) + (90° - \alpha) \tag{7.5}$$

We can see how declination changes through the year in Figure 7.18 (top panel) and has an average of zero. It goes through zero at day number n_e when $n - n_e = 0$ and the sine function value is zero. For example, for the Dallas-Fort Worth (DFW), Texas, airport approximate latitude of 32.90°N (Decimal degrees) or 32°54′0″ N (degrees, minutes, seconds), we can evaluate sun elevation angle to obtain the graph in Figure 7.18 (bottom panel).

We tilt the panel such that it receives as much sunlight as possible, which means looking in the direction of the sun, or perpendicular to the maximum sun elevation for the day, which varies according to latitude (Figure 7.19). Note from this figure the tilt must be $Tilt(n) = 90° - \beta(n)$. Because sun elevation is related to declination and latitude as given in equation (7.5)

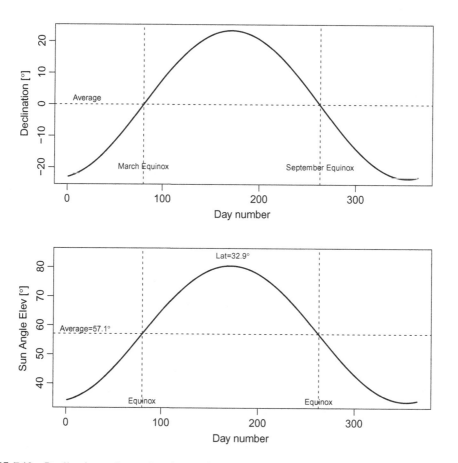

FIGURE 7.18 Declination and sun elevation angle during the year. Sun elevation corresponds to latitude 32.9°.

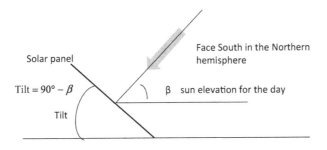

FIGURE 7.19   PV panel tilt.

$$Tilt(n) = 90° - \beta(n) = 90° - \delta(n) - (90° - \alpha) = \alpha - \delta(n) \qquad (7.6)$$

In other words, the tilt for day n is simply latitude minus declination. The best fixed tilt is for average declination, and since average declination is zero, then fixed tilt is just equal to latitude. This is called a *polar mount*, and because it is fixed, it would not be optimal for most days, and we would observe large deviations for extreme values of the declination during the year.

For greater efficiency, the tilt can be varied according to the day of the year and even track the sun during its daily ascent and descent for each day. While this tracking is important in utility-scale, and commercial PV installations, it is not typical of small remote environmental monitoring stations.

ATMOSPHERIC EFFECTS

The extraterrestrial solar radiation I_0 or solar radiation received by Earth outside the atmosphere, measured as power per unit area kW/m², varies with the day n of the year according to

$$I_0 = SC \times \left(1 + 0.034 \times \sin\left(\frac{2\pi}{365}(n - 81) \right) \right) \qquad (7.7)$$

where SC, the solar constant, is the average extraterrestrial radiation and has a value of 1.377 kW/m². As the solar radiation flux goes through the atmosphere, a good part of it is absorbed by atmospheric gases and scattered by particles. Direct radiation reaching the surface of the Earth can be as high as ~70% of the extraterrestrial solar radiation I_0, or solar radiation received by Earth outside the atmosphere. For practical applications, the solar radiation reaching the Earth surface is 1 kW/m² or slightly larger than 70% of I_0. This value is referred to as the *1-sun* or full sun equivalent.

For a given site, solar radiation "insolation" S as energy can be given in kWh/m²day, or equivalently as hours/day of 1-sun or hours of "peak sun". For example, if the average solar radiation is 7.0 kWh/m², then we have the equivalent of 7 hours of 1-sun. We can then use area A of the module (in m²) and system efficiency η to evaluate energy produced by the system at a given site $E = S \times A \times \eta$ either as kWh or hours of 1-sun.

There are two major components of the radiation reaching the Earth's surface: direct and diffuse radiation. A model for radiation flux through the atmosphere follows the Bougher-Lambert-Beer exponential attenuation

$$I_{bn} = I_0 \exp(-\tau_b \times m^{a_b}) \qquad (7.8)$$

$$I_d = I_0 \exp(-\tau_d \times m^{a_d}) \qquad (7.9)$$

where I_{bn} is the *direct beam* normal portion, I_d is the diffuse horizontal portion of clear-sky radiation reaching the Earth's surface, m is the *air mass*, τ_b, τ_d are atmosphere pseudo *optical-depths*, and a_b and a_d are coefficients. Values of τ_b, τ_d are location-specific and vary through the year (Gueymard and Thevenard 2013). The ASHRAE handbook provide coefficients for each month for thousands of locations (ASHRAE 2017). The power coefficients a_b and a_d also vary, but they are related to the optical depths by empirically derived equations

$$a_b = k_b - k_{bb}\tau_b - k_{bd}\tau_d - k_{bbd}\tau_b\tau_d \tag{7.10}$$

$$a_d = k_d - k_{db}\tau_b - k_{dd}\tau_d - k_{dbd}\tau_b\tau_d \tag{7.11}$$

The air mass ratio (m) is a ratio of two path lengths as the sun rays go through the atmosphere. The simplest calculation assumes a flat Earth surface; denote by h_1 path length if the sun were overhead or an elevation angle of 90°, and h_2 path length when the sun is at elevation angle β (Figure 7.20).

$$m = \frac{h_2}{h_1} = \frac{1}{\sin\beta} \tag{7.12}$$

Note that h_2 is larger than h_1 and therefore the air mass ratio will be larger than 1, except when the sun is overhead for which $m = 1$. This special condition is denoted as AM1. For other values of m, say $m = 1.5$ the notation is AM1.5. The extraterrestrial air mass ratio is zero and denoted as AM0. Likewise, we can calculate the effect of other atmospheric processes which allows to understand relative contribution of direct, diffuse, and reflected radiation to the total. The R package renpow has functions to evaluate these components and dataset examples for some locations (Acevedo 2018).

SUN PATH

Figure 7.21 shows Earth rotating at 15°/hour. We define hour angle H in hn (Hours before Noon) as $H = 15 \times hn$. Trigonometric equations relate sun azimuth φ (Figure 7.22) and sun elevation β to hour angle:

$$\sin\phi = \frac{\cos\delta\sin H}{\cos\beta}$$
$$\sin\beta = \cos L\cos\delta\cos H + \sin L\sin\delta \tag{7.13}$$

For example, we calculate these with an R script as shown in the lab 7 guide of the Lab Manual companion to this textbook at the latitude of DFW, Texas and obtain results in Figure 7.23a. Similarly, for azimuth, we obtain results in Figure 7.23b. We can combine these results in an azimuth-elevation plane where the hour is implicit by the graph marker. See example in Figure 7.24 for latitude 32.9 N and day 20.

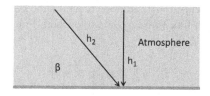

FIGURE 7.20   Air mass ratio.

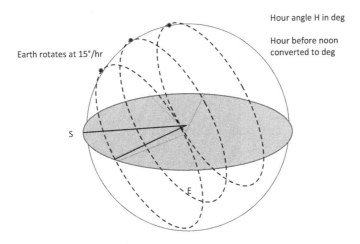

FIGURE 7.21 Sun elevation angle changes during the day. Shaded disc is the horizon plane.

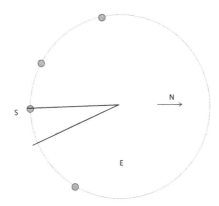

FIGURE 7.22 Projection on horizon plane is azimuth angle.

By using selected days of the year, we obtain a sun path diagram (Figure 7.25). This diagram can be used to analyze potential shading of the solar panel at the installation site using simple tools such as a clinometer and a compass. Locate potential shading features and note their azimuth using compass and height using clinometer. Then place the features on the sun path diagram (Figure 7.25) to determine potential shading by these features. These equations are implemented in functions of the R package renpow (Acevedo 2018), which are covered in lab 7 of the companion Lab Manual (Acevedo 2024).

IMPACT OF TEMPERATURE ON SOLAR PANEL

The temperature of a PV cell depends on ambient temperature and solar radiation, the cell gets hotter as the air gets warmer and radiation increases. The cell temperature T_c in °C can be calculated using the equation

$$T_c = T_a + \left(\frac{\text{NOCT} - 20°C}{0.8} \right) S \tag{7.14}$$

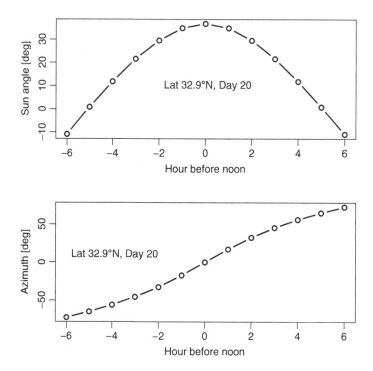

FIGURE 7.23   Sun angle and azimuth as a function of hour angle.

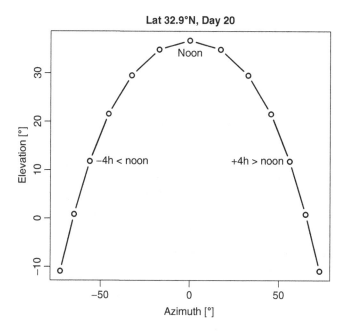

FIGURE 7.24   Position of the sun in Azimuth-Elevation plane.

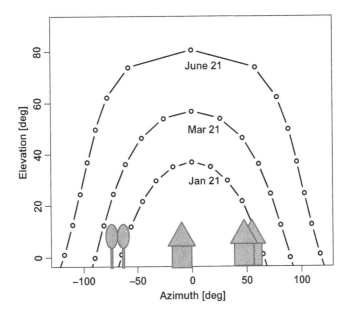

FIGURE 7.25 Sun path diagram and illustration of shading survey.

where, T_a is the ambient temp (°C), S is the solar irradiation (in kW/m²), and NOCT is the Nominal Operating Cell Temp, which is the expected cell temp when the ambient temp is 20°C, the irradiation 0.8 kW/m², and wind speed 1 m/s. The NOCT is given by manufacturer, for many panels this number is in the 45°C–47°C range; see, for example Table 5.3 of Masters (2013b).

Other temperature effect specifications include nominal V_{oc} at 25°C (NV_{oc}), a negative voltage temp coefficient TC_V in %/°C, and a negative power temp coefficient TC_P in %/°C. For example, for voltage

$$V_{oc} = NV_{oc} \times \left(1 + TC_V \times (T_c - 25)\right) \tag{7.15}$$

Using the NOCT, we can estimate cell temperature, and with this value, we can calculate change in open circuit voltage and power of a module made with these cells as ambient temperature and irradiation conditions change. For example, assume a module with NOTC=45°C, NV_{oc} of 38 V, rated power 100 W, TC_v of –0.30%/°C, and TC_p of –0.4%/°C. At full sun (say 1-sun) and ambient 32°C, we get the module temperature $T_c = 32 + \left(\dfrac{46 - 20°C}{0.8}\right) \times 1 = 64.5°C$. With this increased temperature, we calculate the effect on open circuit voltage $V_{oc} = 38 \times \left(1 - 0.003 \times (64.5 - 25)\right) = 33.5$ V

and power $P = 100 \times \left(1 - 0.004 \times (64.5 - 25)\right) = 84.2$ W.

POWER BUDGET AND POWER SYSTEM SIZING

In this section, we present an example of calculating how much power will an environmental monitoring station demand so that we can size the solar panel and battery. For example, suppose a simple case of having a datalogger-based wired sensor system and an SBC that serves a WSN. Required battery capacity and solar panel power are determined through power budget analysis; average power consumption of each power load device is determined by measuring or estimating the average current draw and the time spent in each of its operating modes (Yang et al. 2010; Williams 2014; Acevedo 2018).

Assuming all DC devices, total daily energy load in $W_{h/d}$ is the sum of the power draw of all devices in the operation period in a day.

$$E = \sum_i V_i \times I_i \times T_i \tag{7.16}$$

where V_i is the voltage in V, I_i is the current draw in A, and T_i is hours of daily operation for device i of the station. For example, suppose we have two devices, device 1, drawing 0.2 A at 12 V for a continuous 24-hour operation every day, and device 2 (e.g., a collection of sensors) drawing 50 mA at 5 V totaling 8 hours of operation every day. Total energy load is $E = 12$ V $\times$ 0.2 A $\times$ 24 h/d + 5 V $\times$ 0.05 A $\times$ 8 h/d = 59.6 Wh/d.

To meet this demand for a few backup days, e.g., 3 days, in the absence of sunlight with a 12-V battery, and setting a maximum allowable depth of discharge $\underline{DOD_{max}}$ of 80% for the battery, it would require a battery capacity of

$$C = \frac{E_{DC}}{V} \times \frac{\text{days}}{DOD_{max}} = \frac{59.6 \text{ Wh/d}}{12 \text{ V}} \times \frac{3 \text{ d}}{0.8} \approx 18.6 \text{ Ah} \tag{7.17}$$

We could select a standard capacity near this value. Considering that the battery will be allowed to recharge in a few days, say 2 days, a 12-V solar panel should provide 12 V $\times \left(59.6 \text{ Ah/d} + 18.6/2 \text{ Ah/d}\right)$ = 826 Wh/d. Now we must consider potential abundance of the solar resource at the site. For example, if we have a site with S of at least 4 h/d of 1-sun, we will need a 12-V panel rated at $P = 826.8$ Wh/d/4 h/d ≈ 200 W. These calculations are based on rated and estimated values, and it will be important to monitor battery voltage and current consumption as a part of the remote system status monitoring service to help with early detection of battery degradation in order to prevent system failure and the loss of sensor data (Yang et al. 2010).

POWERING WSN NODES

As WSNs are added to remote environmental monitoring systems, we face the challenge of how to power the nodes of the WSN. The simplest solution is to employ batteries, e.g., rechargeable batteries, and replace these periodically. This task becomes labor-intensive for a large number of nodes when replacement is frequent; nodes that employ more power would require more frequent battery replacement. PV cells can be used to recharge the battery, and this works well in open areas, however become limited in shaded environments, such as a forest understory.

In Chapter 6, we described an example of WSN nodes using low-power Moteino devices with RFM69 radio and powered by a Lithium Polymer battery and a 6-V solar panel for battery charging (Iyiola 2017). In this WSN, the nodes used a Moteino shield by Low Power Lab (2016) to recharge the battery. Power budget calculations are made for various operation modes: sensor reading, radio transmitting with retries, node sleeping assisted to determine reliability of power supply to the nodes.

Alternatives to PV panel charging of batteries include harvesting or scavenging ambient energy, such as vibration, heat, and RF waves (Kahrobaee and Vuran 2013; Kim et al. 2014). For example, harvesting ambient RF allows operation of a WSN (Kim et al. 2014). Although available ambient RF energy is lower than that of other sources, in some sites it is sufficient to power a WSN node, depending on the duty cycle of the node. An advantage of this form of harvesting is that the source is constantly available, in contrast to the variability of sunlight. A system to harvest ambient RF consists of high-gain antenna and a rectifier circuit. The combination of antenna and rectifier is termed a *rectenna*, or rectifying antenna. The rectenna would be designed according to the frequency bands of the available ambient RF field. In addition, it is also possible to combine solar and RF harvesting forming a hybrid energy harvesting system that suffices for many WSN environmental monitoring applications. One technique to harvesting vibration is using piezoelectric sensors acting as resonators; for this purpose, the vibration resource should match the resonant frequency of the harvester (Gibus et al. 2020).

One approach to power storage in WSN nodes is to substitute batteries for supercapacitors, which compared to batteries offer shorter charge and discharge duration (1–10 seconds), longer discharge cycle life (500,000), longer lifetime (15–20 years), and tolerate full discharge. Gurung (2020) experimented with WSN nodes that substituted a battery for supercapacitors, recharged from solar panel, demonstrating that this technology provides an effective manner to power WSN nodes. For detailed analysis of power consumption and the dynamics of charging and discharging, the nodes included current sensors, supercapacitor voltage readings, and a custom-made pyranometer.

ENVIRONMENTAL MONITORING OF RENEWABLE POWER SYSTEMS

Environmental monitoring provides data useful to the design and operation of renewable power systems, for example solar, wind, and hydropower. A major contribution of monitoring is understanding the resource used by these forms of renewable power, e.g., solar radiation, wind speed, and water flow.

SOLAR RADIATION

Solar radiation is measured at many monitoring stations using a *pyranometer* or radiation sensor, which is based on a silicon PV detector mounted on a cosine-corrected head; its output is current, which is converted to voltage by a potentiometer in the sensor head. The resistance of the potentiometer is adjusted when the sensor is calibrated so that all sensors have the same output sensitivity.

Analysis of solar radiation data collected over time allows to understand the solar resource available at a site for electricity production, both for supporting the design of a PV or concentrated solar power facility or powering a remote monitoring station itself. Solar radiation data are typically reported as maximum, minimum, average, and standard deviation over a time interval in W/m². In addition, some dataloggers would integrate over the measurement interval to report energy in kJ/m².

In lab session 7 of the companion Lab Manual (Acevedo 2024), the reader has an opportunity to practice data analysis using an example of data collected by a weather station. One of the results of these analyses is shown in Figure 7.26; which is an example of 10-min average solar radiation data analyzed for a week in December 2018. Note how December 4 and 5 most have been clear days with irradiance reaching 550 W/m², whereas December 6–8 were likely cloudy with irradiance not exceeding 100 W/m², except for a spike occurring on December 7. Three consecutive days like these can lead to low power production by PV panels, and this is significant for battery bank design of off-grid remote stations.

We can calculate energy E_i in kWh/m² accumulated during the ith measurement interval of Δt (in minutes) by $E_i = P_i \times \Delta t / (60 \times 1000)$ where P_i is the solar radiation in W/m² for the ith measurement interval. For each day d, summing E_i for all the intervals within that day $E_d = \sum_{i \in d} E_i$ would yield the daily values of energy in kWh/m² which are the same as hours of 1-sun equivalent for that day d. As an example, Figure 7.27 illustrates the results of that calculation for the week in December analyzed above (Figure 7.26). Note how December 6–8 have less than 1 hour of 1-sun.

Capacity factor (CF) can be evaluated by dividing insolation S in kWh/d or h/day of peak sun by 24 hour/day. Thus, CF for PV installations depends on location, and it is a way of expressing insolation S (Acevedo 2018). For example, an average of 3 hours of 1-sun per day would calculate out to $CF = 3$ h/24 h $= 0.125$ or 12.5%. This calculation can be done for various time periods,

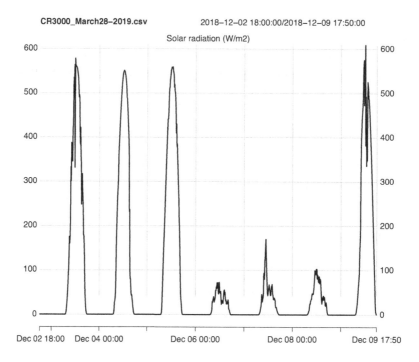

FIGURE 7.26   Example of 10-min average solar radiation data analyzed for a week in December.

such as monthly, seasonally, or annually. For instance, monthly energy E in kWh/mo can be estimated from installed capacity P (in kW), and monthly CF using the number of hours in a month $h_{mo} = 30 \text{ d} \times 24 \text{ h/d} = 720 \text{ h}$ for 30-day months or use $h_{mo} = 31 \text{ d} \times 24 \text{ h/d} = 744 \text{ h}$ for those months with 31 days.

$$E = P \times CF \times h_{mo} \tag{7.18}$$

A histogram of solar radiation energy in hours of 1-sun for an entire year will display similar probability mass above and below an average due to the seasonality of solar radiation. The same would hold as well for data over distinct seasons. As an example, taking the data analyzed above over a period from the end of October 2018 to middle of March 2019 will yield the results shown in Figure 7.28. The highest values (~6 h) would correspond to the higher radiation days in March, many of the lowest values would correspond to lower radiation days in December, and many of the values around the mean would correspond to higher radiation days from November through February. This result illustrates how measuring the number of hours of 1-sun for a site provides guidance on PV-panel–installed capacity design.

WIND SPEED

A fluid of density ρ flowing with velocity v leads to an equation for fluid *specific power $p(t)$* at time t

$$p(t) = \frac{1}{2}\rho(t)v(t)^3 \tag{7.19}$$

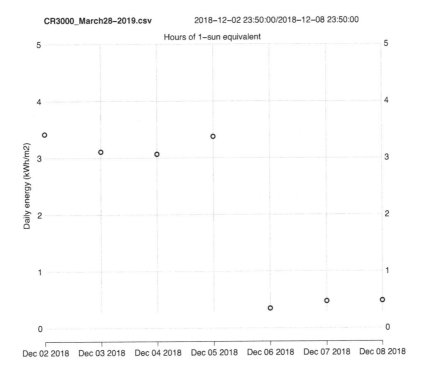

FIGURE 7.27   Example of solar insolation in hours of 1-sun equivalent.

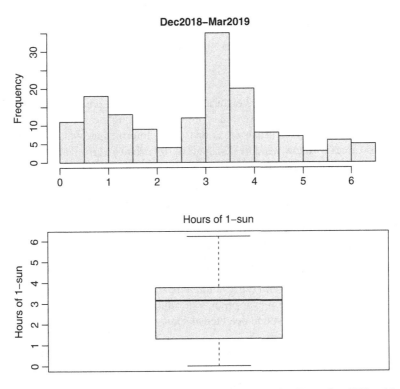

FIGURE 7.28   Distribution of daily hours of 1-sun for a period spanning December 2018 to March 2019.

Specific power is power per unit of cross-sectional area (normal to the direction of flow). Note that specific power is proportional to the density and to the cube of velocity of the fluid, and that this is an instantaneous value since both density and velocity may be varying with time. For wind power, the fluid is air, the velocity is wind speed, and the density is that of air, which changes according to elevation above sea level and temperature. Wind turbine performances are specified for reference conditions, which are pressure 1 atm, temperature 15°C, and density 1.225 kg/m^3. For instance, the instantaneous specific power for wind velocity of 10 m/s at reference conditions will be

$p(t) = \frac{1}{2}\rho v^3 = \frac{1}{2}1.225 \times 10^3 = 612.5$ W/m^2. Functions of the R package renpow facilitate calculations for a variety of conditions (Acevedo 2018).

As we move away from the ground, wind speed increases because friction decreases. Since power in the wind increases with v^3, a wind turbine installed on a taller tower would experience higher wind speed, and thus more power. The increase in velocity due to height is calculated from the empirical relation (Masters 2013a)

$$\frac{v}{v_0} = \left(\frac{H}{H_0}\right)^\alpha \tag{7.20}$$

where H_0 is a *reference height* (10 m), v is the wind speed at height H, v_0 is the wind speed at the reference height, and α is a friction coefficient, which characterizes the terrain conditions considering whether this is on water (the lowest friction), or on land with various types and heights of vegetation and land use given by several tables. An alternative relation, inspired on aerodynamics in the atmospheric boundary layer, is based on the *roughness length l* in m and assumes that the air flow varies logarithmically with elevation (Masters 2013a)

$$\frac{v}{v_0} = \frac{\ln\left(\frac{H}{l}\right)}{\ln\left(\frac{H}{l}\right)} \tag{7.21}$$

The parameter roughness length is the height above the ground at which wind speed is zero for atmosphere thermal conditions with lapse rate of −9.8°C/km. Roughness length varies from 0.0002 for water surface to 1.6 for dense urban areas or forests. Intermediate values are 0.03 (for open areas), 0.1 (for crop areas with few windbreaks), and 0.4 (for urban and rural areas with windbreaks). Since power increases as the cube of wind speed, assuming equal air density, we see that the ratio

of specific power goes as the cube of speed ratio. $\frac{P}{P_0} = \left(\frac{v}{v_0}\right)^3$, and therefore, we can get the ratio of

specific power for both models by raising them to power 3 (Figure 7.29).

Estimation of the friction coefficient or roughness length can be accomplished by monitoring wind speed using *anemometers* installed at two different heights on the tower of a meteorological station and performing a regression of the wind speed measured at those two heights. An example is shown in Figure 7.30. There are several types of *anemometers*, one type is electro-mechanical based on counting pulses on a Hall effect sensor product of rotating cups or a propeller, and another type is sonic based on relative speed of sound detected at sonic sensors. We cover more details in the lab guide 7 of the companion lab manual.

In the same manner as we discussed for solar radiation, it is important to understand wind speed variability to estimate how much power we may harvest during a particular week, month, or season. Very importantly, average wind speed for a location does not alone indicate the energy a wind turbine could produce; frequency of wind speeds at various intervals is also needed. All this

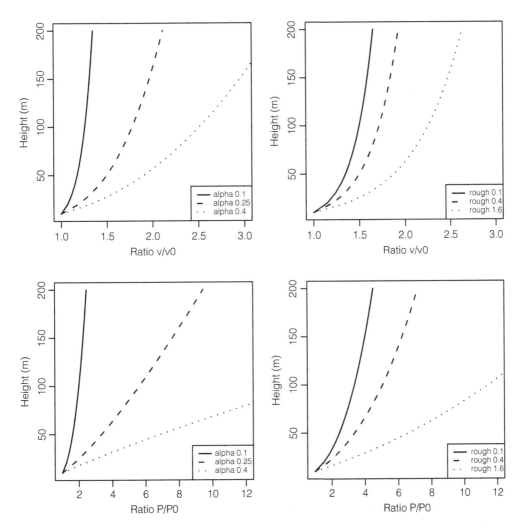

FIGURE 7.29 Wind speed and power ratio as a function of height with respect to a reference. Left: using empirical exponential function. Right: using logarithmic aerodynamics function.

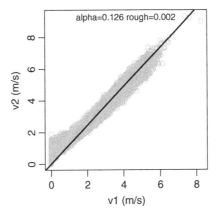

FIGURE 7.30 Calibration of wind speed vs. height function from data collected at two different heights.

information is determined for a location from a probability distribution function fitted to the long-term observed data.

The *Weibull* distribution is often used to model wind speed v, since many wind speed data fit this distribution well. The Weibull PDF has two parameters: scale c and shape k, and it is given by (Masters 2013a)

$$p(v) = \frac{k}{c}\left(\frac{v}{c}\right)^{k-1} \exp\left[-\left(\frac{v}{c}\right)^k\right] \tag{7.22}$$

Here $p(v)$ denotes probability density for a given v. A Weibull with $k=2$ has the special name of *Rayleigh* distribution

$$p(v) = \frac{2}{c}\left(\frac{v}{c}\right)^1 \exp\left[-\left(\frac{v}{c}\right)^2\right] = \frac{2v}{c^2} \exp\left[-\left(\frac{v}{c}\right)^2\right] \tag{7.23}$$

A good fit to wind speed is often found for $k=2$, or a Rayleigh distribution as given in Equation (7.23). The mean or expected value of a Rayleigh PDF is related directly to the scale c (Masters 2013a).

$$\mu_v = E(v) = \int_0^\infty v p(v)\, dv = \frac{\sqrt{\pi}}{2} c \tag{7.24}$$

A first step to find a fit to Rayleigh is to assume shape $k=2$, use Equation (7.24) to determine the scale from the mean wind speed, and estimate the mean of wind speed by its average

$$c = \frac{2\mu_v}{\sqrt{\pi}} \simeq \frac{2\bar{v}}{\sqrt{\pi}} \tag{7.25}$$

where $\bar{v}$ is the wind speed sample mean. Once we have a fitted model, we can calculate the probability of having a range of given wind speed.

For example, Figure 7.31 shows a Rayleigh distribution for a location with average wind of 2.5 m/s, and shape $k=2$, then use $c = \frac{2\bar{v}}{\sqrt{\pi}} = \frac{2 \times 2.5}{1.77} = 2.82$. The vertical lines on the CDF correspond to $v=2$ and $v=3$ to find the probability that wind speed is above 2 and 3 m/s respectively; the probability of these events are $\Pr[v > 2] = 1 - \Pr[v \le 2] = 0.605$ and $\Pr[v > 3] = 1 - \Pr[v \le 3] = 0.323$. Concluding that the wind will exceed 2 m/s 60% of the time and 3 m/s 32.3% of the time.

Once we have an initial estimate for scale and shape, we can verify the fit to the data by calculating the histogram or its density approximation and the empirical cumulative distribution (ECDF) and compare the theoretical values to the data values (Figure 7.32). The fit can be then improved by trial and error or by optimization methods.

It should be evident by now that one cannot apply the specific power equation $p(t) = \frac{1}{2}\rho v^3$ to the average wind speed because the cube term is non-linear. However, to emphasize further, compare the specific energy of two wind regimes with the same average at standard conditions. Suppose we have wind of 3 and 9 m/s each 50% of the time; the average power is $\bar{P} = 0.5 \times 1.225 \times (3^3 + 9^3)/2 = 231.5$ W/m^2 and the average wind speed is $(9+3)/2 = 6$ m/s. If we calculate power using the average wind, we get $P(\bar{v}) = 0.5 \times 1.225 \times 6^3$ W/m$^2 = 132$ W/m^2. We can see how the results are completely different $\bar{P} > P(\bar{v})$. In this case, the average power is almost double the power of the average wind.

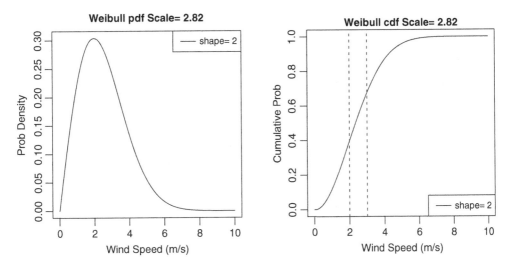

FIGURE 7.31   Example of Rayleigh distribution.

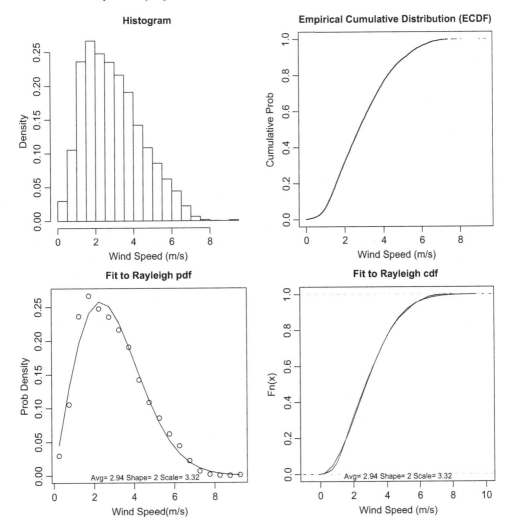

FIGURE 7.32   Wind speed data fit to a Rayleigh distribution.

However, if the wind speed fits a Rayleigh distribution, we can use the average wind speed times a factor $\dfrac{6}{\pi} \approx 1.91$ to calculate average power. To see why we can make this simplification, we would calculate power by integrating using the Rayleigh PDF which is a complicated integration that we will not develop here. We will simply use the known result (Masters 2013a) that

$$\mu_P = E[P] = \frac{1}{2}\rho\frac{6}{\pi}\mu_v^3 \ \text{W/m}^2 \tag{7.26}$$

This is a convenient result in as much as it gives us a rule to calculate mean wind power from mean wind speed as long as wind speed fits a Rayleigh PDF. In practical terms, we use the sample average $\bar{v}$ of wind speed as an estimator of the mean wind speed and multiply by $\dfrac{6}{\pi} \approx 1.91$

$$\mu_P \simeq \frac{1}{2}\rho \times 1.91 \times \bar{v}^3 \ \text{W/m}^2 \tag{7.27}$$

For example, suppose wind speed data can be fit to a Rayleigh PDF with $\bar{v} = 2.94$ m/s and $c = 3.32$. The estimated average power in the wind at standard conditions would be $\mu_P \simeq \dfrac{1}{2}1.225 \times 1.91 \times 2.94^3$ W/m^2 = 29.73 W/m^2. This is relatively low power density since the average wind speed is low.

HYDROELECTRIC

There are several aspects of monitoring that relate to evaluating the resource for hydroelectricity production by surface water, directly as river flow and indirectly as rainfall on the watershed draining to the river that contribute to that flow. In the latter case, rainfall-runoff modeling can be used to predict river flow based on precipitation data and watershed configuration.

Harnessing power in moving water is also related to Equation (7.19) which applies to this situation using water density and water velocity. By employing Bernoulli's equation, in conjunction with Equation (7.19), it is common to summarize power production by the simple equation $P = \rho g h Q$ stating that power is the product of weight density ρg, water flow Q, and head difference h between points 1 and 2, which are the source and the turbine locations in hydroelectric generation. For reservoir-based hydroelectric, this flow Q is produced by stored water, and it is more relevant to monitor reservoir water level, and river flow at the inputs of the reservoir. However, for run-of-river hydroelectric, river flow upstream is of direct importance and more relevant for monitoring.

As an example, consider a run-of-river power plant, and the daily streamflow time series shown in the top panel of Figure 7.33. The flow-duration curve (bottom panel of Figure 7.33) is built using daily streamflow, in such a way that streamflow is sorted into descending order, with the highest levels toward the left of the curve. The horizontal axis is the exceedance in probability units or percent of the time that the value of flow is larger than that value. In the graph, we identify several levels of exceedance probability considered of interest; for instance, 0.5 and 0.95 that correspond to flows Q_{50} and Q_{95} values and can be interpreted as a low potential power and median potential power. In this example, the flow Q_{95} is 17.09 m^3/s, meaning that we exceed this flow 95% of the time. The Q_{50} is 25.65 m^3/s means that the river exceeds this flow 50% of the time.

The Q_{95} flow is used as a low-flow condition that must be maintained in the river when diverting flow through the hydroelectric plant. In addition, we can consider the average flow, called Q_{mean}, and compare to Q_{95} to determine how variable the river is. In this example, the Q_{95} is ~50% of the Q_{mean} or a ratio $Q_{95}/Q_{mean} \approx 0.5$. This high percent is typical of a high-baseflow river, meaning that the watershed has storage of runoff that is released gradually to the stream. A low Q_{95}/Q_{mean} ratio would indicate a flashy river that would change flow fast in response to rain, meaning little storage of runoff in the watershed.

River flow is typically monitored using a rating curve relating stream stage or level to flow, because it is easier to perform real-time monitoring of stream stage than velocity at various points

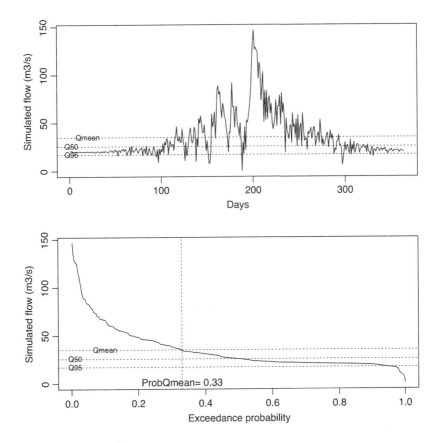

FIGURE 7.33 Flow and exceedance curve.

of the stream cross-section. Water level can be measured by submersed sensors as well as by sonic sensors located above the surface. We will discuss monitoring of river flow with more detail in Chapter 12 that is related to water monitoring.

EXERCISES

Exercise 7.1

Calculate the voltage and current for a 10-Ω load for each curve of the solar module given in Figure 7.10. What is the power given to the load for each one of these curves? Hint: superimpose a straight line with slope 1/10, determine intersection points, and calculate power for each point.

Exercise 7.2

Calculate the voltage and current at which the MPP is established for all the curves of Figure 7.11. What would be the power for each curve if an MPP tracker adjusts the voltage output to the above values? Hint: look at the power curve for each light condition, determine the peak, and look for I and V values at which the peak occurs.

Exercise 7.3

Calculate tilt required for a solar panel at four distinct days: spring equinox (March 21), fall equinox (September 22), winter solstice (December 21), and summer solstice (June 21) for Dallas (latitude 32.9°N). Hint: determine declination for each one of these days and subtract from latitude.

Exercise 7.4

Calculate the sun elevation angle β that will yield an air mass ratio of $m = 1.5$ (AM1.5).

Exercise 7.5

Consider a site located at the DFW, Texas latitude. Determine minimum spacing between a polar mount solar panel and a 2-m tall obstacle located to the south of installation site, in order to avoid shading for most of the day (8 am–4 pm) under worst conditions.

Exercise 7.6

Assume a panel with NOTC = 46°C, V_{oc} of 38 V at 25°C, rated power 100 W at 25°C a voltage temp coefficient of –0.30%/°C, and a power temp coefficient of –0.4%/°C. Estimate module temperature, open circuit voltage, and maximum power under conditions of 1-sun and 35°C.

Exercise 7.7

Design a battery capacity and solar-panel–installed capacity for a monitoring station drawing 0.25 A for 8 hours a day plus 0.1 A during 24 hours every day. Assume that the site has an average of 4 hours of 1-sun energy.

REFERENCES

Acevedo, M. F. 2018. *Introduction to Renewable Electric Power Systems and the Environment.* Boca Raton, FL: CRC Press. 439 pp.

Acevedo, M. F. 2024. *Real-Time Environmental Monitoring: Sensors and Systems.* Second edition – Lab Manual. Boca Raton, FL: CRC Press, Taylor & Francis Group. 463 pp.

ASHRAE. 2017. *ASHRAE Climate Data Center.* accessed November 2017. https://www.ashrae.org/resources–publications/bookstore/climate-data-center#std169.

Gibus, D., P. Gasnier, A. Morel, F. Formosa, L. Charleux, S. Boisseau, G. Pillonnet, C. A. Berlitz, A. Quelen, and A. Badel. 2020. Strongly coupled piezoelectric cantilevers for broadband vibration energy harvesting. *Applied Energy* 277:115518.

Gueymard, C. A., and D. Thevenard. 2013. Revising ASHRAE climatic data for design and standards: Clear-sky solar radiation model. *ASHRAE Transactions* 119:194–209.

Gurung, S. 2020. *Efficient Solar Energy Harvesting and Management for Wireless Sensor Networks Under Varying Solar Irradiance Conditions.* PhD Dissertation, Electrical Engineering. Denton, TX: University of North Texas. 163 pp.

Iyiola, S. 2017. *Moteino-Based Wireless Data Transfer for Environmental Monitoring.* MS Thesis, Electrical Engineering. Denton, TX: University of North Texas. 69 pp.

Kahrobaee, S., and M. C. Vuran. 2013. Vibration energy harvesting for wireless underground sensor networks. In *Communications (ICC), 2013 IEEE International Conference on*, 9–13 June 2013.

Kim, S., R. Vyas, J. Bito, K. Niotaki, A. Collado, A. Georgiadis, and M. M. Tentzeris. 2014. Ambient RF energy-harvesting technologies for self-sustainable standalone wireless sensor platforms. *Proceedings of the IEEE* 102 (11):1649–1666.

Low Power Lab. 2016. *PowerShield.* accessed November 2016. https://lowpowerlab.com/guide/powershield/features/.

Masters, G. M. 2013a. *Renewable and Efficient Electric Power Systems.* Hoboken, NJ: Wiley-IEEE Press. 690 pp.

Masters, G. M. 2013b. *Renewable and Efficient Electric Power Systems, Second Edition.* Hoboken, NJ: Wiley-IEEE Press.

Williams, J. 2014. *An Application of Digital Video Recording and Off-Grid Technology to Burrowing Owl Conservation Research.* MS Thesis, Electrical Engineering. Denton, TX: University of North Texas. 100 pp.

Yang, J, C. Zhang, X. Li, Y. Huang, S. Fu, and M. Acevedo. 2010. Integration of wireless sensor networks in environmental monitoring cyber infrastructure. *Wireless Networks* 16 (4): 1091–1108.

8 Remote Monitoring of the Environment

INTRODUCTION

There is a great variety of airborne and spaceborne platforms and instruments to monitor environmental systems remotely, i.e., the sensors are located at a distance above the ground and allow covering a broader spatial range. *Remote sensing* includes taking images of the land or ocean for specific purposes as needed, for example by airplanes and unmanned aerial vehicles (UAVs), as well as repetitive collection of imagery of the same area, for example by satellites orbiting the Earth. In this chapter, we focus on the use of Landsat imagery responding to different electromagnetic (EM) reflection and absorption of sunlight by land, soil, vegetation, and surface water. For this purpose, we describe major platforms, reference systems, bands, analysis using indices, such as normalized difference indices for vegetation and water, reclassification, and the use of multivariate analysis and machine learning to perform image classification. Images shown in the printed version of this book are in gray scale; full color version is available from the online resources of this book.

REMOTE SENSING OF THE ENVIRONMENT

There is a great variety of airborne and spaceborne platforms and instruments to monitor environmental systems remotely, i.e., the sensors are located at a distance above the ground and allowing to cover a broader spatial range (Huete 2004). *Remote sensing* includes taking images of the land or ocean for specific purposes as needed, for example by airplanes and UAV, or *drones*, as well as repetitive collection of imagery of the same area, for example by satellites orbiting the Earth and responding to different EM reflection and absorption of sunlight by land, soil, vegetation, and surface water. Atmospheric effects, due to reflection and scattering of particulate matter or absorption by atmospheric gases, modify the reflected signal from the ground, which needs to be corrected to account for those effects.

Remote sensors may be *passive* when the source of energy exciting the sensors comes from the solar radiation incoming to Earth and reflected from the Earth (as in optical sensing), or *active* when the sensors require energy, for instance when detecting the response to a signal shot from the platform to the ground, rather than sunlight. These include LiDAR (Light Detection and Ranging) and Radar (RAdio Detection and Ranging). LiDAR uses laser pulses to map objects or the ground in 3D, for example digital elevation models that can be used to monitor land erosion. Radar uses EM in the radio and microwave part of the spectrum, and the backscatter can be used to monitor surface water, forest biomass, and many other environmental systems. Both LiDAR and Radar are also used at ground level to monitor a variety of processes.

Remote sensing of the environment is an extensive topic and there is a wealth of information, books, and journals on these monitoring approaches. In this chapter, we limit ourselves to discuss optical remote sensing by spaceborne platforms, and specifically analysis of Landsat imagery.

OPTICAL REMOTE SENSING

One important class of remote sensing involves the use of remotely placed sensors (air- or spaceborne) responding to different reflection and absorption of sunlight by land, soil, vegetation, and surface water. Atmospheric effects, due to reflection and scattering of particulate matter or

DOI: 10.1201/9781003425496-8

TABLE 8.1

Uses of Various Parts of the EM Spectrum for Remote Environmental Monitoring

Spectral Band	Region	Wavelength	Examples Variables Monitored
Shortwave radiation	UV	300–400 nm	Gases, air quality
	VIS	400–700 nm	Vegetation, soil, and surface water
	NIR	700–1300 nm	Vegetation, biomass
	MIR	1.3–3 μm	Vegetation, leaf moisture
Long-wave radiation	TIR	3–14 μm	Vegetation stress, surface temperature
	Microwave	0.3–300 cm	Vegetation, soil moisture

absorption by atmospheric gases, modify the reflected signal from the ground, which must be considered to correct for these effects. For this purpose, spectrometers are used at ground level to acquire data that can serve to *ground-truth* the remote sensors. Drones for environmental parameter sensing are in the middle between ground data collection and remote sensing from space (Wallerman et al. 2018).

As we discussed in Chapter 7, *incoming* solar radiation reaching Earth is distributed by wavelength, increasing for short wavelengths from ultraviolet (UV) to visible (VIS), reaching a peak at in the VIS range, and then decreasing as wavelength increases. A fraction of the shortwave incoming radiation is reflected and scattered back by reflective surfaces like clouds, snow, and particles. The coefficient representing this fraction is termed *albedo*. As Earth's surface warms, it emits *outgoing* radiation in longer infrared (IR) waves. Earth's average temperature results as a balance of incoming and outgoing radiation. A fraction of the outgoing long-wave radiation is reradiated to Earth due to the greenhouse effect.

Table 8.1, adapted from Huete (2004), summarizes wavelength bands of the EM spectrum used for remote sensing. *Shortwave radiation* encompasses UV, VIS, near IR (NIR), and middle IR (MIR) wavelengths. The shorter of these, UV (300–400 nm), are used for detecting gases in the atmosphere, highlighted by ozone monitoring (further discussed in Chapter 11). VIS (400–700 nm) wavelengths are broadly applicable to monitor vegetation, soil, and surface water (ocean, lakes, and rivers). NIR (700–1300 nm) wavelength is applicable to measure vegetation particularly biomass, and MIR (1.3–3 μm) radiation helps detect surface temperature and moisture in the leaves of vegetation. Long-wave radiation includes thermal IR (TIR, 3–14 μm) wavelengths (applicable to thermal pollution and vegetation stress monitoring) and microwave (0.3–300 cm) applicable to monitor soil moisture.

PIXEL, RASTER, AND IMAGE

Many remote sensing datasets are observations of reflectance data measured at multiple wavelengths of the EM spectrum; in this case, data are *multispectral* or *hyperspectral* if there are many separate wavelengths. By having multiple wavelengths, the data can be used to analyze different characteristics of land surface, soil, and vegetation. Remote sensing datasets are normally stored as *raster* files or *images*, meaning a grid of *pixels*, i.e., a picture element representing a square area of the ground (Figure 8.1). Because there are measurements in multiple wavelengths, a single image has multiple observations for each pixel that are stored in separate raster *layers*. These layers, or variables, are named *bands* (abbreviation for bandwidths), and an image is referred to as a *scene*.

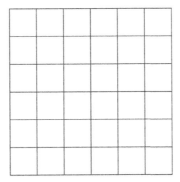

FIGURE 8.1 A raster file is composed of a grid of pixels.

IMAGERY SPECIFICATIONS: RESOLUTION AND QUALITY

The concept of resolution applies to remote sensing imagery in terms of spatial, spectral, temporal, and radiometric resolution. A pixel of remote sensing image represents a square area of the ground, the smallest the pixel the better the *spatial resolution*. For example, 30×30 m size pixels, as used in the Operational Land Imager (OLI) instrument on-board Landsat 8, have lower spatial resolution than pixels of 10 m $\times 10$ m size, as used in SPOT-4.

The values of a pixel correspond to the sensor response in the image corresponding to a given wavelength band. Having more bands means having higher *spectral resolution*. For example, the early Landsat missions had only four bands, while Landsat 9 has 11 bands, making it a higher spectral resolution system.

Frequency of coverage of a location on Earth, or a satellite's return time to image the location, determines the *temporal resolution*. For example, Landsat 8 has a 16-day repeat cycle. Furthermore, available archived images also determine temporal resolution when assembling time series of images over many years.

Radiometric resolution refers to the sensitivity and number of bits of the analog-to-digital converter (ADC) of the sensor for each wavelength band. For instance, Landsat 8 has 12-bit ADC compared to 8-bit ADC of Landsat 1–7. Recall from previous chapters that 12 bits allow representing integers from 0 to $2^{12} - 1 = 4095$, signal levels or a voltage resolution proportional to $1/(2^{12} - 1) = 244 \times 10^{-6}$, whereas 8 bits yield only 256 levels or a much coarser resolution proportional to $1/(2^8 - 1) \approx 3922 \times 10^{-6}$. It should be noted that Landsat 8 products in bands 1–7 and 10 are delivered as 16-bit unsigned integers, which means from 0 to 65,535 (USGS 2022b). Landsat 9 increased radiometric resolution by using 14-bit ADC, or 16,384 levels and voltage resolution proportional to $1/(2^8 - 1) \approx 0.610 \times 10^{-6}$ (USGS 2022c).

Naturally, the purpose and objectives of the monitoring program dictate the level of resolution needed for each type of resolution. For example, low spatial and spectral resolution may suffice to monitor coarse vegetation types over large areas, however, following up details of vegetation change at species level at a particular site may require high spatial and spectral resolution. Together with resolution, image quality also plays a role when selecting imagery. For example, percent coverage by clouds or of artifacts can limit the usefulness of an image.

SPACEBORNE REMOTE SENSING: TYPES OF ORBITS

Three types of orbits are considered for spaceborne remote sensing, geostationary, equatorial, and sun-synchronous (Zhu et al. 2018). A *geostationary* orbit rotates at a period equal to Earth's rotation period (24 hours), therefore the satellite always stays over the same location on Earth; this orbit type is useful for communication and weather satellites. A satellite in an *equatorial* orbit circles the Earth with a small inclination angle (angle between the orbital plane and the equatorial plane); in contrast, *sun-synchronous* satellites have orbits with high inclination angles, passing nearly over the poles, and going over a given location of Earth with a repetition period that varies up to 16 days. Remote sensing satellites tend to in sun-synchronous orbits, so that they can have repeatable sun illumination conditions for specific seasons. In addition, the orbit varies in altitude, typically from 600 to 1000 km.

PLATFORMS AND IMAGERY

By now, there are more than a 1000 remote sensing satellites launched, and many updated with newer technology and hyperspectral sensors, thereby improving spatial and spectral resolutions, as well as periodicity of imaging. Significantly, remote sensing data are becoming available as open data sources (Zhu et al. 2018). In this book, we will cover just some examples of frequently used platforms.

Landsat started in the early 1970s carrying the early multispectral scanner (MSS) with just a few bands VIS green, VIS red, NIR (700–800 nm), NIR (800–1100 nm), and 80-m spatial resolution. Landsat has evolved to Landsat 9 (launched in 2021) with 30-m spatial resolution, carrying the OLI-2, which measures in the VIS, NIR, and SWIR (shortwave IR), and the TIR Sensor 2 (TIRS-2) which measures temperature using two thermal bands. Figure 8.2 shows spectral bands of all Landsat satellites. MSS bands 1–4 in this figure were called bands 4–7 in Landsat 1–3. To illustrate, we consider the 11 bands of Landsat 8 in Table 8.2. Landsat is a great example of long-term and continuous measurement that characterizes environmental monitoring. From Landsat 1 to Landsat 9, we have more than 50 years of data.

GEODETIC DATUM OR SYSTEM

A *geodetic datum* or system is a reference frame to specify any location on the Earth, and it is of great importance for satellite imagery. A horizontal datum is given in latitude and longitude or another coordinate system; a vertical datum is given in elevation or depth relative to a reference, e.g., mean sea level. The World Geodetic System (WGS) and its revision of 1984, known as WGS84, is intended for global use and has become prevalent in many applications; for instance, WGS84 is used for the Global Positioning System. In WGS84, the origin of coordinates is the Earth center of mass (with uncertainty <2 cm), the prime meridian (0°0′0″ longitude) is the Greenwich meridian

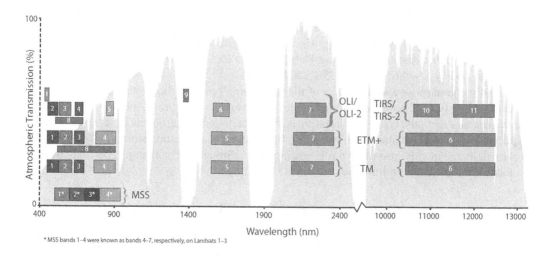

FIGURE 8.2 Spectral bands for all Landsat satellites. Figure from USGS Landsat Missions (2022b).

TABLE 8.2
Landsat 8 Bands (USGS 2022b)

Instrument	Band Number		Wavelength (nm)	Spatial Resolution (m)
OLI	1	Coastal aerosol	433–453	30
	2	Blue	450–515	30
	3	Green	525–600	30
	4	Red	630–680	30
	5	NIR	845–885	30
	6	SWIR	1560–1660	30
	7	SWIR	2100–2300	30
	8	PAN	500–680	15
	9	Cirrus	1360–1390	30
TIRS	10	Thermal	10,600–11,200	100
	11	Thermal	11,500–12,500	100

(just 5.3″ off), the Equator is 0° 0′ 0″ latitude, and the vertical or surface datum is an oblate spheroid calculated from several models, including Earth gravitation.

WORLD REFERENCE SYSTEM (WRS)

The Landsat WRS (NASA 2022) consists of *Path* and *Row* numbers that enable to find an image for any part of the world. The center of the scene is designated by the Path number (given first) followed by the Row number. For instance, 127–043 refers to Path 127 and Row 043. As platforms changed, so did the WRS; the WRS-1, corresponding to Landsat 1–3, sequences 251 path numbers from east to west, with the first crossing the equator at 65.48°W longitude.

WRS-2 corresponding to Landsat 4, 5, 7, 8, and 9 is an extension of WRS-1 using a Path/Row notation as well but with differences in repeat cycles (16-day), coverage, swaths, and number designation due to the large orbital differences. WRS-2 sequences 233 paths east to west, with the first crossing the equator at 64.60°W longitude and partitioning each path in 248 row intervals. Rows 60 and 184 coincide with the equator during the descending (daytime) and ascending (nighttime) parts of the orbit, respectively; Row 1 is at 80°47′ N latitude, row numbers increase southward and then northward ending at Row 248 located at 81°22′ N. For illustration, Figure 8.3 shows a sector of the WRS-2 map for the descending part of the orbit. The slanted sequences of dots are the paths, whereas the horizontal sequences of dots are the rows. The vertical lines correspond to Universal Transverse Mercator (UTM) Zones, which we will describe in the next section. For better understanding of the WRS, the reader is referred to the full maps available at USGS Landsat Missions (2022a).

In the companion lab manual (Acevedo 2024), we indicate how to obtain Landsat imagery which is available via the EarthExplorer USGS website (USGS 2022a). In that lab guide, we work with images of a scene downloaded from EarthExplorer and corresponding to Path 27, Row 37 of the Landsat WRS-2. The coordinates of the center of this scene are 33.176999° and -97.102621° latitude and longitude, respectively. This example is illustrated in Figure 8.4 which shows the relationship between Paths, Rows, coordinates of the center of the scene, and UTM Zones. The example covers an area of the North Central Texas region, State of Texas, USA. Th extracted scene and boundaries are illustrated in Figure 8.5 with datum WGS84.

UTM COORDINATE SYSTEM

In the UTM coordinate system, the world is divided into 60 zones, running in the North South direction. Thus, 360° of longitude divided into 60, means that the width of each zone is 6° of longitude;

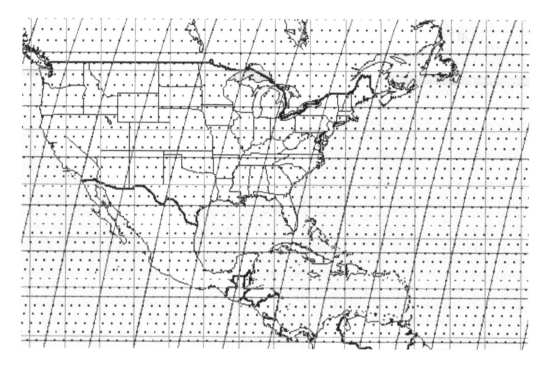

FIGURE 8.3   Section of path/row map showing descending (daytime) part of the orbits. From USGS Landsat Missions (2022a).

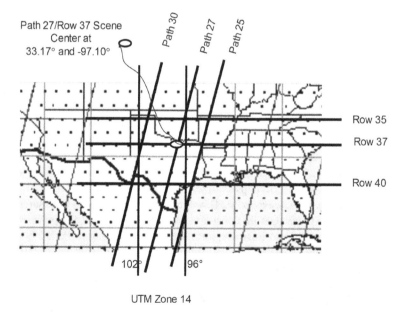

FIGURE 8.4   Example of a scene for Path 27, Row 37. Its center is located at 33.176999° and –97.102621° (latitude and longitude). Vertical lines are UTM Zone boundaries. Shown here is Zone 14.

FIGURE 8.5   Polygon for Path 27 and Row 37 scene.

this width in km is maximum at the Equator. UTM Zones have integer numbers beginning with Zone 1, at longitude 180° West, increasing to the East, and ending in Zone 60, which spans from 174° to 180° East longitude. Taking for instance the USA, the westernmost part of Alaska is in Zone 1, and Maine is in Zone 19.

For each zone, coordinates are *northing* and *easting* in meters. Northing values are referenced to the Equator (given a northing value of 10,000,000 m), and easting values are referenced to a meridian (given an easting value of 500,000 m) running through the middle of each zone.

For example, in the lab session 8 of the Lab Manual companion to this book (Acevedo 2024), we will work with a portion of the Landsat image discussed in the previous section. This portion is shown in Figure 8.6 (using false color composite which we will discuss in a later section), and is in UTM Zone 14, has easting extending from 664,995 to 700,005 m (this means it is located east of the central meridian which is at 500,000 m) and northing extend from 3,671,805 to 3,706,815 (this means is north of the Equator which is at 10,000,000 m). The total extent in easting is to 700,005−664,995=35,010 m or 35.01 km, similarly the total extend in northing is 3,706,815−3,671,805=35,010 m or 35.01 km. Therefore, this image has the same range in northing as in easting.

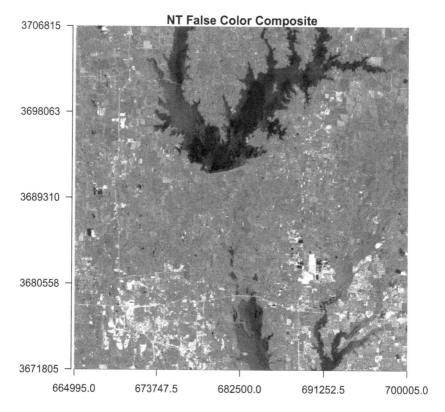

FIGURE 8.6 Portion of the scene for Path 27 and Row 37 selected for analysis and illustrating UTM coordinates. Shown here is a gray tone recolored version of the false color composite.

We end up with a raster image of 1167 rows and 1167 columns of 30 m × 30 m cells or pixels, for a total of 1,361,889 pixels.

BANDS AND IMAGE DISPLAY

Now we will look at the analysis of bands, using as an example a Landsat 8 scene captured on June 18, 2021 for Path 27, Row 37 that we have used as an example in the previous sections. To simplify the presentation, we will focus on bands 1–7 of Landsat 8. The data for each band are contained in a raster file, and as already mentioned when discussing radiometric resolution, the pixel values are given in 16-bit unsigned integers, that is from 0 to $2^{16} - 1 = 65535$. For further simplicity, we will frame the subsequent discussion using values scaled from 0 to 1, which are obtained dividing the original values into 65,535.

Displaying the contents of only one band on a computer monitor (Figure 8.7) does not yield a good visualization of many features of the terrain; this is evident in the green and red bands for the entire image. Interestingly, the NIR and SWIR1 bands do show differences between water and land. However, we can do a *true or natural color composite*, meaning we send the red, green, and blue bands to the red, green, and blue channels of a computer monitor (Figure 8.8) and now we can visualize more features.

To help the reader be acquainted with the area of the image, Figure 8.9 composed using Goggle Earth, shows the cities, towns, creeks, lakes, roads, and rural landscapes. These features can be identified in the natural color image of Figure 8.8 where we can visualize the lakes (Ray Roberts to the North and Lewisville to the south), including differences in water quality, roads and highways, urban areas in the southwest part of the image (corresponding to the City of Denton).

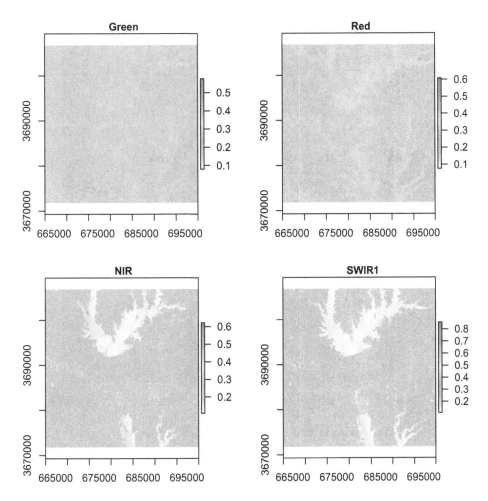

FIGURE 8.7   Green, red, NIR, and SWIR1 bands.

Often differences can be better visualized in a *false color composite* that sends the NIR, red, and green bands to the red, green, and blue channels of the monitor (Figure 8.6). The contrast between the lakes and the surrounding land areas is enhanced, as well as the contrast between the urban areas and surrounding rural areas.

We gain insight regarding the contents of the bands by looking at the distribution of the values (Figure 8.10). Indeed, the values of the green and red bands do not show much variability and opposite skewness, whereas NIR shows bimodality due to the differences between water and land, with higher values due to terrestrial vegetation. That difference is attenuated in the SWIR1 band.

We can compare the relationship between pairs of the four bands given in Figure 8.10 by plotting pair-wise scatter plots on comparable scales Figure 8.11. We see that green and red are highly correlated, but their relationship with NIR shows an interesting pattern. An important relationship to analyze vegetation is the one between the red and NIR (upper right-hand panel of Figure 8.11), since vegetation reflects more in the NIR than in the red. Note that the scatter plot for these two bands has a unique triangular shape. Its top corner is due to pixels with high NIR (vertical axis) and low red (horizontal axis), indicating vegetation. The bottom corner has low reflectance for both bands, indicating water, whereas the furthest corner corresponds to high reflectance in both bands, indicating exposed surface such as bright soil or concrete (Ghosh and Hijmans 2019). The scatter plot between green and NIR has a similar triangular shape, with a bottom corner that has low reflectance

FIGURE 8.8   North Texas area image bands as true or natural color composite.

FIGURE 8.9   North Texas area image from Google Earth.

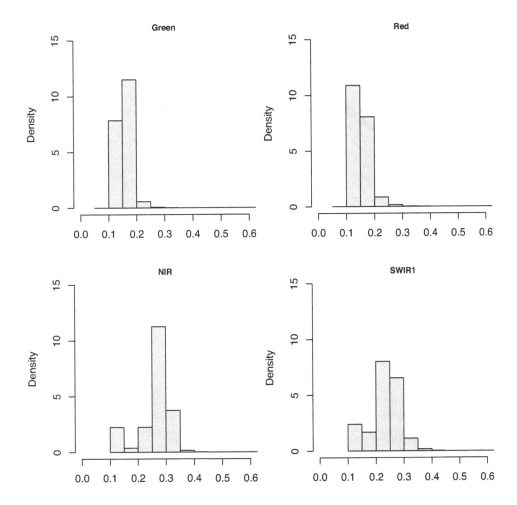

FIGURE 8.10   Histograms of the green, red, NIR and SWIR1 bands.

for both bands, indicating water. The unique relationship of NIR with red and green with respect to vegetation and water will be employed in the next section to construct vegetation and water indices.

ANALYSIS USING INDICES

Combining bands into *indices* allows to extract features from remote sensing imagery. As examples we use indices that can be calculated from various bands of Landsat 8 and Landsat 9. Several indices, such as those for vegetation and water, use a *normalized difference* that consists of taking the pixel-by-pixel difference between the values for two bands k and i and dividing by the sum of the values for these two bands

$$nd = \frac{b_k - b_i}{b_k + b_i} \tag{8.1}$$

These indices vary between −1 and +1. We can use it for instance, to calculate the *Normalized Difference Vegetation Index* (NDVI), which uses band 5 (or NIR) for b_k and band 4 (or red) for b_i.

$$NDVI = \frac{b_5 - b_4}{b_5 + b_4} \tag{8.2}$$

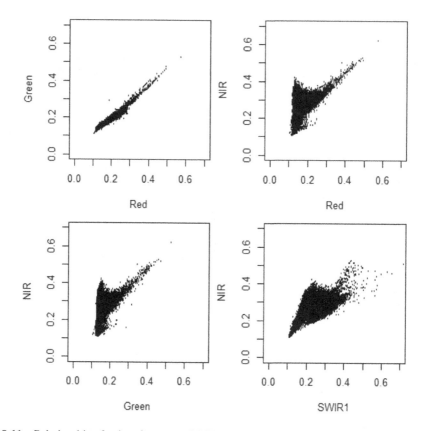

FIGURE 8.11   Relationship of pairs of green, red, NIR, and SWR1 bands.

Applying it to the example we have been using produces the image shown in Figure 8.12. Higher values of NDVI, darker (or green in the color version) areas (above ~0.3) indicate vegetation. In this case, we see, for example, higher values along Clear Creek that runs Southeast toward Lake Lewisville, and particularly an area named the Greenbelt Corridor, just to the north of the Lake Lewisville headwaters (see Figure 8.9 for locations).

NDWI, or the *Normalized Difference Water Index*, allows detecting pixels with water from the image; NDWI can be calculated using the expression

$$NDWI = \frac{b_3 - b_5}{b_3 + b_5} \tag{8.3}$$

which uses band 3 (or green) for b_k and band 5 (or NIR) for b_l. We can see the image in Figure 8.13. Higher values of NDWI, darker (or blue in the color version) areas (above ~0.2), indicate water, and these pixels show clearly for the lakes.

It is helpful to look at the statistics of NDVI and NDWI (Figure 8.14). We see how these two indices give us contrasting information, the pixels with low negative values of NDVI centered around −0.05, correspond to water, that coincides with the positive values of NDWI ~0.1. Contrastingly, high positive values of NDVI for vegetation correspond to negative values of NDWI. We can use this information to extract areas in vegetation or water using *reclassification*.

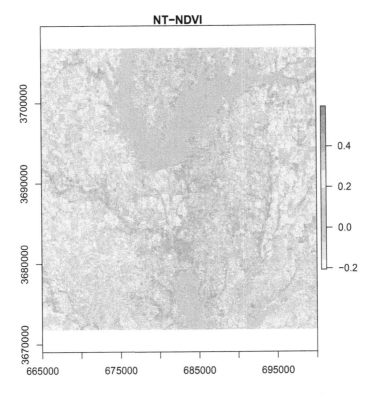

FIGURE 8.12 NDVI for the example image. Darker pixels correspond to vegetation.

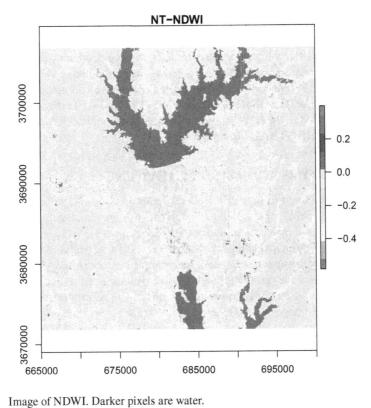

FIGURE 8.13 Image of NDWI. Darker pixels are water.

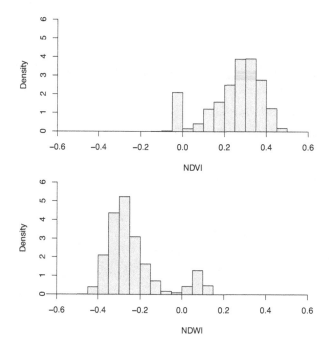

FIGURE 8.14 Histogram of the NDVI and NDWI values.

RECLASSIFICATION

A useful analysis of remote sensing imagery is reclassification, which can select pixels above a threshold or in an interval, to have a better idea of what they correspond to in the image. We can use it to mask out those pixels with a value lower than the threshold, which will show as white when we plot the image. As an example, consider reclassifying the NDVI image to mask out those values lower than 0.3 (Figure 8.15). We confirm the vegetation pattern we had determined previously. As another example, we can query for those pixels with NDVI around −0.05 (Figure 8.16). We confirm that those pixels correspond to the lakes and some smaller bodies of water. We could perform similar reclassification using NDWI to confirm the results.

MULTIVARIATE ANALYSIS AND MACHINE LEARNING

Besides the methods explained in previous sections, analysis of remote sensing data may require using Machine Learning (ML) algorithms. Although ML is a broad subject, in the case of remote sensing, this type of method includes techniques to reduce dimensionality, such as singular value decomposition (SVD) and principal components analysis (PCA), as well as ML *classification*. The latter can be *supervised* or *unsupervised*, based, for example on *cluster analysis*, of which there are two major types, agglomerative (grouping) and divisive (splitting into groups). In general, remote sensing data expressed as raster files require techniques of image processing. There are specialized programs to analyze imagery, but Geographic Information Systems (GIS) software typically includes remote sensing analysis modules; for instance, GRASS GIS (GRASS 2022) and QGIS (QGIS 2022). Analysis can also be conducted with programming tools, such as R (Ghosh and Hijmans 2019), and Mathematica (Haneberg 2004). There is a trend to seek solutions of remote sensing data classification problems and biophysical parameters estimation by additional ML techniques which we will cover in Chapter 9.

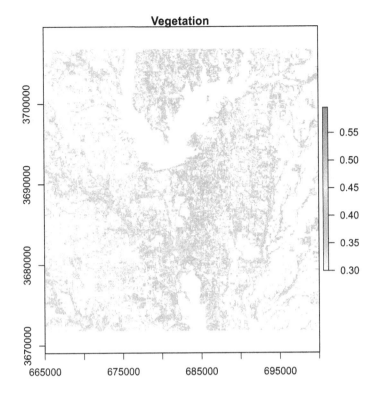

FIGURE 8.15   NDVI image reclassified to values above 0.3.

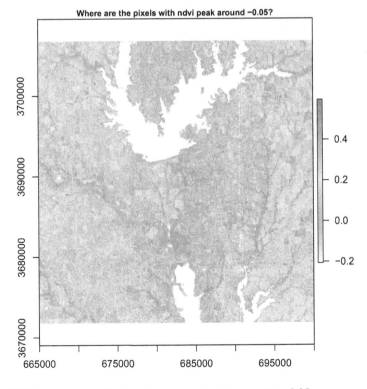

FIGURE 8.16   NDVI image reclassified to values approximately equal to −0.05.

REDUCING DIMENSIONALITY

PCA

In remote sensing applications, Principal Component Analysis (PCA) can be used to explain multispectral data contained in the images from new uncorrelated variables called *principal components* that are linear combination of the original variables or bands. A reduction in dimensionality is achieved when a few components can explain most of the variance observed in the images. These principal components are the eigenvectors of the covariance matrix and are orthogonal to each other.

Consider X to be an $n \times m$ data matrix, with columns corresponding to values of m bands $[X_1, X_2, ..., X_m]$ and the rows to the n observations or pixels. We assume that $n \gg m$, or that we have many more pixels than bands. The covariance matrix $C = cov(X)$ is $m \times m$, with a trace (the sum of main diagonal terms) equal to the total variance and in turn equal to the sum of m eigenvalues. Therefore, each eigenvalue is a percent of the total variance. When selecting components, a typical approach is to select those components that explain at least 90% of total variance. Alternatively, there are other criteria, such as the Kaiser rule, that retain a component if it explains as much as one of the original variables. Instead of C, we can also use the correlation matrix $R = cor(X)$ to scale uniformly.

The new variables $[Z_1, Z_2, ...Z_m]$ make up the new matrix Z which is also $n \times m$. Note that Z is the linear function of X, i.e., $Z = XA$ where A is a matrix $m \times m$ composed of the eigenvectors of the covariance matrix.

$$A = \begin{bmatrix} v_{11} & v_{21} & ... & v_{31} \\ v_{12} & v_{22} & ... & v_{32} \\ ... & ... & ... & ... \\ v_{1m} & v_{2m} & ... & v_{mn} \end{bmatrix} \tag{8.4}$$

The entries of A are the *loadings*, whereas the entries of Z are *scores* or new coordinate values. Observations are plotted as points in this new set of coordinates. Points close to each other are similar. Matrix A satisfies

$$C = ALA^T \tag{8.5}$$

where L is a diagonal matrix made up with the eigenvalues of C arranged in descending order.

$$L = \begin{bmatrix} \lambda_1 & 0 & ... & 0 \\ 0 & \lambda_2 & ... & 0 \\ ... & ... & ... & 0 \\ 0 & 0 & ... & \lambda_m \end{bmatrix} \tag{8.6}$$

here $\lambda_1 \geq \lambda_2 \geq ... \geq \lambda_m$. The trace of L is equal to the total variance.

For example, in a similar manner to the example in Carr (1995, p96), consider the data matrix of $n = 5$ observations of two variables

$$X = \begin{bmatrix} 3 & 14 \\ 4 & 13 \\ 5 & 6 \\ 6 & 5 \\ 7 & 0 \end{bmatrix} \tag{8.7}$$

First, center the columns by subtracting the means

$$x = \begin{bmatrix} -2 & 6.4 \\ -1 & 5.4 \\ 0 & -1.6 \\ 1 & -2.6 \\ 2 & -7.6 \end{bmatrix} \tag{8.8}$$

Now the covariance matrix is

$$C = \frac{1}{n-1} x^T x = \frac{1}{5-1} \begin{bmatrix} 10 & -36 \\ -36 & 137.2 \end{bmatrix} = \begin{bmatrix} 2.5 & -9 \\ -9 & 34.3 \end{bmatrix} \tag{8.9}$$

The total variance is the trace of C or tr(C) $= 34.3 + 2.5 = 36.8$. The eigenvalues are

$$\lambda_1 = 36.67 \quad \text{and} \quad \lambda_2 = 0.13 \tag{8.10}$$

Their sum is also equal to the total variance ($36.67 + 0.13 = 36.8$). The first eigenvalue 36.67 is a large fraction of total variance 36.8. Indeed, the percentage of total variance explained by first eigenvalue is $100 \times (36.67/36.8) = 99.6\%$. Therefore, only one component explains most of the variance. The eigenvectors are

$$v_1 = \begin{bmatrix} -0.25 \\ 0.97 \end{bmatrix} \quad \text{and} \quad v_2 = \begin{bmatrix} -0.97 \\ -0.25 \end{bmatrix}$$

are arranged as columns in a matrix A to form the loadings

$$A = \begin{bmatrix} -0.25 & -0.97 \\ 0.97 & -0.25 \end{bmatrix}$$

The new or transformed data matrix Z is obtained by post-multiplying the data by the loadings

$$Z = XA = \begin{bmatrix} 3 & 14 \\ 4 & 13 \\ 5 & 6 \\ 6 & 5 \\ 7 & 0 \end{bmatrix} \begin{bmatrix} -0.25 & -0.97 \\ 0.97 & -0.25 \end{bmatrix} = \begin{bmatrix} 12.77 & -6.47 \\ 11.55 & -7.18 \\ 4.52 & -6.36 \\ 3.30 & -7.08 \\ -1.78 & -6.77 \end{bmatrix}$$

We can verify that L is diagonal with eigenvalues in main diagonal

$$L = A^T CA = \begin{bmatrix} -0.25 & -0.97 \\ 0.97 & -0.25 \end{bmatrix} \begin{bmatrix} 2.5 & -9 \\ -9 & 34.3 \end{bmatrix} \begin{bmatrix} -0.25 & -0.97 \\ 0.97 & -0.25 \end{bmatrix} = \begin{bmatrix} 36.67 & 0 \\ 0 & 0.13 \end{bmatrix}$$

Note that in this case, the transpose of matrix A is the same as matrix A. Graphically we can compare the original data X to the transformed scores Z as shown in Figure 8.17. Here we label the points with observation numbers. Proximity of pairs of observations indicates similarity between those observations. The first principal component (horizontal axis) explains most of the variance and observations sort themselves along this axis; observation 1 has the highest vale and observation 5 the lowest. However, all observation pairs differ little along the vertical axis (second principal component).

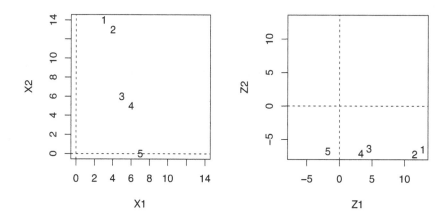

FIGURE 8.17 Original data X (left) and transformed data (scores) Z (right) using observation number for labels.

When the observations are not all on the same scale, which could happen for example, when the variables are not in the same units, it is important to first standardize the observations or equivalently use the correlation matrix. In the example at hand, the second column has much larger values than the first, therefore it is important to standardize or use the correlation matrix.

To standardize the observations, divide the centered columns by the standard deviation of the column. For example, matrix X above is standardized

$$x_s = \begin{bmatrix} -1.26 & 1.09 \\ -0.63 & 0.93 \\ 0 & -0.27 \\ 0.63 & -0.44 \\ 1.26 & -1.30 \end{bmatrix}$$

We can either calculate the correlation matrix from the covariance matrix

$$R = \begin{bmatrix} 2.5/2.5 & \dfrac{-9}{\sqrt{2.5}\sqrt{34.3}} \\ \dfrac{-9}{\sqrt{2.5}\sqrt{34.3}} & 34.3/34.3 \end{bmatrix} = \begin{bmatrix} 1 & -0.97 \\ -0.97 & 1 \end{bmatrix}$$

or calculate it directly from the standardized observations themselves

$$R = \frac{1}{n-1} x_s^T x_s = \begin{bmatrix} 1 & -0.97 \\ -0.97 & 1 \end{bmatrix}$$

The total correlation (scaled variance) is the trace of R or $\mathrm{tr}(R) = 1+1 = 2$. The eigenvalues are

$$\lambda_1 = 1.97 \text{ and } \lambda_2 = 0.03$$

Their sum is also equal to the total correlation ($1.97 + 0.03 = 2$). The first eigenvalue 1.97 is a large fraction of total correlation. Indeed, the percentage of total correlation explained by the first eigenvalue is $100 \times (1.97/2) = 98.5\%$. Therefore, only one component explains most of the variance, but note that the standardization has made the value smaller. The eigenvectors

$$v_1 = \begin{bmatrix} -0.71 \\ 0.71 \end{bmatrix} \text{ and } v_2 = \begin{bmatrix} -0.71 \\ -0.71 \end{bmatrix}$$

have all the same magnitudes due to the standardization. These are arranged as columns in a matrix A_s to form the loadings

$$A_s = \begin{bmatrix} -0.71 & -0.71 \\ 0.71 & -0.71 \end{bmatrix}$$

The new or transformed data matrix Z_s is obtained by post-multiplying the data by the loadings.

$$Z_s = x_s A_s = \begin{bmatrix} -1.26 & 1.09 \\ -0.63 & 0.93 \\ 0 & -0.27 \\ 0.63 & -0.44 \\ 1.26 & -1.30 \end{bmatrix} \begin{bmatrix} -0.71 & -0.71 \\ 0.71 & -0.71 \end{bmatrix} = \begin{bmatrix} 1.67 & 0.12 \\ 1.10 & -0.20 \\ -0.19 & 0.19 \\ -0.76 & -0.13 \\ -1.81 & 0.02 \end{bmatrix}$$

Plots of the data and scores allow comparison as before (Figure 8.18). On the right-hand side panel, sorting along the horizontal axis is nearly the same, but now we can appreciate differences among observation pairs along the vertical axes (second component). Observations 2 and 4 are relatively similar, and so are 1 and 3. However, these two pairs differ among themselves; 2 and 4 are negative, whereas 1 and 3 are positive.

SVD AND BIPLOTS

SVD is a method to obtain a diagonal form for the covariance matrix C, similar to equation (8.5), but it is more general and applied directly to the rectangular data matrix X. SVD relates to the Eckart-Young theorem and applies to many multivariate techniques: PCA, correspondence analysis, and canonical correlation analysis.

SVD can work directly with the X data matrix, or the centered observations x, or the standardized observations x_s. In the following, we work with the centered column data matrix x. The SVD of matrix x is

$$x = U\Gamma V^T \tag{8.11}$$

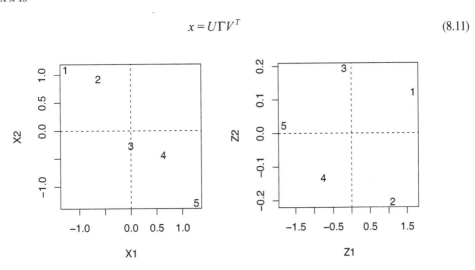

FIGURE 8.18 Original data (standardized) Xs (left panel) and transformed data (scores) Zs (right panel) using observation number for labels.

where U is $n \times m$, Γ is $m \times m$, and V is $m \times m$. U and V are matrices of singular vectors, the columns of U are eigenvectors of xx^T (referred to as *left* eigenvectors), and the columns of V are eigenvectors of x^Tx (referred to as *right* eigenvectors). Note that xx^T can be a large matrix because it is $n \times n$ and that x^Tx is smaller because it is $m \times m$. Matrix Γ is a diagonal matrix of singular values, the square root of eigenvalues of x^Tx, arranged in non-decreasing order. The non-zero eigenvalues of xx^T are the same as the eigenvalues of x^Tx.

$$\Gamma = diag(\sqrt{\lambda_i} \quad i = 1,...,m) \tag{8.12}$$

A least squares approximation of the matrix x is obtained using the first few (say k) dominant singular values and vectors

$$\hat{x}_k = U_k \Gamma_k V_k^T \tag{8.13}$$

In this equation, the subscript k-means we take the first k columns of U, Γ, and V^T. A useful graphical visualization of the reduced dimensionality is called a *biplot* (Gabriel 1971). This plot shows both the observations and the variables simultaneously. The prefix *bi* refers to the simultaneous display of both rows and columns of the transformed data matrix given by Equation (8.13), and not to the fact that the plots are bi-dimensional. Biplots are usually drawn in 2D ($k = 2$) for ease of interpretation, but conceptually they can be done in 3D and multi-dimensional space ($k > 2$).

A factorization of equation (8.13) is

$$\hat{x}_k = U_k \Gamma_k^\alpha \Gamma_k^{1-\alpha} V_k^T = GH^T \tag{8.14}$$

where G and H^T are

$$G = U_k \Gamma_k^\alpha$$
$$H^T = \Gamma_k^{1-\alpha} V_k^T \tag{8.15}$$

and the value of the *scale* parameter α in the range 0 to 1 ($0 \leq \alpha \leq 1$) determines whether emphasis is placed on the rows or columns of x. The biplot display in 2D is the plot of the row "markers" G and column "markers" H given in equation (8.15) for $k = 2$.

$$G = U_2 \Gamma_2^\alpha$$
$$H = V_2 \Gamma_2^{1-\alpha} \tag{8.16}$$

In other words, the biplot is a plot of the coordinates associated with G or columns of G, superimposed over the coordinates associated with H or columns of H.

Although any values of α are possible to accomplish the factorization, three are most used, 1, ½, and 0. When $\alpha = 1$ is selected, which is the original value used in Gabriel (1971), the result is a *row metric preserving* biplot. This display is useful for studying relationships among the observations. When the value 0 is selected, the result is a *column metric preserving* biplot. This display is useful for interpreting relationships among variables (for example, interpreting a covariance or correlation matrix). The other value of α, 1/2, gives equal scaling or weight to the rows and columns. It is useful for interpreting interaction in two factor experiments (Gower and Hand 1996).

We can see that PCA is a special case of SVD, since in PCA X^TX is transformed (centered and scaled) to be the covariance or correlation matrix. This assumption is not generally made in SVD. Therefore, the singular values of PCA are eigenvalues of a covariance or a correlation matrix.

The singular values of an SVD of the centered observations x must be squared and divided by $n-1$ to obtain variances. The V matrix is equivalent to the matrix of loadings A.

Let us look at the numerical example given in the previous section and apply the SVD to the centered observations given in Equation (8.8) and the x^Tx matrix given in Equation (8.9)

$$x^Tx = \begin{bmatrix} 10 & -36 \\ -36 & 137.2 \end{bmatrix}$$

with eigenvalues

$$\lambda_1 = 146.68 \qquad \lambda_2 = 0.52$$

take the square roots to obtain the singular values

$$s_1 = \sqrt{146.68} = 12.11$$

$$s_2 = \sqrt{0.52} = 0.72$$

and the singular values matrix has the diagonal elements equal to the singular values

$$\Gamma = \begin{bmatrix} 12.11 & 0 \\ 0 & 0.72 \end{bmatrix}$$

In this case, the matrix x is of rank 2 (only two variables) and therefore the least squares approximation for $k=2$ is exact. The singular values matrix is just 2×2 and has no zeros in the main diagonal. Note that $12.11^2/4 = 36.67$ and $0.72^2/4 = 0.13$ are the eigenvalues of the covariance matrix.

The eigenvectors of the x^Tx matrix are arranged as columns in a matrix V

$$V = \begin{bmatrix} -0.25 & -0.96 \\ 0.96 & -0.25 \end{bmatrix}$$

Now calculate the xx^T matrix, note that $n=5$ and therefore this matrix will be 5×5

$$xx^T = \begin{bmatrix} 44.96 & 36.56 & -10.24 & -18.64 & -52.64 \\ 36.56 & 30.16 & -8.64 & -15.04 & -43.04 \\ -10.24 & -8.64 & 2.56 & 4.16 & 12.16 \\ -18.64 & -15.04 & 4.16 & 7.76 & 21.76 \\ -52.64 & -43.04 & 12.16 & 21.76 & 61.76 \end{bmatrix}$$

with eigenvalues

$$\lambda_1 = 146.68$$

$$\lambda_2 = 0.52$$

$$\lambda_3 = 0.0$$

$$\lambda_4 = 0.0$$

$$\lambda_5 = 0.0$$

Note that the first two are non-zero eigenvalues and are the same as the eigenvalues of $x^T x$. This fact is part of the Eckart-Young theorem and SVD.

The eigenvectors of the xx^T matrix can be arranged as columns of matrix U

$$U = \begin{bmatrix} -0.55 & -0.42 & 0.72 & 0.00 & 0.00 \\ -0.45 & 0.56 & -0.01 & -0.68 & -0.06 \\ 0.12 & -0.56 & -0.23 & -0.49 & -0.60 \\ 0.22 & 0.42 & 0.43 & 0.25 & -0.72 \\ 0.64 & -0.00 & 0.50 & -0.47 & 0.33 \end{bmatrix}$$

We can verify that the SVD $x = U\Gamma V^T$ yields the correct results. Let us calculate the projection matrices G and H for the biplot in 2D ($k = 2$). Use scale equal to 1.

$$G = U_2 \Gamma_2^\alpha = U_2 \Gamma = \begin{bmatrix} -0.55 & -0.42 \\ -0.45 & 0.56 \\ 0.12 & -0.56 \\ 0.22 & 0.42 \\ 0.64 & 0.00 \end{bmatrix} \begin{bmatrix} 12.11 & 0 \\ 0 & 0.72 \end{bmatrix} = \begin{bmatrix} -6.70 & -0.30 \\ -5.47 & 0.41 \\ 1.54 & -0.41 \\ 2.76 & 0.30 \\ 7.86 & -0.00 \end{bmatrix}$$

$$H = V_2 \Gamma_2^{1-\alpha} = V_2 = V = \begin{bmatrix} -0.25 & -0.96 \\ 0.96 & -0.25 \end{bmatrix}$$

Now we can construct the biplot in 2D using G and H as in Figure 8.19. In this plot, the observations are labeled 1–5 and the variables V1 and V2. The bottom and left axis labels and scale correspond to the observations (coordinates related to U), whereas the top and right axis labels and scale correspond to the variables (coordinates of V). Proximity between pairs of observations denotes degree of similarity between those observations, whereas the cosine of the angle in between variables represents the correlation between those variables. In the example, we see that observation pairs 1, 2 are similar and 3, 4 are similar.

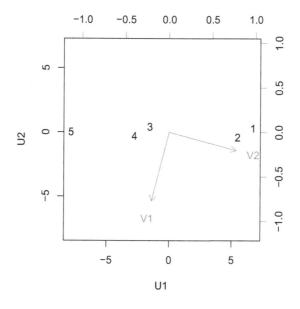

FIGURE 8.19 Biplot for the example.

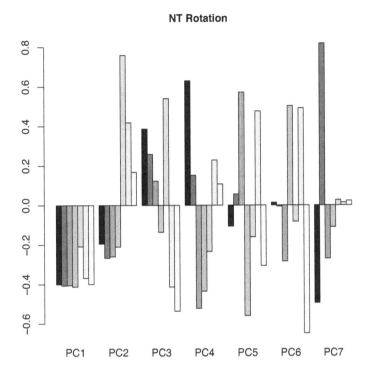

FIGURE 8.20   PCA rotation plot of the example image.

The biplot concept can be applied to PCA, the observations are plotted on the two principal components PC1 and PC2. Their coordinates given by G are the scores. The variables are given by coordinates of H, which are equivalent to the loadings or matrix A.

PCA APPLIED TO REMOTE SENSING IMAGES

We will demonstrate the application of PCA to remote sensing images using the images worked upon in the previous sections. Since the number of pixels is very large (1,361,889 in this example), we can take a random sample, e.g., size 10,000 before performing PCA. We obtain that the first three components explain more than 98% of the variance, the loadings or rotation indicate that the first component loadings are of the same sign and nearly similar magnitude for the first four bands (VIS), whereas the second component loadings are different sign for the last three bands (NIR and SWIR) (Figure 8.20). A biplot of the first two components shows that all bands 1–7 follow the PC1 axis but bands 5–7 (NIR and SWIR bands) separate from bands 1–4 (VIS) for the PC2 axis (Figure 8.21). We conclude that we can explain the images from just the first three components, PC1, PC2, and PC3 and create an image assigning the first three components to the RGB channels (Figure 8.22). We obtain a clear distinction of pixels by types of land use, highlighting the lakes, vegetated areas, rural areas, and urban areas.

UNSUPERVISED CLASSIFICATION: CLUSTER ANALYSIS

One major goal of remote sensing applications is to group or classify pixels of similar spectral characteristics. In *unsupervised classification*, we do not supply information indicating that pixels belong to a particular class or group. This is useful when we do not have knowledge of the area under monitoring nor ground-truth data. *Supervised classification* implies training the algorithm by a set of pixels for which we know the correct class.

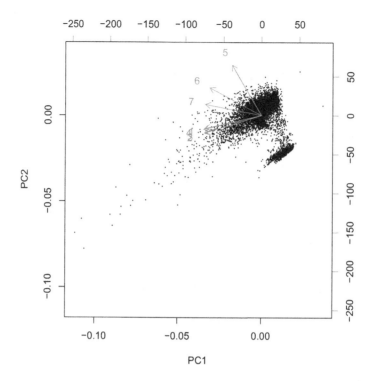

FIGURE 8.21   PCA biplot for PC1 and PC2 of the example image.

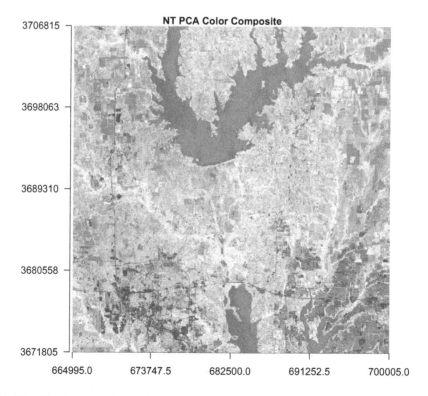

FIGURE 8.22   RGB image of first three PCA components.

Cluster analysis is a method to classify objects or observations into groups; this analysis helps to identify the possible groups and reveal relations among groups. Besides remote sensing image classification, it has many applications such as numerical taxonomy. There are several types of cluster analysis. For example, *hierarchical agglomerative clustering* successively joins or merges the most similar observations to form clusters, and divisive such as *K-means* that successively divides the pixels into groups.

HIERARCHICAL CLUSTER ANALYSIS

Assume we have n observations and m attributes or variables that make up an $n \times m$ data matrix. First, the distance (dissimilarity) matrix D is $m \times m$. Entries can be computed in various ways, for example calculating the Euclidian distance in m-space between observations or data points i and j

$$d_{ij} = \sqrt{\sum_{k=1}^{m}(x_{ik} - x_{jk})^2} \tag{8.17}$$

The sum is over $k = 1, ..., m$ attributes. Here x_{ik} is the kth variable value of point i, and x_{jk} is the kth variable value of point j.

A small distance between two points implies that the observations are similar. Matrix D is symmetrical. The hierarchical agglomerative cluster analysis process consists of the following steps: (1) Initially set $i = n$ clusters, this is to say only one object in each cluster, compute D. (2) Join two clusters to form a new cluster; at this point, we have i–1 clusters. To decide what clusters to join there are several options; for example, the Sum of Squares method: merges those two with the smallest increase in within-cluster Sum of Squares, the Link method: merges two with the smallest distance. There are several options to evaluate this distance between clusters: for example, nearest neighbor, complete linkage or farthest neighbor, centroid, and average. (3) Recompute D, which is now $(i-1) \times (i-1)$; the new cluster is represented by a "centroid". (4) Repeat steps 2 and 3 until $i = 2$, when the set reduces to only two clusters. (5) Use the distance determined for joining clusters as a vertical axis (height) to draw a tree or dendrogram.

We will work out a simple example using the data matrix given in Equation (8.7), consisting of $n = 5$ observations of two variables, and used to illustrate PCA

$$X = \begin{bmatrix} 3 & 14 \\ 4 & 13 \\ 5 & 6 \\ 6 & 5 \\ 7 & 0 \end{bmatrix}$$

We will use Euclidian distances and the complete linkage or furthest neighbor method for merging clusters. First calculate distances between observations to obtain matrix D. For illustration, take rows 1 and 2, and calculate distance $d_{12} = \sqrt{(3-4)^2 + (14-13)^2} = \sqrt{2} = 1.414$

After calculating all unique pairs of observations, we get the following matrix D, which is symmetrical and thus we have not written the values in the upper right half because they are equal to the lower left half.

$$D = \begin{bmatrix} 0 \\ 1.414 & 0 \\ 8.246 & 7.071 & 0 \\ 9.487 & 8.246 & 1.414 & 0 \\ 14.560 & 13.342 & 6.325 & 5.099 & 0 \end{bmatrix}$$

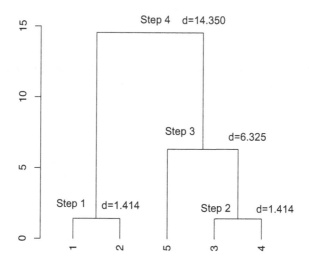

FIGURE 8.23   Dendrogram generated by cluster analysis.

Observations 1 and 2, as well as 3 and 4, are the closest at $d=1.414$ and then at step 1 we form one cluster merging 1 and 2, and another cluster merging 3 and 4 (step 2). Next, we examine the remaining single observation 5 and we see that the shortest distance 5.099 is to observation 4 which is part of previous cluster 3–4, so now to merge 3–4, and 5 we use furthest neighbor which implies using the larger distance 6.325 at step 3. At this point we have 2 clusters, 1–2 and 3–4–5, and to merge them at the final one cluster we select the largest distance 14.560 at step 4.

We can put this information in a dendrogram as shown in Figure 8.23 where the vertical axis (height) corresponds to distance, and the steps mentioned above are identified together with the cluster merging distances. Three clusters could be kept at a distance (height) of 5. Increasing the cutoff distance to 10, we have only two clusters.

It is worth noting the location of these observations as points in two-dimensional space as shown in the left-hand side of Figure 8.17, which suggests that observations 3, 4, and 1,2 are the closest, with 5 most closely related to 3-and 4. These provides additional insight on the results of the hierarchical cluster analysis.

Model-based clustering uses a statistical model for the clusters; for example, Gaussian density for each cluster with mean μ and covariance C as parameters. The eigenvectors of C determine orientation of the cluster, and the eigenvalues of C determine the shape. If we make all clusters have same ratios of eigenvalues to the dominant eigenvalue, then all clusters have the same shape which is parameterized by the ratios; for example, when ratio$=1$ we have hyper spherical clusters, but small ratio values lead to highly elongated hyper ellipsoids.

A difficult part of cluster analysis is to decide how many clusters to include at the end of the process. This often becomes a judgment call made by looking at the dendrogram. However, some help is available as a Bayes factor $B_k = \dfrac{P(mc_k)}{P(mc_1)}$ where mc_k is the event that the model has k clusters and mc_1 that the model has 1 cluster. Then $B_1=1$ and larger B_k indicates more evidence for the existence of k clusters. The Bayes factor can be converted into the Approximate Weight of Evidence for k clusters AWE_k as $AWE_k = 2 \times \log(B_k)$. Now $AWE_1=0$. The rate of change of AWE and max (AWE) are used to decide on the number of clusters.

K-Means

The K-means method is based on partitioning the data into non-overlapping clusters of similar observations and with centroids calculated as the mean distance of observations assigned to each cluster. K-means is used in many areas and is one important tool in unsupervised ML.

The letter "K" in the name K-means refers to the number k of clusters (given a priori by the user) to divide the data into, whereas the "means" part of the name refers to using the mean to calculate the centroids. Mathematically, the intent is to classify the observations into k clusters such that the total within-cluster sum of squares is as small as possible. There are several algorithms commonly employed to implement this method. For example, Hartigan and Wong (1979), MacQueen (1967), and Lloyd (1982) or its equivalent Forgy (1965).

Assume we have n observations and m attributes or variables that make up an $n \times m$ data matrix. Denote by c_{ij} the centroid of cluster i where $i = 1,...,k$. for variable $j = 1,...,m$ Observations are assigned to a cluster based on the shortest distances from the points to the centroid. For example, using the Euclidian distance

$$d_{il} = \sqrt{\sum_{j=1}^{m}(x_{ij} - c_{ij})^2}$$

Between a point l in cluster i, with coordinates x_{ij} For all points in cluster i we want to minimize the within-cluster sum of squares SS

$$SS = \frac{1}{w}\sum_{p,r \in i}\sum_{q=1}^{m}(x_{pq} - x_{rq})^2 \tag{8.18}$$

Where p, r denote pair of points in the cluster and w is the total number of points in the cluster. K-means proceeds following these steps: (1) Assign k centroids randomly in m-space, (2) partition the observations based on their distance from the centroids; points relatively closer to a centroid will get assigned to that cluster. (3) Then, the cluster centroids are calculated based on the means of the cluster points and relocated, (4) based on the newly assigned centroids, assign each observation falling closest to the new centroids, check for lower within-cluster sum of squares. (5) Repeat steps 3 and 4 until the cluster centroids do not change or the stopping criterion is reached.

As a simple example, we will use the data matrix given in equation 8.7, consisting of $n = 5$ observations of two variables, which was also used to illustrate PCA and hierarchical clustering. As mentioned in the previous section, the location of these points in two-dimensional space shown in the left-hand side of Figure 8.17, suggests that observations 3, 4, 5 are most likely a cluster separated from another cluster composed of observations 1, 2. Our hierarchical cluster work of the previous section also supports this hypothesis. Therefore, we will work a k-means example postulating two clusters.

As mentioned above, initial centroid coordinates are assigned at random, and then distances are calculated to decide which observations go with each centroid. However, in this simple example we can calculate the coordinates of the centroids of the hypothetical clusters averaging the coordinates of all points in the cluster. For cluster 1–2 the means are $\begin{matrix} x1 = (3+4)/2 = 3.5 \\ x2 = (14+13)/2 = 13.5 \end{matrix}$ and for cluster 3–5 the means are $\begin{matrix} x1 = (5+6+7)/3 = 6 \\ x2 = (6+5+0)/3 = 3.66 \end{matrix}$. You can visually estimate where these centroids will appear in the plane of the left-hand side of Figure 8.17, for further insight into the process.

Let us calculate the within sum of squares for each cluster, For this purpose we can use matrix D from the previous section. For cluster 1–2, we already know the distance is 1.414 and its square is 2, since we have two points $w = 2$, therefore the within SS should be SS=2/2=1. We can see this also from equation 8.18 $SS = \frac{1}{2}[(3-4)^2 + (14-13)^2] = 1$. For cluster 2, using equation 8.18

$$SS = \frac{1}{3}\left\{[(5-6)^2 + (6-5)^2] + [(5-7)^2 + (6-0)^2] + [(6-7)^2 + (5-0)^2]\right\} =$$

$$= \frac{1}{3}\{2 + 40 + 26\} = 68/3 = 22.66$$

we obtain within SS of 22.66, which can also be obtained from matrix D. Note also that total within SS is 1+22.66 = 23.66.

It is also of interest to calculate "between SS", which are calculated using points across clusters and "total SS" which is obtained from all data points. A ratio of these quantities expressed in percentage represents an evaluation of the results; the larger the ratio the better the separation between the clusters. This is somewhat lengthier calculation to be made by hand but applying a computer program to this example will yield a result in one iteration due to the simplicity of the data matrix. In this case the between SS is 123.53 and the total SS is 147.2, yielding a ratio of 123.53/147.2=83.9% indicating relatively good results for this model. We learn how to use a k-means R package in lab 8 of the companion Lab Manual.

UNSUPERVISED CLASSIFICATION OF REMOTE SENSING IMAGES USING K-MEANS

Hierarchical agglomerative clustering is not very practical to classify remote sensing images due to the large number of pixels, which implies calculating a very large distance matrix. However, it can be applied to a relatively small sample of the image. We will not examine this method here but rather focus on divisive clustering using K-means. For this purpose, we perform unsupervised classification

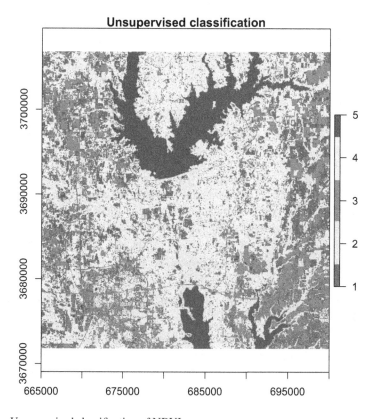

FIGURE 8.24   Unsupervised classification of NDVI.

of the NDVI image obtained earlier and targeting five clusters. The computational process is described in the companion lab guide. The five clusters have means 0.235, 0.392, 0.136, 0.311, and −0.025. From our knowledge of the NDVI, cluster 5 that is centered around −0.025 must correspond to the lakes, cluster 2 must be dense vegetation, cluster 4 must relate to vegetation, and clusters 1 and 3 must relate to open areas or urban areas. Armed with these classes, we assign cluster number to the pixels, create colors to correspond to these classes, and display the image (Figure 8.24).

EXERCISES

Exercise 8.1

Compare radiometric resolution using 12 bits and 14 bits.

Exercise 8.2

Determine the longitude of the center meridian for UTM Zone 14.

Exercise 8.3

Calculate the NDVI of a pixel with NIR band equal to 0.15 and red band equal to 0.1.

Exercise 8.4

Consider the following data matrix, with rows corresponding to observations and columns to the variables

$$X = \begin{bmatrix} 3 & 14 \\ 4 & 13 \\ 5 & 6 \\ 6 & 5 \\ 7 & 0 \end{bmatrix}$$

Apply PCA to this simple matrix. Calculate the centered values and the covariance matrix of the centered values, the percentage of variance explained by the components, loadings matrix, and the transformed data.

Exercise 8.5

Consider the data matrix of the previous exercise with rows corresponding to observations and columns to the variables. Apply hierarchical cluster analysis to this simple matrix. Calculate the Euclidian distance for all observations pairs, merge clusters according to the complete linkage method, and draw a dendrogram.

Exercise 8.6

Consider the data matrix of the previous exercise with rows corresponding to observations and columns to the variables. Apply k-means analysis to this simple matrix using clusters 1–2 and 3–5. Calculate coordinates of the centroids and within SS for both clusters.

REFERENCES

Acevedo, M. F. 2024. *Real-Time Environmental Monitoring: Sensors and Systems, Second Edition. – Lab Manual*. Boca Raton, FL: CRC Press, Taylor & Francis Group. 463 pp

Forgy, E. 1965. Cluster analysis of multivariate data: Efficiency versus interpretability of classifications. *Biometrics* 21: 768–780.

Gabriel, K. R. 1971. The biplot-graphic display of matrices with application to principal component analysis. *Biometrika* 58: 453–467.

Ghosh, A., and R. J. Hijmans. 2019. *Remote Sensing Image Analysis with R*. 48 pp. accessed January 2023. https://rspatial.org/raster/rs/index.html

Gower, C., and D. J. Hand. 1996. *Biplots*. London: Chapman & Hill. 277 pp.

GRASS. 2022. *GRASS GIS*. accessed March 2022. https://grass.osgeo.org/.

Haneberg, W. 2004. *Computational Geosciences with Mathematica*. Berlin, Heidelberg: Springer. 381 pp.

Hartigan, J. A., and M. A. Wong. 1979. Algorithm AS 136: A K-means clustering algorithm. *Journal of the Royal Statistical Society* 28 (1):100–108.

Huete, A. R. 2004. "Remote Sensing for Environmental Monitoring." In *Environmental Monitoring and Characterization*, edited by J. F. Artiola, I. L. Pepper and M. L. Brusseau, 183–206. Burlington, MA: Elsevier Academic Press.

Lloyd, S. 1982. Least squares quantization in PCM. *IEEE Transactions on Information Theory* 28 (2):129–137.

MacQueen, J. 1967. Some methods for classification and analysis of multivariate observations In *Proceedings of the Fifth Berkeley Symposium on Mathematical Statistics and Probability*, edited by L. M. Le Cam and J. Neyman, Berkeley, CA: University of California Press.

NASA. 2022. *The Worldwide Reference System*. accessed December 2022. https://landsat.gsfc.nasa.gov/about/the-worldwide-reference-system/.

QGIS. 2022. *QGIS A Free and Open Source Geographic Information System*. accessed December 2022. https://www.qgis.org/en/site/.

USGS. 2022a. *EarthExplorer*. accessed December 2022. https://earthexplorer.usgs.gov/.

USGS. 2022b. *Landsat 8*. accessed March 2022. https://www.usgs.gov/landsat-missions/landsat-8.

USGS. 2022c. *Landsat 9*. accessed March 2022. https://www.usgs.gov/landsat-missions/landsat-9.

USGS Landsat Missions. 2022a. *Landsat Shapefiles and KML Files*. accessed December 2022. https://www.usgs.gov/landsat-missions/landsat-shapefiles-and-kml-files.

USGS Landsat Missions. 2022b. *Spectral Bandpasses for all Landsat Sensors*. accessed December 2022. https://www.usgs.gov/media/images/spectral-bandpasse-all-landsat-sensors.

Wallerman, J., J. Bohlin, M. B Nilsson, and J. E. S Franssen. 2018. Drone-based forest variables mapping of icos tower surroundings. In *IGARSS 2018-2018 IEEE International Geoscience and Remote Sensing Symposium*, 22–27 July 2018.

Zhu, L., J. Suomalainen, J. Liu, J. Hyyppä, H. Kaartinen, and Haggren. H. 2018. "A Review: Remote Sensing Sensors." In *Multi-Purposeful Application of Geospatial Data*, edited by R. B. Rustamov, S. Hasanova and M. H Zeynalova, 19–42. London: Intech Open.

9 Probability, Statistics, and Machine Learning

INTRODUCTION

This chapter covers probability, statistics, and machine learning (ML) applications to environmental monitoring. Starting with basic probability theory, we formulate the basics of ML classification, interpreted from the point of view of Bayes' theorem, emphasizing the concepts of false negative errors, false positive errors, and confusion matrix. Discrete random variables (RVs) are introduced to support the analysis of counts and proportions, as well as contingency analysis. Analysis of error or confusion matrix is covered using Kappa statistics and contingency tables. Then, we introduce multiple linear regression, including approaches to select the explanatory variables, emphasizing collinearity issues and stepwise regression. This chapter ends with an introduction to classification and regression trees (CART) and its application to supervised classification. These concepts are expanded by computer exercises in Lab 9 of the companion Lab Manual (Acevedo 2024).

PROBABILITY

Probability theory is the basis for the analysis of uncertainty in science and engineering and it also plays a role in many ML algorithms. The concept of probability ties to a numerical measure of the likelihood of an *event*, i.e., an outcome of an experiment or measurement and which belongs to the *sample space* that is the set of all possible outcomes (Acevedo 2013). Probability is defined as a real number between zero and one (0 and 1 included) assigned to the likelihood of an event. As a shorthand for the probability of an event, we can write Pr[event] or P[event]. For example, for event A then Pr[A] or $P[A]$ where $0 \leq P[A] \leq 1$.

A simple approach to assign a probability value to an event is related to the proportion of time that an event could occur when compared to others. Consider picking a pixel at random from a remote sensing image classified to six vegetation classes, 1, ..., 6, and thus the sample space U has six possible outcomes. Suppose there are 100 out of 1000 pixels in class 1 and define event A = pixel is class 1, then $P[A] \sim 100/1000 = 0.1$ or a likelihood of 1 out of 10.

ALGEBRA OF EVENTS

For didactic purposes, events are illustrated using Venn diagrams and set theory. Events are represented by shapes or areas located in a box or domain. The *universal* event is the sample space U (includes all possible events) and therefore occurs with absolute certainty $P[U] = 1$. In the previous example, U = a pixel is one of class 1–6. See Figure 9.1.

The *null* event is an impossible event or one that includes none of the possible events, and therefore its probability is 0, $P[\phi] = 0$. For example, ϕ = the pixel is class 7. An oval shape represents an event A within U as shown in Figure 9.1. We also refer to B as the *complement* of A, i.e., the only other event that could occur. Therefore, the only outcomes are that either A or B happens. In addition, A and B are *mutually exclusive* and collectively exhaustive. The complement is an important concept often used to simplify solving problems. It is the same as B is NOT A which in shorthand is $B = \overline{A}$ where the bar on top of the event denotes *complement*, i.e., the logical operation NOT. In

DOI: 10.1201/9781003425496-9

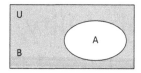

FIGURE 9.1   Universal event U or sample space (left), event A in U (right) showing B as the complement of A.

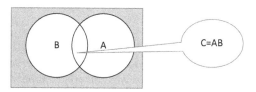

FIGURE 9.2   Intersection of two events; C is the sliver shared by events A and B.

the Venn diagram of Figure 9.1, B is shaded. The box represents U and the clear oval represents A. The key numeric relation is

$$P[B] = 1 - P[A] \tag{9.1}$$

Also, note that the complement of U is the null event.

Take the previous pixel example, define $B =$ pixel is any class except 1, $A =$ pixel is class 1, and determine $P[B]$. Since $B = \overline{A}$ use $P[B] = 1 - 1/10 = 9/10 = 0.9$. We did not have to enumerate B with detail, just subtracted from 1.

When two events share common outcomes, we define the *intersection* of two events as the common or shared events. In other words, the intersection of A and B is the event C that is in both A and B. Denote the intersection by $C = AB$, then the probability of the intersection is $P[C] = P[AB]$. In the popular diagram illustrated in Figure 9.2, AB is contained in A and in B. It corresponds to the AND logical operation.

Returning to the pixel example, define $A =$ pixel is class 1 or 2, $B =$ pixel is class 2 or 3. Obviously event $C =$ pixel class 2 is common to A and B, therefore $C = AB$. Suppose there are 200 class-2 pixels out of 1000, then $P[C] = 0.2$. When A and B do not intersect, then AB is the null event $AB = \phi$ and therefore $P[AB] = 0$. For example: $A =$ pixel is class 1 or 2, $B =$ pixel is class 3 or 4, $C = AB = \phi$ null and $P[C] = 0$.

The *Union* of A and B is the event C defined as A happens or B happens. It is the OR logical operation and is denoted by $A + B$. In reference to Figure 9.2, it would the addition of the two circles, but we avoid double counting the sliver of the intersection. Therefore, we discount the intersection AB once.

$$P[C] = P[A + B] = P[A] + P[B] - P[AB] \tag{9.2}$$

For example, $A =$ pixel is class 1 or 2, $B =$ pixel is class 2 or 3, then $C = A + B$ is $C =$ pixel is class 1, 2, or 3. Suppose there are 400 pixels of class 3 out of 1000. Then probabilities, $P[A] = 0.1 + 0.2 = 0.3$, $P[B] = 0.2 + 0.4 = 0.6$, $P[AB] = 0.2$, $P[A + B] = 0.3 + 0.6 - 0.2 = 0.7$.

An event B is included in A when event B is a subset of A, in set notation $B \subset A$ and therefore $P[B] < P[A]$. See Figure 9.3.

COMBINATIONS

When we complicate the experiment, for example, tossing a coin three times in a sequence, rolling a die five times in a sequence, we can combine the probabilities from the simpler components of the experiments to obtain probabilities of the more complex outcomes. The number n of independent

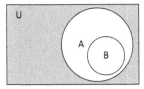

FIGURE 9.3   Event B is included in event A.

repetitions (trials) and the number k of outcomes of each repetition determine the total number of possible outcomes (Acevedo 2013). In general

$$N = k^n \tag{9.3}$$

For example, consider tossing a coin twice and denote H for head and T for tail. We have two outcomes $k = 2$ for each trial and $n = 2$ trials. The outcome of a toss is independent of the other. Possible combinations lead to four events $N = 2^2$ and this constitutes the sample space $U = HH, HT, TH, TT$. Each outcome is equally likely with probability 1/4. To see this, we reason that the probability of getting H in first toss, $P[H] = 1/2$ and to get H in the second toss is the same because of independence, therefore $P[HH] = 1/4$.

The combinations of n items taken r at a time are of great interest

$$\binom{n}{r} = \frac{n!}{r!(n-r)!} \tag{9.4}$$

The exclamation point "!" symbol is a factorial operation defined as $n! = n \times (n-1) \times (n-2) \times \ldots \times 2 \times 1$. For example, how many events have exactly one tail in two tosses of a coin? What is the probability of obtaining event A = exactly one tail in tossing a coin twice? Using equation (9.4) yields $\binom{2}{1} = \frac{2!}{1!(2-1)!} = \frac{1 \times 2}{1 \times 1} = 2$. Thus, there are two possible combinations of one head in two tosses.

This makes sense because from the previous example, we know that we have four possible events. Only a set of these would have one tail in two trials and we can count them HT, TH. We can calculate the probability as $P[A] = P[HT] + P[TH] = 0.25 + 0.25 = 0.5 = 1/2$.

PROBABILITY TREES

A useful visual aid in probability theory is a *tree*. The basic unit is a node from which we branch in arcs denoting events. Next to the arc, we write the probability of the event and the name or code event at the tip of the arc (Acevedo 2013; Drake 1967). For example, in the coin toss, we have the basic branch shown in Figure 9.4.

This basic unit can be iterated and combined to visualize situations that are more complex. For example, the two-toss coin experiment shown in Figure 9.5. Here the end branches of the tree correspond to the four outcomes. Multiplication of the probabilities of all the arcs traversed yields the probability of each path. Thus,

$$P[HH] = 0.5 \times 0.5 = 0.25$$

$$P[HT] = 0.5 \times 0.5 = 0.25$$

$$P[TH] = 0.5 \times 0.5 = 0.25 \tag{9.5}$$

$$P[TT] = 0.5 \times 0.5 = 0.25$$

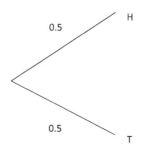

FIGURE 9.4 Basic element of a probability tree: node and set of arcs.

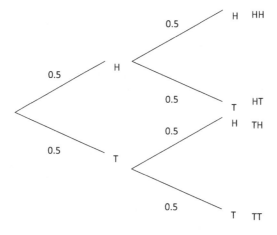

FIGURE 9.5 Example of a sequence of branches: tossing a coin twice.

As we can see, the probability of each path is the same and the sum of all probabilities is equal to 1.

CONDITIONAL PROBABILITY

Consider an experiment performed in two steps. At the first step, an event can occur with probability $P[A]$ and then at the second step the occurrence of an event C is *dependent* on whether A occurred. We use a vertical bar to denote occurrence of one event conditioned on another

$$Pr\left[C \text{ occurs given } A\right] = Pr\left[C \mid A\right] = P[C \mid A]$$

The key relation here is that A and C must have non-null intersection $P[AC] \neq 0$

$$P[C \mid A] = \frac{P[AC]}{P[A]} \tag{9.6}$$

$P[A]$ is also called *prior* probability, and $P[C|A]$ is also called *posterior* probability. A Venn diagram (Figure 9.6) and a tree (Figure 9.7) help illustrate this relation. The events at the end of the second set of arcs of Figure 9.7 are intersections and their probabilities can be found using equation (9.6) rewritten as

$$P[AC] = P[C \mid A]P[A] \tag{9.7}$$

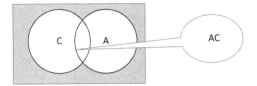

FIGURE 9.6   Intersection of events C and A: shows that C is conditioned on A.

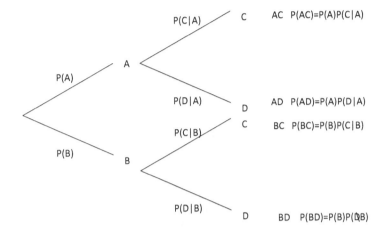

FIGURE 9.7   Second set of arcs shows conditional probabilities.

For example, consider a sequence of two days. In the first day define event: A = rains the first day, with $P[A] = 0.5$ and event C = rains the second day. Assume that the probability of raining the second day given that it rains the first is 0.7. What is the probability that it rains both days? In this case, the intersection event AC = rains the first day AND rains the second day. Let us calculate $P[AC]$ using conditional probability. Applying equation (9.7) $P[AC] = P[C \mid A]P[A] = 0.7 \times 0.5 = 0.35$

If events A and C are independent, then $P[C \mid A]$ is just $P[C]$ and therefore

$$P[AC] = P[A \mid C]P[C] = P[A]P[C] \tag{9.8}$$

Example: consider a sequence of two tosses of a fair coin. In the first toss, define event A = heads in first toss, with $P[A] = 0.5$ and event C = heads in second toss. What is the probability of getting a head in both tosses? The probability of getting a head in the second toss is independent of whatever we got in the first toss, thus $P[C \mid A] = P[C] = 0.5$. In this case, the event AC = head the first toss AND head the second toss. Then, calculate $P[AC]$ using conditional probability using equation (9.8) $P[AC] = P[C]P[A] = 0.5 \times 0.5 = 0.25$.

BINARY (2-CLASS) CLASSIFICATION

Let us consider an example based on testing from water quality (Carr 1995). A water quality test is conducted to decide whether water of a site is contaminated, event A, or not, event B. Assume that 20% of sampling sites are contaminated, then $P(A) = 0.2$; this is a *prior* probability also called a *base rate*. Define C = the test result is negative, D = the test result is positive. Suppose that the test yields a *false negative*, i.e., fails to determine contaminated water, 3% of the time; this means that its *sensitivity* or true positive rate is $1 - 0.03 = 0.97$ or 97%. Suppose that the test yields a *false positive* 7% of the time; this means that its *specificity* or true negative rate is $1 - 0.07 = 0.93$ or 93%.

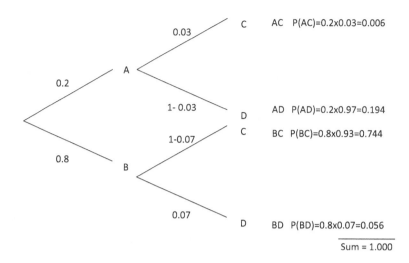

FIGURE 9.8 Water quality test: positive and negative results.

TABLE 9.1
Binary Classification Confusion Matrix

	Contaminated	Not Contaminated	Row Total
Test positive	True positive 97	False positive 7	Total positive 104
Test negative	False negative 3	True negative 93	Total negative 96
Column total	Total contaminated 100	Total not contaminated 100	Total wells 200

We can think of this scenario as a binary (2-class) classification: the results of the test, positive or negative, are used to classify the wells into contaminated or not. Note that 0.03 and 0.07 are probabilities for classification test errors. We build the tree shown in Figure 9.8 using this information. In the tree, the probability that the test is negative if the water is contaminated is $P[C|A]=0.03$, and that the water is indeed contaminated $P[A]=0.2$; then the probability that the test yielded negative results AND that the water was contaminated is $P[AC]=0.03\times0.2=0.006$. Note that events C and D are dependent on both A and B.

CONFUSION MATRIX

The *error matrix* or *confusion matrix* is a method to specify the errors and determining accuracy of the classification. See Table 9.1 for this example, where we can see that the diagonal terms for true positive and negative cases (97+93) are correct classifications and the diagonal terms for false positive and negative are total error rates (3+7). The sensitivity is calculated as true positive cases divided by total contaminated cases $97/(97+3)=0.97$, specificity is calculated as true negative cases divided by total not contaminated cases $93/(93+7)=0.93$. The *positive prediction value* (PPV) is calculated as true positive cases divided by total positive cases $97/(97+7)=97/104=0.932$. The negative prediction value (NPV) is given by true negative cases divided by total negative cases $93/(93+3)=93/96$. In the next section, we relate the binary classification to Bayes' theorem.

BAYES' THEOREM AND CLASSIFICATION

Bayes' rule connects the conditional probability of an event A given C to its opposite, or the conditional probability of C given A; in other words, it provides a link between $P[A|C]$ and $P[C|A]$

(Acevedo 2013). It is quite simple to derive Bayes' rule. From the conditional probability equation, we have

$$P[AC] = P[C \mid A]P[A] \tag{9.9}$$

In addition, A depends on C, and thus we can write

$$P[A \mid C] = \frac{P[AC]}{P[C]} \tag{9.10}$$

Or in the same form as equation (9.9)

$$P[AC] = P[A \mid C]P[C] \tag{9.11}$$

Equating the two relations (9.9) and (9.11) for $P[AC]$

$$P[A \mid C]P[C] = P[C \mid A]P[A] \tag{9.12}$$

and solving for $P[C \mid A]$

$$P[C \mid A] = \frac{P[A \mid C]P[C]}{P[A]} \tag{9.13}$$

This relation is Bayes' theorem or rule for two events and is of great use, as we will see in the next examples.

For example, consider the tree of Figure 9.8 with known prior probabilities for A and B, and posterior for $P[C \mid A]$ and $P[C \mid B]$. Can we back-calculate? e.g., what is the probability that we have contaminated water given a positive test result? In other words, what is the probability of correctly classifying the water as contaminated if we have a positive test result? $P[A \mid D]$? Use Bayes' theorem to calculate

$$P[A \mid D] = \frac{P[AD]}{P[D]} = \frac{P[D \mid A]P[A]}{P[D]} \tag{9.14}$$

the probability of D should be obtained from either A or B happening

$$P[D] = P[D \mid A]P[A] + P[D \mid B]P[B] \tag{9.15}$$

That is adding the two paths leading to a D at the end of the tree. Now substitute equation (9.15) in the denominator of equation (9.14) to obtain

$$P[A \mid D] = \frac{P[AD]}{P[D]} = \frac{P[D \mid A]P[A]}{P[D \mid A]P[A] + P[D \mid B]P[B]} \tag{9.16}$$

For example, in the tree of Figure 9.8, $P[AD] = 0.2 \times 0.97 = 0.194$, $P[BD] = (1-0.2) \times (0.07) = 0.056$, and $P[D] = 0.194 + 0.056 = 0.25$. Then, using equation (9.16), we get $P[A \mid D] = 0.194/0.25 = 0.776$. That is, if the test result is positive, there is a 77.6% probability that the water is contaminated. We can run a similar calculation for a correct classification given a negative test, $P[B \mid C] = \dfrac{P[BC]}{P[C]} = \dfrac{P[C \mid B]P[B]}{P[C \mid B]P[B] + P[C \mid A]P[A]}$. Or

P[B|C] = 0.744/(0.744 + 0.006) = 0.992 or 99.2%. Table 9.2 maps the confusion matrix of the binary classification given in Table 9.1 to the conditional probabilities used in Bayes' theorem, and Table 9.3 maps the Bayes' rule calculations to the binary classification.

TABLE 9.2

Binary Classification Confusion Matrix Mapped to Conditional Probabilities

	Contaminated	Not Contaminated	Predictive Values
Test result positive	True positive rate (sensitivity) P(D|A) 0.97	False positive rate P(D|B) 0.07	Positive predictive value 97/104 = 0.932
Test result negative	False negative rate P(C|A) 0.03	True negative rate (specificity) P(C|B) 0.93	Negative predictive value 93/96 = 0.968

TABLE 9.3

Binary Classification Confusion Matrix Mapped to Bayes' Rule Probabilities

	Contaminated	Not Contaminated	Correct Prediction
Base rate or prior probability	P(A) 0.2	P(B) 0.8	
Test result positive	True positive prediction P(DA)=P(D|A) P(A)=0.97×0.2=0.194	False positive prediction P(DB)=P(D|B) P(B)=0.07×0.8=0.056	Correct positive prediction P(DA)/(P(DA)+P(DB)) =0.194/(0.194+0.056) =0.776
Test result negative	False negative prediction P(CA)=P(C|A) P(A)=0.03×0.2=0.006	True negative prediction P(CB)=P(C|B) P(B)=0.93×0.8=0.744	Correct negative prediction P(CB)/(P(CA)+P(CB)) =0.744/(0.744+0.006) =0.992

GENERALIZATION OF BAYES' RULE TO MANY EVENTS

Suppose there are several events $B_1, B_2, \ldots$ that condition event C. For each event B_i where $i = 1, n$, we can write an equation like (9.9)

$$P[CB_i] = P[C \mid B_i]P[B_i] \tag{9.17}$$

and using Bayes' rule for each event B_i, we have

$$P[B_i \mid C] = \frac{P[C \mid B_i]P[B_i]}{P[C]} \tag{9.18}$$

Note that if events B_i account for all ways in which we can get event C, then by adding equation (9.17) for all events B_i we get

$$P[C] = \sum_{i=1}^{n} P[CB_i] = \sum_{i=1}^{n} P[C \mid B_i]P[B_i] \tag{9.19}$$

And by substituting this last expression in the denominator of Bayes' rule equation (9.18), we can derive an extension to Bayes' rule like equation (9.16)

$$P[B_i \mid C] = \frac{P[C \mid B_i]P[B_i]}{P[C]} = \frac{P[C \mid B_i]P[B_i]}{\sum_{i=1}^{n} P[C \mid B_i]P[B_i]} \tag{9.20}$$

BIOSENSING OF WATER QUALITY

Let us consider an application of Bayes' rule to a biosensing method developed to detect contamination or other water quality issues by electronically monitoring the gape of clams (Allen et al. 1996). The working concept is that organisms integrate many signals from the environment and therefore are excellent environmental sentinels. Because of variability in behavioral response to stress, it is necessary to setup the monitor using more than one individual, e.g., ten or more clams.

For the sake of a simple scenario, suppose we use two clams to sense water quality and that we are only considering valves completely shut or open. At any one measurement time, there are three events from measuring clam gape, B_1=two animals have valves shut, B_2=one animal has valves shut and the other has them open, and B_3=both animals have open valves. Define C as the event of contaminated water. Take an interval of 100 measurements and assume that 70% result in event B_1, 20% in B_2 and 10% in B_3. The probabilities of false positive errors are 0.1, 0.2, and 0.9 for B_1, B_2, and B_3, respectively. What is the probability that the water is contaminated? What is the probability that two animals have valves shut if the water is contaminated? The probability of contaminated water is $P[C]$, thus first apply equation (9.19) and use the complement for all error probabilities because we are giving the false positive

$$P[C] = \sum_{i=1}^{n} P[C \mid B_i]P[B_i] = \sum_i \left(1 - P[\bar{C} \mid B_i]\right)P[B_i]$$

$$= (1 - 0.1) \times 0.7 + (1 - 0.2) \times 0.2 + (1 - 0.9) \times 0.1 = 0.8$$

The probability that two animals have valves shut if the water is contaminated is $P[B_1 \mid C]$, and then apply equation (9.20) $P[B_1 \mid C] = \dfrac{P[C \mid B_1]P[B_1]}{P[C]} = \dfrac{(1 - 0.1) \times 0.7}{0.8} = 0.79$, which means the probability that two animals shut the valves if the water is contaminated is 79%.

BAYES' RULE AND ML

Looking at Bayes' equation (9.13) in terms of ML classification, we think of C as a *hypothesis* or what we target to learn (e.g. a hypothetical event) and A as the *data*, (e.g., the result of an observation event) then $P[C]$ is the *prior* probability of the event before we know the result of A; we can refer to $P[A]$ as the *marginal* probability, $P[C|A]$ as the *posterior* probability of the hypothetical event given the observation, and $P[A|C]$ as the likelihood probability of the observation based on the hypothetical event.

For emphasis, we change the notation C to h (for hypothesis) and A to d (for data) and restate Bayes' rule applied to ML as

$$P[h \mid d] = \frac{P[d \mid h]P[h]}{P[d]} \tag{9.21}$$

In other words, we calculate the posterior probability of a hypothesis given the data, based on its prior probability $P[h]$, the probabilities of observing the data given the hypothesis $P[d/h]$, and the observed data $P[d]$. Selecting the best among several hypotheses h_1, h_2, ... that can be explained by the data can be posed as a *maximum a posteriori* problem,

$$\max_i(P[h_i \mid d]) = \max_i \left(\frac{P[d \mid h_i]P[h_i]}{P[d]} \right) \tag{9.22}$$

To simplify, it is assumed that $P[d]$ is the same for all hypotheses and that there is no prior $P[h_i]$ and therefore this last equation reduces to

$$\max_i(P[d \mid h_i]) \tag{9.23}$$

Which means simply that we choose the hypothesis that best explains the data, and note that this is equivalent to building an empirical model from the data by optimization, e.g., linear least squares (LLS) regression. Thinking now specifically of classification, h_i is a label for a class, and the ML algorithm would find the classes that best explain the data.

When using ML, we divide the dataset into a subset for *training* the classifier and a subset to *test* or *evaluate* the classifier. In the lab guide 9 of companion Lab Manual, we study how to split the dataset, train a classifier, and make predictions based on the trained classifier and the evaluation subset.

NAÏVE BAYES CLASSIFIER

For multidimensional data, d consists of a set of features or attributes X_1, X_2, ..., X_n that depend on each other and therefore it becomes very complicated to compute the equation for the posterior of a class given the data, $P[h_i \mid d] = \dfrac{P[d \mid h_i]P[h_i]}{P[d]}$. For practical purposes, it is assumed that the features X_j are independent of each other, then we can write

$$P[h_i \mid d] = \frac{\left(\prod_{j}^{n} P[X_j \mid h_i] \right) P[h_i]}{P[d]} \tag{9.24}$$

This is called a Naïve Bayes classifier, which is often used for ML classification. The symbol Π denotes product in a similar way in which Σ denotes sum. Just for illustration, if you had two variables $P[h_i \mid d] = \dfrac{P[X_1 \mid h_i]P[X_2 \mid h_i]P[h_i]}{P[d]}$. The independence assumption makes the naïve classifier easier and faster to compute.

DECISION TREES

Probability theory is one of the bases of decision theory. We can frame a decision problem as selecting the most promising alternative *action* or *option* given the uncertainties. The major concept is that events occurring with given probabilities follow the options. Say, we choose option A_1 and event E_1 occurs with probability p, whereas event E_2 occurs with probability $1-p$. Therefore, if we associate a cost or a loss to a combination of action and event (say A_1E_1), we can weigh the cost by its probability to calculate an *expected* cost or loss for this branch of the decision. We do this for all options and then select the option with minimum expected loss or cost. Depending on the situation, we would rather formulate the decision in terms of benefits. For example, assign a profit or gain to the outcomes and tackle the decision by selecting the alternative with the maximum profit or gain.

A very simple example is one with two alternative actions A_1 and A_2 and two events E_1, E_2 as shown in the tree of Figure 9.9. This is a *decision tree* and is similar to a probability tree, except that some nodes (marked with squares) are decision nodes and others are event nodes (marked with circles). Suppose the alternative actions are to invest in an environmental protection at an investment cost of I_1 or at a lower level I_2 to prevent the detrimental ecosystem effect of a potential spill. Suppose the events are E_1 (contaminant spilled) with probability $p = P[E_1]$ and E_2 (no spill).

DISCRETE RVs

We can define an RV once we have defined the sample space, based on the possible events and their probabilities (Acevedo 2013). An RV is a rule, or a function, or a map associating a number to each event in the sample space. See Figure 9.10.

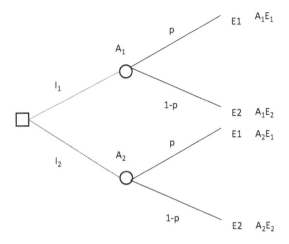

FIGURE 9.9   Decision tree: a simple example from environmental protection.

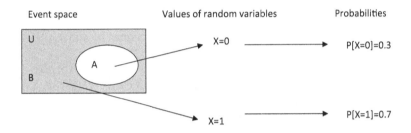

FIGURE 9.10   Constructing an RV.

Example: In the roll of six-sided die, the events are Ai=side facing up is i where i=1,2, ..., 6 with $P[Ai]$=1/6. We can make a "discrete" RV, denoted by X, associating each event with the value of the RV; e.g., X taking values X= 1, 2, 3, 4, 5, 6 each with $P[x_i]$=1/6. We refer to the type of RV described in the example as *discrete* because its values are discrete, i.e., a set of numbers, in this case integers 1,2, ..., 6.

However, the events for a discrete RV can also be defined by quantizing a range of *real* values. For example, suppose we measure concentration X of a mineral (in ppm) at a given location and it can take values between 0 and 2000 ppm. Two events could be defined as A is $0 \le X < 1000$ ppm and B is $1000 \le X \le 2000$ ppm.

PROBABILITY MASS FUNCTION (PMF)

A *discrete* distribution or probability mass function (PMF, for short) $p(X)$ is a set of probabilities, one for each value of X. More precisely, denoting x_i as the values of X

$$p(x_i) = P[X = x_i] \tag{9.25}$$

for all values x_i of X

$$0 \le p(x_i) \le 1 \text{ for all } i \tag{9.26}$$

$$\sum_i p(x_i) = 1 \tag{9.27}$$

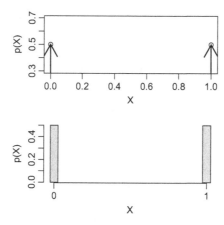

FIGURE 9.11   PMF of a discrete RV represented as a spike graph and as bar graph.

the last equation says that the total probability (sample space) must be equal to 1 when all probabilities are summed over all i.

Example: Toss a coin and assign 0 to T and 1 to H. For a fair coin, $p(0)=0.5$, $p(1)=0.5$. This is an example of a *uniform* discrete RV for which the probabilities of each event are the same.

We can represent the probabilities as a graph as illustrated in Figure 9.11 for the example above. Vertical thick arrows or bars represent a spike or impulse with intensity given by the height of the spike and equals the probability of that value. Alternatively, it can be represented as a bar graph where the height of each bar represents the probability.

Example: Roll a six-side die and assume it is not loaded; then the PMF is $p(xi)=1/6$ where $xi=1$, 2, ..., 6. This is also a uniform discrete RV.

CUMULATIVE MASS FUNCTION (CMF)

The cumulative mass function (CMF) at a given value are defined by "accumulating" all probabilities up to that value. Accumulation is simply a summation in the case of discrete RV

$$F(x) = P[X \le x] = \sum_{i=1}^{x} p(x_i) \tag{9.28}$$

Please note that the value at which we evaluate the cumulative is the upper limit of the accumulation. The value of the cumulative for the largest value of X is the largest value of the cumulative and should be equal to 1.

For example, toss of a coin and roll of a die, for which the CMF is a stepwise function. At each value, we add the intensity of a spike to obtain a staircase for the CMF. See Figure 9.12.

FIRST MOMENT OR MEAN

The *first moment* of X is the *expected value* of X denoted by the operator E applied to X, this is $E[X]$. The first moment is the same as the *mean* of X. When X is discrete, the *mean* is

$$\mu_X = E[X] = \sum_i x_i p(x_i) \tag{9.29}$$

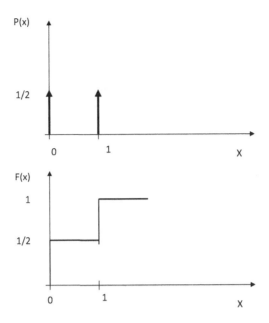

FIGURE 9.12 Integration of PMF to obtain CMF.

where the sum is over all values of X. It is equivalent to the location of the center of mass of the PMF.

Example: toss a coin $E(X) = 0 \times 0.5 + 1 \times 0.5 = 0.5$ this is the value of X in between the two spikes or bars in Figure 9.11. Example: Roll of a die

$$E(X) = 1 \times \frac{1}{6} + 2 \times \frac{1}{6} + 3 \times \frac{1}{6} + 4 \times \frac{1}{6} + 5 \times \frac{1}{6} + 6 \times \frac{1}{6}$$

$$E(X) = (1 + 2 + 3 + 4 + 5 + 6) \times \frac{1}{6} = \frac{21}{6} = 3.5$$

This is the value of X in between the two spikes or bars in the PMF.

Example: Suppose we perform a coin toss ten times and that we assign $H=1$, $T=0$ as in example one. Suppose we get six heads and four tails. Then the average is $\bar{X} = \frac{1}{10} \times 6 = 0.6$

This sample mean is different from the population mean that we calculated to be $\mu_X = 0.5$.

SECOND CENTRAL MOMENT OR VARIANCE

The second *central* (i.e., with respect to the mean) moment is the *variance* or the expected value of the square of the difference with respect to the mean

$$\sigma_X^2 = E[(X - \mu_X)^2] = \sum_i (x_i - \mu_X)^2 p(x_i) \tag{9.30}$$

The variance can also be calculated using the simplified expression $\sigma_X^2 = E[X^2] - E[X]^2$. Taking the square root of the variance, we obtain the standard deviation $\sigma_X = \sqrt{\sigma_X^2}$. For example, suppose the RV for the toss of a coin $x_i = 0$, 1 and $p(x_i) = 0.5$ for all i. Using the definition in equation (9.30) $\sigma_X^2 = E[(X - \mu_X)^2] = (0 - 0.5)^2 \times 0.5 + (1 - 0.5)^2 \times 0.5 = 0.25$. The standard deviation is the square root of the variance $\sigma_X = \sqrt{\sigma_X^2} = \sqrt{0.25} = 0.5$

Take another example, suppose we get six heads in ten tosses of a coin. The sum of squares is six and the sum of the x_i is also six, then the sample variance is $s_X^2 = \frac{1}{9} \times \left[6 - \frac{1}{10} \times 6^2 \right] = 0.26$ and the sample standard deviation is $s_X = \sqrt{.26} = 0.509$. Note that the population variance is 0.25 and standard deviation 0.5 according to calculation in previous exercise. Therefore, the statistic sample variance has overestimated the variance.

BINOMIAL DISTRIBUTION

As an example of discrete RV, consider the *binomial* model that describes the probability $P[X = r] = p(r)$ of r "successes" in n independent trials, given that the probability of a success is p. The values of r are the integers 0, 1, 2, .., n and the PMF is

$$p(r) = \binom{n}{r} p^r (1 - p)^{n-r} \tag{9.31}$$

The mean is $\mu_X = np$ and the variance is $\sigma_X^2 = np(1-p)$.

For example, assume a binomial RV ((Davis 2002) pages 13–16) with probability of success $p = 0.2$ and $n = 3$. Use a calculator and evaluate the binomial at $r = 0, 1, 2, 3$ to get $p(0) = 0.512$, $p(1) = 0.384$, $p(2) = 0.096$, $p(3) = 0.008$. The mean is $3 \times 0.2 = 0.6$ and the variance is $0.6(1-0.2) = 0.6 \times 0.8 = 0.48$.

BIVARIATE DISCRETE RANDOM VARIABLES

The joint PMF of two RVs X and Y is

$$p_{XY}(x, y) = P[X = x, Y = y] \tag{9.32}$$

where the comma stands for a logical AND. This function must satisfy $p_{XY}(x, y) \geq 0$ for all pairs x, y, that is all probabilities are non-negative and $\sum_x \sum_y p_{XY}(x, y) = 1$ that is the sum of the probabilities of all pairs must be 1.

For example, suppose $X = 0, 1$, $Y = 0, 1$, and a uniform joint PMF, shown in Table 9.4. This can also be written as a matrix

$$p_{XY}(x, y) = \begin{pmatrix} \frac{1}{8} & \frac{1}{4} \\ \frac{3}{8} & \frac{1}{4} \end{pmatrix} \tag{9.33}$$

With rows representing values of X and columns values of Y. Note that the sum of all joint probabilities is equal to 1.

TABLE 9.4
Joint PMF Example

	Y=0	Y=1
X=0	1/8	1/4
X=1	3/8	1/4

The marginal PMF of X is the PMF of X given one value of Y, say $p_X(x) = \sum_y p_{XY}(x,y)$ and similarly, the marginal PMF of Y is $p_Y(y) = \sum_x p_{XY}(x,y)$ In this example,

$$p_X(0) = p_{XY}(0,0) + p_{XY}(0,1) = \frac{1}{8} + \frac{1}{4} = \frac{3}{8}$$

$$p_X(1) = p_{XY}(1,0) + p_{XY}(1,1) = \frac{3}{8} + \frac{1}{4} = \frac{5}{8}$$

Therefore, the marginal PMF of X is $p_X(x) = \begin{cases} 3/8 \text{ for } X = 0 \\ 5/8 \text{ for } X = 1 \end{cases}$

Similarly, for Y we can find

$$p_Y(0) = p_{XY}(0,0) + p_{XY}(1,0) = \frac{1}{8} + \frac{3}{8} = \frac{1}{2}$$

$$p_Y(1) = p_{XY}(0,1) + p_{XY}(1,1) = \frac{1}{4} + \frac{1}{4} = \frac{1}{2}$$

$$p_Y(y) = \begin{cases} 1/2 \text{ for } Y = 0 \\ 1/2 \text{ for } Y = 1 \end{cases}$$

Dividing values of the joint PMF by values of the marginal PMF yields the conditional probability for Y taking value y given that X has value x

$$p_{Y|X}(y \mid x) = \frac{p_{XY}(x,y)}{p_X(x)} \tag{9.34}$$

Similarly,

$$p_{X|Y}(x \mid y) = \frac{p_{XY}(x,y)}{p_Y(y)} \tag{9.35}$$

Using the example above, $p_{Y|X}(1 \mid 0) = \dfrac{p_{XY}(0,1)}{p_X(0)} = \dfrac{1/4}{3/8} = 2/3$ and $p_{Y|X}(0 \mid 0) = \dfrac{p_{XY}(0,0)}{p_X(0)} = \dfrac{1/8}{3/8} = 1/3$, and of course these two conditional probabilities add to 1.

INFORMATION THEORY

Based on Shannon's information theory, the *entropy* of an RV X is defined as

$$H(X) = -\sum_{i=1}^{n} p(x_i) \log\big(p(x_i)\big) \tag{9.36}$$

where $p(x_i)$ is the value of the PMF for value $X = x_i$, $i = 1, \ldots, n$. The base of the logarithm determines the units. For example, if we select base 2, we obtain binary units or bits. Entropy defined this way represents information in the sense that higher entropy means more uncertainty.

As an application example, species diversity is measured by species richness n and by some index of the relative abundances of the species. Richness is simply the number of coexisting species. In the case of three-species communities, richness can be 1, 2, or 3. For an index of relative abundances, we need the species composition, given by the relative abundances of the species. That is the proportions p_i of each species with respect to the total.

Diversity of species composition is calculated by indices based on functions of the species distribution p_i. Several indices are common, among them the Simpson diversity index, and the Shannon diversity index. The latter is popular and used frequently and termed evenness. It is derived from the concept of information

$$E = -\sum_{i=1}^{n} p_i \ln(p_i) \tag{9.37}$$

Here we have used natural logarithms; however, this is flexible since the base of the logarithms determines the units. For example, if we select base 2, we obtain binary units. As an example, let us calculate evenness for three species, uniformly distributed. In this case, $p_i = 1/3$. equation (9.37) yields $1.098 \sim 1.1$.

We will see another example of using information theory later in the chapter when we study the Akaike information criterion (AIC) applied to regression analysis, and measures of purity when calculating a classification tree.

COUNTS AND PROPORTIONS

Consider the event A to be a "success" with a given probability of occurrence denoted here as p_s in a series of trials. An estimate of p_s is the ratio of number of successes to number of trials. For one sample, we formulate the following question: is P[A] different from a presumed value? For two samples, is P[A] different between the samples? The probability of obtaining r successes in n trials follows the binomial distribution

$$P(X = r) = \binom{n}{r} p_s{}^r (1 - p_s)^{n-r} \tag{9.38}$$

For which, the mean is $\mu_X = np_s$ and the variance is $\sigma_X^2 = np_s(1 - p_s)$. The binomial test uses the binomial distribution as a test statistic to determine the hypothetical value of the probability of obtaining the observed ratio r/n. For one sample, the null H0 is that $p_s = p_0$ where p_0 is the hypothetical probability of success, and two-sided alternative $p \neq p_0$. A p-value is the sum of cumulative probabilities of obtaining less than r successes and more than r successes. These calculations are easily conducted in R as explained in the lab manual.

For example, suppose the hypothetical p_0 is 0.5 and that we observe a ratio of $r = 3$ successes in $n = 10$ trials. The p-value is 0.34 and we decide against rejection of the null and allowing that p_s may be 0.5. However, if we were to observe a ratio of $r = 1$ in $n = 10$ trials, the p-value is 0.02 or 2% which is low enough to reject H_0.

χ^2 (CHI-SQUARE) TEST

The Pearson's χ^2 (chi-square) statistic is the squared difference between observed counts, in bins or intervals, and theoretical counts, in the same bins, from the hypothesized distribution. Denote n = sample size, k = number of bins or classes, then the chi-square statistic is calculated as

$$\chi^2 = \sum_{j=1}^{k} \frac{(c_j - e_j)^2}{e_j} \tag{9.39}$$

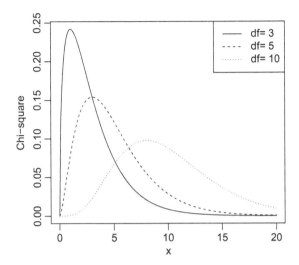

FIGURE 9.13 χ^2 PDF for $df=3$, 5, and 10.

the sum is over $j=1, ..., k$ where k is the number of classes or bins, c_j is the number of observations counted in the jth bin, e_j is the number of observations or counts expected in bin j. See Figure 9.13 that illustrates the shape of the χ^2 density for various degrees of freedom.

The chi-square test consists of checking if the χ^2 statistic is too large (for a given level of significance) because this indicates substantial departure from the hypothesized distribution. In this test, the null hypothesis H_0 is that the sample is drawn from a given distribution, and thus when you reject the null, you conclude that the sample is not drawn from the hypothesized distribution (with given alpha error). When you fail to reject, the sample may come from the hypothesized distribution.

To calculate the degrees of freedom (df) in a χ^2 test you would use o the number of bins minus one (k–1), and not to the number of observations minus one (n–1). Increasing the number of bins may increase the df but decrease the number of points per bin, which must be kept above five counts per bin. If additional parameters are estimated, then df decreases; for example, when testing for a fit to the standard normal $N(0, 1)$, $df=k$–3, both mean and variance are estimated from the sample to generate standard values.

Example: Suppose that we want to check if 100 values come from a uniform distribution. We got the following counts in $k=5$ categories (bins or intervals) 22, 13, 18, 27, and 20. The expected value in each bin is 100/5=20. Now we calculate the statistic

$$\chi^2 = \sum_{j=1}^{k} \frac{(c_j - e_j)^2}{e_j} =$$

$$= \frac{(22-20)^2 + (13-20)^2 + (18-20)^2 + (27-20)^2 + (20-20)^2}{20} =$$

$$= \frac{4+49+4+49+0}{20} = \frac{106}{20} = 5.30$$

The df value is 5–1=4. The probability of getting this value or higher is the tail of the χ^2 for a $df=4$ which is calculated as $1 - F(5.30) = 0.258$ where F is the cumulative density function (CDF, see Chapter 1) of the χ^2 for $df=4$. Therefore, the p-value is 0.258 which is too high to reject the null with a reasonable α that the sample is drawn from a uniform distribution.

CONTINGENCY TABLES AND CROSS-TABULATION

Another application of the χ^2 distribution is the analysis of contingency tables. For bivariate cases, we count the number of occurrences of observations falling in joint *categories* and thus we develop a *contingency* table that can help answer the question: is there association between the two variables? This is an *independence* test, which is different from a goodness of fit test, although the calculations are similar.

Contingency analysis applies to categorical data. Use factors with several levels, count the number of cases in each level, and calculate the frequency of two values occurring in joint or combined levels. To build a contingency table, we use all the levels of one factor (categorical variable) as *rows* and the levels of the other variable as *columns*. We then count occurrences in each cell and find the frequency of joint levels or cell frequency for each cell in the table. This is a *cross-tabulation*.

Now to perform a test, first we sum the cell frequencies across both rows and columns and place the sums in the margins, which are the *marginal* frequencies. The lower right-hand corner value contains the sum of either the row or the column marginal frequencies, which both must be equal to the total number of observations n.

The test assesses whether one factor has effects on the other. Effects are relationships between the row and column factors; for example, are the levels of the row variable distributed at different levels according to the levels of the column variables? Rejection of the null hypothesis (independence of factors) means that there may be effects between the factors. Non-rejection means that chance could explain any differences in cell frequencies and thus factors may be independent of each other.

For example, there are three possible land cover types in a remote sensing image (crop field, grassland, forest). Using a digital elevation model, we classify the area in two elevation classes: lowlands and uplands. We divide the total area so that we have 80 squares of equal size and classify each square in one of the cover types and terrain classes. Then we count the number of squares in each joint class. The resulting cross-tabulation is Table 9.5.

The first step in computing the χ^2 statistic is the calculation of the row totals and column totals of the contingency table as shown in Table 9.6. The next step is the calculation of the expected cell count for each cell. This is accomplished by multiplying the marginal frequencies for the row and column (row and column totals) of the desired cell and then dividing by the total number of observations

$$e_{ij} = \left(\frac{\text{Row}_i\, \text{Total} \times \text{Column}_j\, \text{Total}}{n} \right) \qquad (9.40)$$

TABLE 9.5
Cross-Tabulation of Topographic Class and Vegetation Cover

	Crop Field	Forest	Grassland
Lowlands	12	20	8
Uplands	8	10	22

TABLE 9.6
Row and Column Totals

	Crop Field	Forest	Grassland	Row Total	Total of Rows
Lowlands	12	20	8	40	
Uplands	8	10	22	40	
Column total	20	30	30		80
Total of columns				80	

TABLE 9.7
Expected Frequencies

	Crop Field	Forest	Grassland	Row Total	Total of Rows
Lowlands	12	20	8	40	
e_{1j}	10	15	15	40	
Uplands	8	10	22	40	
e_{2j}	10	15	15	40	
Column total	20	30	30		80
	20	30	30		
Total of columns				80	

TABLE 9.8
Deviation from Expected

	Crop Field	Forest	Grassland	Row Total	Total of Rows
Lowlands	12	20	8	40	
e_{1j}	10	15	15	40	
$o_{1j}-e_{1j}$	+2	+5	−7	0	
Uplands	8	10	22	40	
e_{2j}	10	15	15	40	
$o_{2j}-e_{2j}$	−2	−5	+7	0	
Column total	20	30	30		80
	20	30	30		
	0	0	0		
Total of columns				80	

For example, computation of the expected cell frequency for cell in row 1 and column 1 is $e_{11} = (20 \times 40)/80 = 10$. Use this expression to calculate all the expected cell frequencies and including these as additional rows in each cell (Table 9.7). Then we subtract the expected cell frequency from the observed cell frequency for each cell, to obtain the deviation or error for each cell. Including these in the preceding table, we obtain the next table shown in Table 9.8. Note that the sum of the row total for expected is the same as the sum of the observed row totals; the same is true for the column totals. Note also that the sum of the Observed − Expected for both the rows and columns equals 0. Now, to get the final table, we square the difference and divide by the expected cell frequency, resulting in the chi-square values (Table 9.9). The χ^2 statistic is computed by summing the last row of each cell; in the example, this would result in $\chi^2 = 0.4 + 0.4 + 1.66 + 1.66 + 3.26 + 3.26 = 10.64$. The number of degrees of freedom is obtained by multiplying the number of rows minus one, times the number of columns minus one $df = (\text{Rows} - 1) \times (\text{Columns} - 1)$. In the example above $df = (2-1) \times (3-1) = 2$. We would then determine the p-value for 10.64 and compare to the desired significance level. If lower, the rows and columns of the contingency may be dependent. In this case, $1 - F(10.64) = 0.00489$ for $df = 2$. Therefore, the p-value ≈ 0.005 suggests rejecting H_0 and that vegetation cover is not distributed evenly across the different elevations.

SUPERVISED CLASSIFICATION: CONFUSION MATRIX

As we discussed in a previous section, the error matrix or confusion matrix is a method to determine accuracy of the classification or of the results of ML. In the case of remote sensing, we need a *reference* class for pixels (from ground truth, maps, or field work) and cross-tabulate with the classification results for those pixels. See Table 9.10.

TABLE 9.9
Final Table with Chi-Square Values

	Crop Field	**Forest**	**Grassland**	**Row Total**	**Total of Rows**
Lowlands	12	20	8	40	
e_{1j}	10	15	15	40	
$o_{1j}-e_{1j}$	+2	+5	−7	0	
$(o_{1j}-e_{1j})^2/e_{1j}$	4/10=0.4	25/15=1.66	49/15=3.26	5.32	
Uplands	8	10	22	40	
e_{2j}	10	15	15	40	
$o_{2j}-e_{2j}$	−2	−5	+7	0	
$(o_{2j}-e_{2j})^2/e_{2j}$	4/10=0.4	25/15=1.66	49/15=3.26	5.32	
Column total	20	30	30		80
	20	30	30	**10.64**	
	0	0	0		
	0.8	3.32	6.52		
Total of columns				80	

TABLE 9.10
Confusion Matrix: Correct Class in Diagonal and Incorrect Classification in Off-Diagonal Entries

		Ground			
		Grass	**Forest**	**Urban**	**Row Total**
Image	Grass	75	10	5	90
	Forest	10	80	5	95
	Urban	8	4	85	97
	Column total	93	94	95	**282**

Denote by z_{ij} the entries of the confusion matrix where i and j are the row and column indices, and by N the total observations studied, or the sum of all entries of the matrix

$$N = \sum_{i,j} z_{ij} \qquad (9.41)$$

In the example, $N = 282$. The *overall accuracy* or *proportion of agreement Po* is given by

$$Po = \frac{\sum_i z_{ii}}{N} \qquad (9.42)$$

That is to say, the sum of diagonal entries (the trace of the matrix) $75+80+85=240$ divided by the total pixels studied; thus, in this example, $Po = 240/282 = 0.851$ or 85.1%.

The row totals rt_i are the sum of the entries for each row i, and their marginal proportions rp_i are obtained dividing by N

$$rt_i = \sum_j z_{ij} \qquad rp_i = \frac{\sum_j z_{ij}}{N} \qquad (9.43)$$

In this example, $rt = 90$, 95, 97 and $rp = 0.32$, 0.34, 0.34. Likewise, the column totals are the sum of the entries of each column j and the marginal proportions dividing by the grand total

$$ct_j = \sum_i z_{ij} \quad cp_j = \frac{\sum_i z_{ij}}{N} \tag{9.44}$$

In this example, $ct = 93$, 94, 95 and $cp = 0.33$, 0.33, 0.34.

Now, define as Pe the sum of all products of a row marginal proportions times the column marginal proportions.

$$Pe = \sum_{i,j} rp_i \times cp_j \tag{9.45}$$

In this example, the proportion Row 1 margin $90/282 = 0.3191$ times column 1 margin $93/282 = 0.329$ is the probability of pixels being grass randomly, that is 0.105. For class forest, the row and column marginal proportions are 0.337 and 0.33 and the product is 0.112, for class urban we have 0.344 and 0.336 with product 0.116. The sum of all these products is 0.333. In other words, $Pe = 0.319 \times 0.330 + 0.337 \times 0.333 + 0.344 \times 0.337 = 0.333$.

The *kappa* coefficient is a measure of how the classification results compare to values assigned at random, and it takes values from 0 to 1; with higher kappa coefficient occurring when the classification is more accurate. Cohen's kappa coefficient is

$$\kappa = \frac{Po - Pe}{1 - Pe} \tag{9.46}$$

In this example, kappa is $\kappa = \frac{0.851 - 0.333}{1 - 0.33} = 0.777$ which is relatively good but not too close to 1.

There are two metrics for the accuracy of each class depending on whether you look at the data from the prediction side (the rows), yielding the *user accuracy*, or the reference side (the columns) yielding the *producer accuracy*. User accuracy calculated as the diagonal term of the class divided by the row total, i.e., it represents the probability that a predicted value is indeed in that class. For each row i

$$ua_i = \frac{z_{ii}}{rt_i} \tag{9.47}$$

In the example, $ua = 0.833$, 0.842, 0.876. In contrast, the producer accuracy is the diagonal term of a class divided by the column total, i.e., represents the probability that a reference value of a class was classified correctly. For each column j

$$pa_j = \frac{z_{jj}}{ct_j} \tag{9.48}$$

In the example, $pa = 0.806$, 0.851, 0.895, and the user accuracy is lower than the producer accuracy for all classes except grass.

Two error metrics are of interest and are calculated directly from the user and producer accuracy. The *error of commission* of a class is the fraction of values predicted to be in a class but do not belong to that class, i.e., false positives, calculated as the sum of off-diagonal values of the row divided by the row total, which is the same as subtracting each user accuracy value from 1.

$$ec_i = \frac{\sum_{j \neq i} z_{ij}}{rt_i} = 1 - ua_i \tag{9.49}$$

In the example, $ec = 0.166, 0.158, 0.124$. The second error metric, *error of omission* of a class, is the fraction of values that belong to a class and yet were predicted to be in a different class, i.e., false negatives; calculated as the sum of off-diagonal values of the column divided by the column total, which is the same as subtracting each producer accuracy value from 1.

$$eo_j = \frac{\sum_{i \neq j} z_{ij}}{ct_j} = 1 - pa_j \tag{9.50}$$

In the example, $eo = 0.193, 0.149, 0.105$ and the error of omission is lower than the error of commission for all classes except grass.

MULTIPLE LINEAR REGRESSION

We learned in Chapter 1 that we can find an optimum estimator of a response or dependent variable from an independent or explanatory variable by linear least-squares regression, and that this is a powerful method to create empirical models from data. In this chapter, we extend the simple linear regression model of Chapter 1 to more than one explanatory variable X; that is, several (m) explanatory variables X_i, influencing one response variable Y (Acevedo 2013).

This method has many applications in monitoring, since we could predict a variable of interest based on measurements of several other variables. In addition, this method can support ML since once we estimate coefficients from a training dataset, we can use them for prediction. As discussed previously in reference to the application of Bayes' rule to ML, maximizing the posterior probability of the hypothesis given the data can be reduced to finding the hypothesis that best explains the data, and this is similar to linear regression.

As we did in Chapter 1, we develop an LLS estimator of Y

$$\hat{Y} = b_0 + b_1 X_1 + b_2 X_2 + \ldots + b_m X_m \tag{9.51}$$

with *intercept* b_0 and *coefficients* b_k, $k = 1, \ldots, m$, and therefore we have $m + 1$ regression parameters to estimate. For each observation i there is a set of data points y_i, x_{ki} and the estimated value of Y at the specific points x_{ki} is

$$\hat{y}_i = b_0 + b_1 x_{1i} + b_2 x_{2i} + \ldots + b_m x_{mi} \tag{9.52}$$

We need to find the values of the coefficients b_k that minimize the square error

$$\min_b q = \min_b \sum_{i=1}^{n} e_i^2 = \min_b \sum_{i=1}^{n} (y_i - \hat{y}_i)^2 \tag{9.53}$$

A starting visual aid in multiple regression is to obtain pairwise scatter plots of all variables involved, this allows one to explore potential relationships among the variables. We encountered an example when we examined scatter plots of pairs of green, red, near IR (NIR), and SWR1 bands of Landsat 8 that we studied in Figure 8.11 of Chapter 8. As another example see Figure 9.14 that refers to air quality and more specifically how ozone relates to meteorological conditions. We will study air quality in Chapter 11.

MATRIX APPROACH

We proceed as in simple linear regression; first, find derivatives of q with respect to b_0 and each one of the coefficients b_k. Then set these derivatives equal to 0, so that q is at a minimum, and find

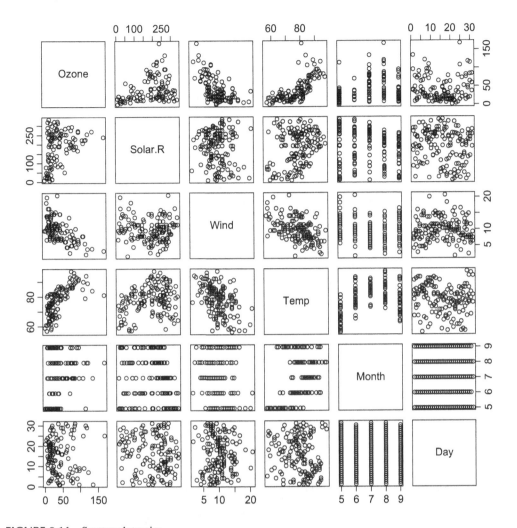

FIGURE 9.14 Scatter plot pairs.

equations to solve for the unknowns (Acevedo 2013). Matrix algebra will help us now. Write vector y of values of the observations y, this is $n \times 1$ (a column vector)

$$y = \begin{bmatrix} y_1 \\ y_2 \\ y_3 \\ \dots \\ y_n \end{bmatrix} \qquad (9.54)$$

and a matrix x which is rectangular $n \times (m + 1)$, n rows for observations, and $m + 1$ columns for the intercept and m variables

$$x = \begin{bmatrix} 1 & x_{11} & x_{21} & \dots & x_{m1} \\ 1 & x_{12} & x_{22} & \dots & x_{m2} \\ \dots & \dots & \dots & \dots & \dots \\ 1 & x_{1n} & x_{2n} & \dots & x_{mn} \end{bmatrix} \qquad (9.55)$$

Now, the unknown coefficient vector b is a column vector $(m+1) \times 1$ with entries $b_0, b_1, \ldots, b_m$

$$b = \begin{bmatrix} b_0 \\ b_1 \\ \ldots \\ b_m \end{bmatrix} \tag{9.56}$$

we need to solve b from

$$y = xb \tag{9.57}$$

Pre-multiplying both sides by x^T

$$x^T y = x^T xb \tag{9.58}$$

For brevity, we can use the notation $S_x = x^T x$ and $S_y = x^T y$. Thus,

$$S_y = S_x b \tag{9.59}$$

The S_x entries are the sum of squares and cross products of entries of x, whereas S_y entries are sum of squares and cross products of entries of y with entries of x. Solve for b by pre-multiplying both sides by the inverse of S_x

$$S_x^{-1} S_y = S_x^{-1} S_x b \tag{9.60}$$

and therefore

$$b = S_x^{-1} S_y \tag{9.61}$$

To gain some insight into the solution, let us develop it for $m = 2$, so that it is easier to find the inverse

$$\begin{bmatrix} n & \sum x_{1i} & \sum x_{2i} \\ \sum x_{1i} & \sum x_{1i}^2 & \sum x_{1i} x_{2i} \\ \sum x_{2i} & \sum x_{2i} x_{1i} & \sum x_{2i}^2 \end{bmatrix} \begin{bmatrix} b_0 \\ b_1 \\ b_2 \end{bmatrix} = \begin{bmatrix} \sum y_i \\ \sum x_{1i} y_i \\ \sum x_{2i} y_i \end{bmatrix} \tag{9.62}$$

Divide both sides by n and use average or sample mean notation (i.e., a bar on top of the variable)

$$\begin{bmatrix} 1 & \overline{X}_1 & \overline{X}_2 \\ \overline{X}_1 & \overline{X_1^2} & \overline{X_1 X_2} \\ \overline{X}_2 & \overline{X_1 X_2} & \overline{X_2^2} \end{bmatrix} \begin{bmatrix} b_0 \\ b_1 \\ b_2 \end{bmatrix} = \begin{bmatrix} \overline{Y} \\ \overline{X_1 Y} \\ \overline{X_2 Y} \end{bmatrix} \tag{9.63}$$

After some algebraic work, we can find that the determinant of Sx is

$$|Sx| = (s_{X_1})^2 (s_{X_2})^2 - (s_{\mathrm{cov}(X_1, X_2)})^2 \tag{9.64}$$

Note that in the special case of perfect correlation between X_1 and X_2, say $X_2 = X_1$, this expression reduces to

$$|Sx| = (s_{X_1})^2 (s_{X_1})^2 - \left((s_{X_1})^2\right)^2 = 0 \qquad (9.65)$$

which means that we cannot find the inverse and there will be no solution. So, let us assume that X_1 and X_2 are not perfectly correlated. After performing the inverse and multiplication operations, we find

$$\begin{bmatrix} b_0 \\ b_1 \\ b_2 \end{bmatrix} = \begin{bmatrix} \bar{Y} - b_1 \bar{X}_1 - b_2 \bar{X}_2 \\ \dfrac{(s_{X_2})^2 s_{\text{cov}(X_1,Y)} - s_{\text{cov}(X_2,Y)} s_{\text{cov}(X_1,X_2)}}{(s_{X_1})^2 (s_{X_2})^2 - (s_{\text{cov}(X_1,X_2)})^2} \\ \dfrac{(s_{X_1})^2 s_{\text{cov}(X_2,Y)} - s_{\text{cov}(X_1,Y)} s_{\text{cov}(X_1,X_2)}}{(s_{X_1})^2 (s_{X_2})^2 - (s_{\text{cov}(X_1,X_2)})^2} \end{bmatrix} \qquad (9.66)$$

Note that the covariance between X_1 and X_2 plays an important role here.

The strength of the linear relationship between independent variables X_1 and X_2 is *collinearity*, and we can see how it affects the results. When we have several independent variables, we can have multicollinearity among these variables, this effect is strong if there is a linear relationship among some of the independent variables.

In the ideal case of absence of collinearity, i.e., when the covariance between X_1 and X_2 is 0, that is if X_1 and X_2 are uncorrelated, then the solution simplifies to

$$\begin{bmatrix} b_0 \\ b_1 \\ b_2 \end{bmatrix} = \begin{bmatrix} \bar{Y} - b_1 \bar{X}_1 - b_2 \bar{X}_2 \\ \dfrac{s_{\text{cov}(X_1,Y)}}{s_{X_1}^2} \\ \dfrac{s_{\text{cov}(X_2,Y)}}{s_{X_2}^2} \end{bmatrix} \qquad (9.67)$$

Also note that b_1, b_2 are *partial* or *marginal* coefficients, i.e., they represent the rate of change of Y with one of the X while holding all the other Xs constant. We can see that in the special case of uncorrelated X_1 and X_2, this marginal change of Y with X_1 or X_2 depends only on the covariance of X_1 or X_2 and Y and the variance of X_1 or X_2. However, when X_1 and X_2 are correlated, then the marginal coefficient for one variable is affected by (1) the variance of the other variable, (2) the covariance of the other variable with Y, and (3) the covariance of X_1 and X_2.

Recall that by using the definition of correlation coefficient

$$r_{(X,Y)} = \frac{s_{\text{cov}(X,Y)}}{s_X s_Y} \qquad (9.68)$$

we can substitute in equation (9.67) to obtain

$$\begin{bmatrix} b_0 \\ b_1 \\ b_2 \end{bmatrix} = \begin{bmatrix} \bar{Y} - b_1 \bar{X}_1 - b_2 \bar{X}_2 \\ \dfrac{r_{(X_1,Y)} s_{X_1} s_Y}{s_{X_1}^2} \\ \dfrac{r_{(X_2,Y)} s_{X_2} s_Y}{s_{X_2}^2} \end{bmatrix} = \begin{bmatrix} \bar{Y} - b_1 \bar{X}_1 - b_2 \bar{X}_2 \\ \dfrac{r_{(X_1,Y)} s_Y}{s_{X_1}} \\ \dfrac{r_{(X_2,Y)} s_Y}{s_{X_2}} \end{bmatrix} \qquad (9.69)$$

In this case, the product of a correlation coefficient and a ratio of standard deviations give the marginal or partial coefficients b_1, b_2. The correlation coefficient is that between the corresponding independent variable X_i and the dependent variable Y. The ratio is that of the standard deviation of the dependent variable Y to the standard deviation of independent variable X_i.

EVALUATION AND DIAGNOSTICS

In a similar fashion to simple regression (see Chapter 1), we use ANOVA, and *t*-tests for each coefficient. For diagnostic of residuals and outliers, we use plots of residuals vs. fitted, Q–Q plots of residuals, and a plot of residuals vs. leverage. As an example, Figure 9.15 illustrates these diagnostic plots when we perform a multiple regression of ozone vs. temperature, wind, and solar radiation.

The interpretation is as in simple regression, observations 117, 62, 30 are identified as potential outliers by most plots, observations 9 and 48 are detected as outliers by the residual vs. leverage plot. High leverage occurs when its value exceeds $2 \times m/n$ or twice the number of coefficients (*m*) divided by the number of observations (*n*), and those values of Cook's distance larger than 1 are considered having large influence. In this example, high leverage would occur for $2 \times 4/153 = 0.052$.

A useful plot is that of ozone estimated by the regression model vs. the ozone observed in the data, together with the hypothetical line where both values would coincide, or the line with 1:1 slope. An example is in Figure 9.16 for the ozone example already described.

However, there is also the need to check for multicollinearity among independent variables. As we concluded in the previous section, correlation among the independent variables X_i makes the marginal coefficient depend on correlations among these variables and among these variables and the response Y. We also saw that in extreme cases of perfect correlation, there is no solution to the

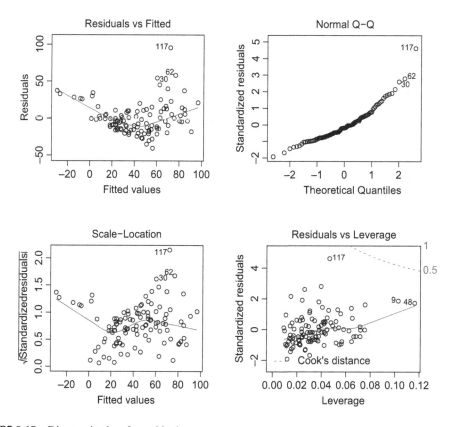

FIGURE 9.15 Diagnostic plots for residuals.

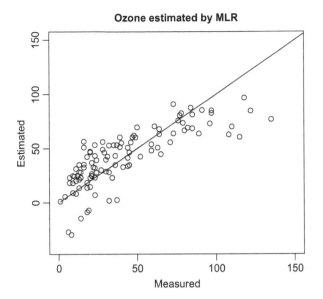

FIGURE 9.16   Ozone measured vs. fitted.

matrix equation. Correlation among independent variables can make a coefficient more important than what it really is.

A practical metric of collinearity is *tolerance*, defined as the amount of variance in an independent variable not explained by the other variables. Its inverse is the *variance inflation factor* (VIF). To calculate this metric for the variable X_j use

$$Tol = 1 - R_j^2 \tag{9.70}$$

where R_j^2 is the R^2 of the regression of variable X_j on all other independent variables. A high value of R_j^2 implies that variable X_j is well explained by the others, and then the tolerance will be low, and because *VIF* is the reciprocal $VIF = 1 / Tol$ we will have high *VIF*. Therefore, low tolerance or high *VIF* imply problems due to multicollinearity. As a rule of thumb $Tol < 0.2$ or $VIF > 5$ indicates potential problems (Rogerson 2001). A less strict rule is $Tol < 0.1$ or $VIF > 10$. Although these thresholds are arbitrary, they provide practical guidance.

One possible approach to remedy multicollinearity problems is to remove some variables from the analysis considering our knowledge of the underlying processes, and the *VIF* of the various variables. This is only one aspect of variable selection, which we consider, with more details in the next section.

VARIABLE SELECTION

Key issue to consider when applying multiple linear regression is how many variables and which X_i to use (Hocking 1976; Rogerson 2001). There are several ways of proceeding. (1) *Backward* selection: start by including all variables at once, and then *drop* variables iteratively without significant reduction of R^2. (2) *Forward* selection: we start with the X_i most likely to affect the Y variable, and then *add* independent variables. (3) *Stepwise* selection: drop and add variables as in forward and backward selection; as we add variables, we check to see if we can drop a variable added before. This process can be automated using metrics that describe how good the current selection is. The *Mallows' Cp* statistic or the *AIC* is used to decide whether an X can be dropped or added and as guide to stop the trial-and-error process.

Mallows' Cp is calculated for a subset of p variables of all m independent variables in the following manner:

$$Cp(r) = \frac{SS_r}{SS_m} + 2r - n = \frac{\sum_{i=1}^{n}(y_i - y_i(r))^2}{\sum_{i=1}^{n}(y_i - y_i(m))^2} + 2r - n \qquad (9.71)$$

Here SS_r and SS_m correspond to the residual mean square error obtained when the regression is calculated for r and m independent variables, respectively. Those mean square errors use $y_i(r)$ and $y_i(m)$, which are the fitted dependent variable for the ith observation. Recognize that if r were to be equal to m, Cp takes the value $1 + 2m - n$. In addition, the ratio of the sum of square errors would tend to be larger than 1. Cp is used to select the set of variables by picking the set that would make Cp less than $2r$ when the sets are ordered according to increasing values of r.

AIC is based on a likelihood function and information content. The likelihood function applies to discrete and continuous RVs. For example, for a discrete RV X, the likelihood function $L(\theta)$ of a parameter θ of its PMF is a continuous function formed by the products of the PMF evaluated at the observed values

$$L(\theta) = \prod_{i=1}^{n} p(\theta, x_i) \qquad (9.72)$$

Note that because the PMF is evaluated at the observed values, then L is only a function of θ. A maximum likelihood estimate (MLE) is a value $\hat{\theta}$ of θ that maximizes $L(\theta)$. An MLE is obtained by calculus taking the derivative $\frac{\partial L}{\partial \theta}$, making it equal to 0, and finding the point where the optimum occurs as we explained in Chapter 1, or by numerical optimization methods. This concept is expanded for more than one parameter.

The AIC is defined as

$$AIC = 2r - 2\ln(L) \qquad (9.73)$$

where r is the number of variables and L is the maximized likelihood function of the regression model for these r variables. This expression comprises two parts: a positive cost $2r$ of increasing the number of variables and a negative benefit $2\ln(L)$ derived from the goodness of fit. Then we select the set of variables such that AIC or the balance of two terms should be as low as possible. The AIC gives a measure of information lost when using a regression model vs. another.

Cp and AIC can be used in an iterative or stepwise process. For example, to augment a model by one variable, we add the X_i for which AIC is lowest among all alternatives. To reduce a model, drop the X_i for which AIC is lower than current. At each step, we run multiple regression. This stepwise regression procedure is automated by programming. It usually commences by augmentation from zero-order regression (intercept only).

CART

Decision trees are used to implement predictive classification and regression methods by iteratively or *recursively* partitioning the data into subsets such that each subset is as homogeneous as possible. At each node of the decision tree, we produce a rule that predicts how we split or partition the data; the rules can consist of inequalities for continuous variables, yielding a *regression tree*, or logical rule for categorical variables yielding a *classification tree*. Because the method includes both types

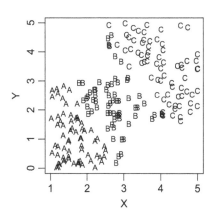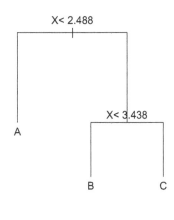

FIGURE 9.17 Left: Continuous variables X, Y, and categorical Z. Right: CART result.

of trees, it is named *classification and regression trees* hence the acronym, CART (Breiman 1984; Berk 2016; Gareth et al. 2013) and typically produces a binary tree for visualization, which is drawn vertically with the *root* on top and *leaves* at the bottom. See, for example Figure 9.17 right, which has two *decision nodes* and three *terminal nodes* or leaves.

The decision nodes are generated by recursive splits of the explanatory variables say X_1, X_2, ..., X_n starting with the variable that has the highest association with the response variable Y, which can be evaluated by one of a variety of metrics, leading to the leaf nodes which are the predicted responses.

CLASSIFICATION TREES

As a simple example, suppose we have a dataset with continuous variables X, Y with values between 0 and 5, and a categorical variable Z, with values A, B, C (Figure 9.17 left). Qualitatively, we can see that when variable X is larger than ~3.5, it can segregate the group of C values, and when X is less than ~2.5, it can separate the A values. Most B values are in X between ~2.5 and ~3.5. Once we perform CART, the resulting tree (Figure 9.17 right) is such that at the top node, which is a decision node is split based on $X < 2.48$; if true, we go to the left and all these observations are at the leaf node labeled A for value of Z. However, if the split of the top node is false, we go to the right and it can be further divided at another decision node into the leaf node labeled B if $X < 3.438$ or C otherwise.

For more details, the tree can be drawn with more information (Figure 9.18 right), particularly the probabilities of having one of the three values A, B, C in each node and the percentage of the observations in that node. For example, the top node has PMF $p = [0.33, 0.32, 0.35]$ and is labeled C because the C value is the majority, and it contains 100% of the observations. For another example, the leaf node labeled C has PMF $p = [0.00, 0.06, 0.94]$, with dominance of C, and it contains 33% of the observations.

The PMF at each node indicates the *purity* (or homogeneity) of the partition using the Gini index (not discussed here) or the entropy (discussed in a previous section on Information Theory) of the partition. In classification, decision rules are derived such that purity or homogeneity is maximized. As you can see from the plot of the left-hand side of Figure 9.18, the lower interval is almost pure A, with a few Bs but no C, and it corresponds to the leaf node labeled A ($p = [0.84, 0.16, 0.00]$). Likewise, the upper interval of the graph is almost pure C, with a few Bs but no A ($p = [0.00, 0.06, 0.94]$). There is some bleeding of As and Cs into the middle interval, which is almost pure B; this corresponds to the PMF $p = [0.00, 0.86, 0.14]$ of the leaf node labeled B.

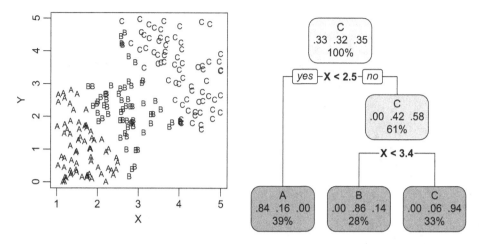

FIGURE 9.18   Left: Continuous variables X, Y, and categorical Z. Right: CART result showing probabilities and percentage of observations in each node.

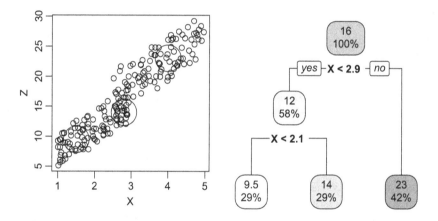

FIGURE 9.19   Left: Z variable is continuous. Right: Regression tree.

REGRESSION TREES

A simple example for a regression tree is shown in Figure 9.19 right. In this case, at the node we display the mean of the response variable for this node, instead of the PMF as in classification trees. For example, the top decision node shows a mean of 16 for 100% of the observations, and the leaf nodes have their mean at 9.5, 14, and 23. For illustration, the large circles shown in Figure 9.19 (left) indicate the approximate location of these mean values, in each one of the intervals defined by the cutoff decision values of 2.1 and 2.9.

For a more practical example, let us perform CART on a training set of the air quality dataset used in the previous section to describe multiple regression. The response variable is ozone, and the explanatory variables are solar radiation, wind, and temperature. In the lab guide, we learn how to run CART using R, which would produce the regression tree of Figure 9.20. At the top node observations are split based on wind values above six leading to a leaf node of high ozone values (mean 108) when false (wind less than six). When true (89% of the observations), we go to another decision node where a cutoff temperature of 84 splits the remaining observations into two terminal nodes, one for low ozone values (mean 24) for low temperature, and another of medium ozone values (mean 71) for higher temperature.

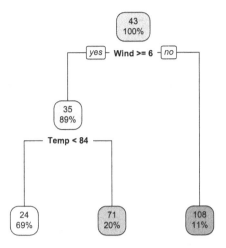

FIGURE 9.20   Regression tree for the air quality dataset.

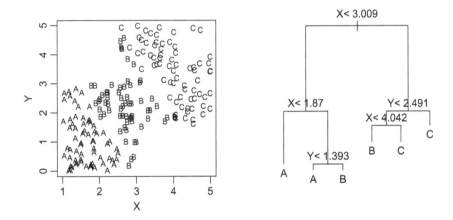

FIGURE 9.21   A classification that fits the data with higher complexity.

MODEL COMPLEXITY

Selecting a good classification or regression model involves a compromise between *overfitting* and *underfitting* the data. When overfitting the data, the model incorporates too much of the variability or noise in the data and therefore while it may fit the training set very well, it may do poorly when predicting new data. However, when the model underfits the data, it may not capture the essence of the data, and the predicted result is also poor.

For illustration, consider the tree in Figure 9.21 (right) obtained from a training subset of the data shown in the left side, and performing the CART algorithm such that it stops when adding additional nodes does not contribute to reducing the error. You will note that the classification has additional decision and terminal nodes. This model allows to further segregate the As from the Bs as well as the Cs from the Bs.

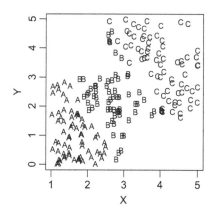 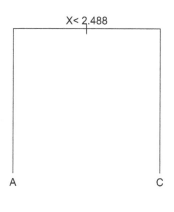

FIGURE 9.22   A classification that underfits the data.

Applying this model to an evaluation set, the confusion matrix analysis $pa = \begin{pmatrix} 1.000 & 0.727 & 0.955 \end{pmatrix}$, $Po = 0.896$, $\kappa = 0.843$ shows lower producer accuracy for class B. While this may fit this dataset well, it may not do well when we apply to a different set.

For illustration of underfitting, consider the tree in Figure 9.22 (right) obtained from a training subset of the data shown in the left-hand side. You will note that the classification consists of only one decision node and two terminal nodes, missing class B altogether. Obviously, this may not do well if we apply to a different set, since we will not be able to predict class B observations.

Indeed, applying to an evaluation set, the producer accuracy $pa = (0.957 \ 0 \ 1)$ indicates nearly perfect producer accuracy for class A, perfect for class C, but zero producer accuracy for class B.

The number of decision and terminal nodes depends on when the CART algorithm stops trying to add more variables and rules, because of lack of improvement of the classification or regression. This stopping criterion is subsumed in a metric named *complexity parameter* denoted by cp, not to be confused with Mallows Cp. The lower the complexity parameter, cp, value, the more terminal nodes and therefore the tree possibly overfits the data; whereas the higher the cp the fewer terminal nodes and with potential to underfit the data. For example, the tree in Figure 9.22 has high $cp = 0.4$, whereas the tree in Figure 9.21 has low $cp = 0.01$; as a middle point, consider the tree of Figure 9.17 which has $cp = 0.15$.

CROSS-VALIDATION

Cross-validation methods are of great importance to estimate errors of prediction of many models (Efron and Tibshirani 1993), in particular for more complete ML model evaluation, we can use multiple training and evaluation datasets. One popular algorithm is *k-fold* cross-validation that consists in splitting the dataset in k subsets and repeating the train-evaluation process k times: for each iteration i, subset $k = i$ is used for testing and the other four subsets for training. One can generate a confusion matrix for each repetition and examine the results, calculate an average of accuracy metrics, or lump all matrices in one and compute the evaluation metrics on the combined matrix. Cross-validation can also be used to find the best model when performing CART. As an example, Figure 9.23 shows producer and user accuracy for a k-fold cross-validation using $k = 5$ for the dataset of Figure 9.21 where each bar corresponds to one of the five iterations.

CART APPLIED TO SUPERVISED CLASSIFICATION FOR REMOTE SENSING

CART methods are useful for remote sensing image classification and can be employed in supervised classification by developing a model based on a training set, predicting based on an evaluation

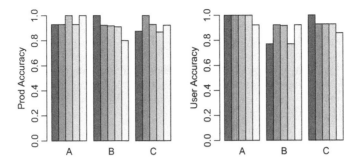

FIGURE 9.23 Cross-validation using *k*-fold.

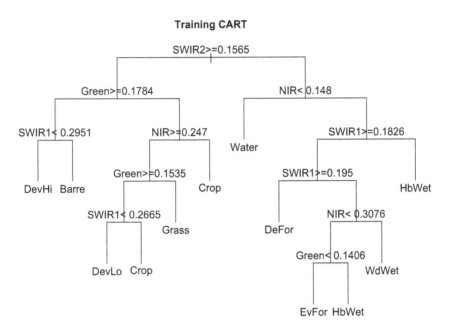

FIGURE 9.24 CART of Landsat 8 image used in Lab guide 9.

dataset, and calculating the confusion matrix for analysis. For example, as we learn in lab guide 9, we can derive a tree from the seven bands of a Landsat 8 image (employed in Chapter 8 and lab guide 8) by sampling an image containing a reference land cover class for each pixel and predict this class from the values of the bands using CART (Figure 9.24).

The nodes are split based on four bands, green, NIR, SWIR1, and SWIR2 (the last two are shortwave). The leaves or terminal nodes have the land cover classes Water, DevHi (developed high intensity), DevLo (developed open, low, and medium intensity), DeFor (deciduous forest, mixed-forest, and shrub), EvFor (evergreen forest), Grass (grass and pasture), Crop, Barre (barren land), WdWet (woody wetlands), and HbWet (herbaceous wetlands).

The confusion matrix for this classification results in $Po = 0.514$, $Pe = 0.204$, $\kappa = 0.389$, which does not indicate much accuracy. The errors of commission are all elevated 20%–90% except for Water at 2.2%. Similarly, the errors of omission are all elevated 30%–77% except for Water at 7%. Other models could be developed to improve this classification.

EXERCISES

Exercise 9.1

Define event A = rain today with probability 0.2. Define the complement of event A and its probability. Determine the sample space and the possible outcomes stating the probabilities of each. Now, define A = rains less than 1 inch, B = rains more than 0.5 inches. What is the intersection event C obtained by intersection of A and B?

Exercise 9.2

A pixel of a remote sensing image can be classified as grassland, forest, or residential. Define A = land cover is grassland, B = land cover is forest. What is the union event C? What is D = the complement of C?

Exercise 9.3

Assume we take water samples from water wells to determine if the well is contaminated. Assume we sample four wells and that they are independent. Calculate the number and enumerate the possible events of contamination results. Calculate the number and enumerate those that would have exactly two contaminated wells in the four trials.

Exercise 9.4

Using the tree of Figure 9.8, calculate the probability that test is positive, and that water was not contaminated $P[BD]$. What is the total probability of the test is in error? Hint: BD or AC. What is the probability that the test is correct? Using Bayes' theorem: what is the probability that the water is contaminated given a positive test result? Hint: calculate $P[A|D]$.

Exercise 9.5

Assume 20% of an area is grassland. We have a remote sensing image of the area. An Image classification method yields correct grass class with probability = 0.9 and correct non-grass class with probability = 0.9. What is the probability that the true vegetation of a pixel classified as grass is grass? Repeat assuming that grasslands are 50% of the area. Which one is higher and why?

Exercise 9.6

Define a discrete RV based on outcomes of classification of a pixel of a remote sensing image as grassland (prob = 0.2), forest (prob = 0.4), or residential (prob = 0.4). Plot the distributions (density or mass) and cumulative. Calculate the mean and variance. Calculate the sample mean, variance, and standard deviation of sampled data consisting of 300 grass pixels, 500 forest, and 200 residential out of 1000 pixels.

Exercise 9.7

Assume 60% of a landfill is contaminated. Suppose that we randomly take three soil samples to test for contamination. We define event C = soil sample contaminated. We define X to be an RV where x = number of contaminated soil samples. Determine all possible values of X. What distribution do we get for X? Calculate the values of PMF and CMF for all values of x. Graph the PMF and CMF. Calculate the mean and the variance.

Exercise 9.8

Calculate statistics of the confusion matrix given in Table 9.10.

Exercise 9.9

Interpret the results of multiple linear regression for air quality given in Figure 9.16.

Exercise 9.10

Interpret the results of regression tree for air quality given in Figure 9.20 and of remote sensing given in Figure 9.24.

REFERENCES

Acevedo, M. F. 2013. *Data Analysis and Statistics for Geography, Environmental Science & Engineering. Applications to Sustainability*. Boca Raton, FL: CRC Press, Taylor & Francis Group. 535 pp.

Acevedo, M.F. 2024. *Real-Time Environmental Monitoring: Sensors and Systems, Second Edition – Lab Manual*. Boca Raton, FL: CRC Press, Taylor & Francis Group. 463 pp.

Allen, J. H., W. T. Waller, M. F. Acevedo, E. L. Morgan, K. L. Dickson, and J. H. Kennedy. 1996. A Minimally-Invasive Technique to Monitor Valve Movement Behavior in Bivalves. *Environmental Technology* 17:501–507.

Berk, R. A. 2016. *Statistical Learning from a Regression Perspective*. Berlin: Springer Nature. 433 pp.

Breiman, L. 1984. *Classification and Regression Trees*. Belmont, CA: Wadsworth International Group. 358 pp.

Carr, J. R. 1995. *Numerical Analysis for the Geological Sciences*. Englewood Cliffs, NJ: Prentice Hall. 592 pp.

Davis, J. C. 2002. *Statistics and Data Analysis in Geology*. Third ed. New York, NY: Wiley. 638 pp.

Drake, A. 1967. *Fundamentals of Applied Probability Theory*. New York: McGraw Hill.283 pp.

Efron, B., and R. Tibshirani. 1994 "Cross-Validation and Other Estimates of Prediction Error." In *An Introduction to the Bootstrap*, edited by B. Efron, and R. Tibshirani. New York: Chapman and Hall/CRC.

Gareth, J., D. Witten, T. Hastie, and R. Tibshirani. 2013. *An Introduction to Statistical Learning: With Applications in R*. New York: Springer.

Hocking, R. R. 1976. A Biometrics Invited Paper. The Analysis and Selection of Variables in Linear Regression. *Biometrics* 32 (1):1–49.

Rogerson, P. A. 2001. *Statistical Methods for Geography*. London: SAGE Publications. 368 pp.

10 Databases and Geographic Information Systems

INTRODUCTION

This chapter provides an overview of database technology and its application to environmental monitoring. It covers relational databases, a mature technology, and Extensible Markup Language (XML) databases, which are becoming popular. The basic notions of schema, entity relation diagrams, and structured query language (SQL) are presented. Then, geographic information systems (GIS) are introduced as databases organized as layers of georeferenced information, focusing on the major types of layers, raster and vector, and their analysis. Examples illustrate the major concepts involved in queries and analysis. Developing database skills are very important to environmental monitoring practitioners because of the need to organize and store data collected by sensors and dataloggers as illustrated in Figure 10.1. This chapter concludes with material to help understand what a database system can do in terms of storage and retrieval of real-time sensor data, understand schemas and metadata of a database and their importance in data sharing through a web service. These concepts will be expanded by computer exercises in Lab 10 of the companion Lab Manual (Acevedo 2024).

DATABASES

Informally, a database (DB) is an organized collection of data, or in other words, a collection of records, or pieces of knowledge. Common examples of a DB would be a telephone directory, mailing list, and a recipe collection. A Database Management System (DBMS) is software designed to create a database as well as to store and query this database; i.e., a DBMS is a collection of software modules that manage data storage, query processing, and data recovery (Silberschatz et al. 2010).

There are several types of DB. Of these, we will briefly cover *relational* DB, which is a mature type and employed in many applications, and *XML* DBs, which are becoming popular. Some well-known standards in DBMS are SQL and Open Database Connectivity (ODBC). Examples of DBMS are MySQL and PostgreSQL, which is an open-source option that offers spatial data capabilities (PostGIS).

Database technology provides standards and schemas, storage methodologies, and helps us design constrains and data quality assurance. In addition, it helps us provide exchange of data through real-time web interface and standard web services, e.g., Open Geospatial Consortium (OGC) sensor web enablement (SWE) (OGC 2022a).

Developing database skills are very important to environmental monitoring practitioners because we need to organize and store data collected by sensors, dataloggers, WSN, and IoT devices as described in Chapters 5 and 6. In this chapter, we cover material to help understand schemas and metadata of a database and their importance in data sharing, as well as define XML and its off-spring e.g., GML and sensorML. We will briefly see how to share data through the standard web service SWE. The first part of this chapter is just a brief outline based on an environmental monitoring mini-course material developed at the University of North Texas (Huang et al. 2008).

SERVER CLIENT: DATALOGGING AND DB

We will distinguish between a DB *server* and a data station or *client* (datalogger or sensor system), and two main modes, *push* and *pull*, to import data from a field station into a database server. In

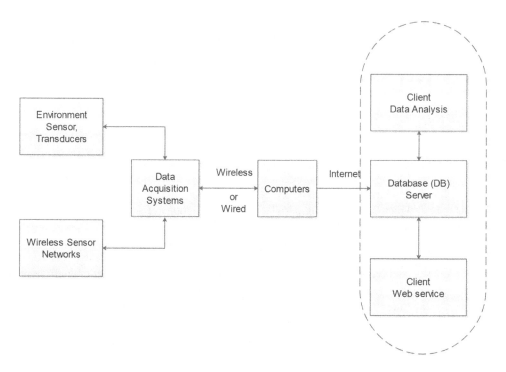

FIGURE 10.1   From sensors to DB systems.

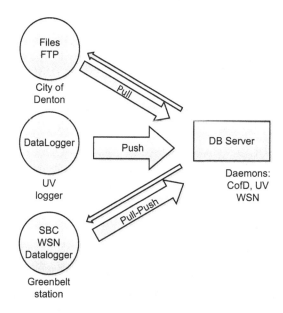

FIGURE 10.2   Push and pull modes. Adapted from Huang et al. (2008).

pull, the server initiates the request for data transaction, whereas in push the request is initiated by the station. In the example of Figure 10.2, the DB server, using FTP, pulls files of rainfall data available at City of Denton computers; files from UV data collection (see Chapter 11) are pushed to the server by the UV datalogger. Push and pull transactions can also be combined.

As you recall from Chapter 6, the MQTT protocol is based on the server and client concept, and in this context a WSN node or IoT device may be a publisher client that sends data to a server broker,

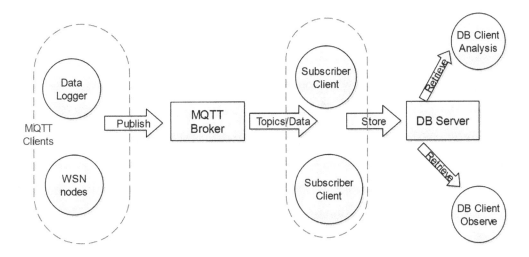

FIGURE 10.3   MQTT protocol and DB.

while a subscriber client can access the broker based on topics. You can see how a subscriber client may access the broker, collect data, and organize it in a DB server, which can then be accessed by other clients querying data from the DB (Figure 10.3).

Data collected by dataloggers are not necessarily ready for DB ingestion. We need to specify how to format the raw data to organize DB files. For example, we can look at some of the data from the TEO system. These include City of Denton FTP Data

```
Directory Structure: YEAR/MONTH/cod_date_time
2007/9/cod_2007-9-21_8.30
File Format: Station_Name/Gate_Name, date, time, rain_gage_reading/
gate_status
RG1, 9/21/2007, 8:30:57 AM, 0.00
HC_N_Gate_STATUS, 9/21/2007, 8:30:57 AM, UP
```

and UNT UV Data

```
Directory Structure: YEAR/MONTH/uv_date_time.txt
2007/9/uv_2007-9-15_12.30
File Format: Date, Time, SUV #1, UVA #2, Temp 1, Temp2
"15.09.2007", "12:30", 1.367, 0.000, 32, 0
```

RELATIONAL DATABASES

Relational DBs are built upon a well-founded theory, consisting of *Relations* (tables), *Keys*, *Functional Dependencies*, and *Normal Forms*. An *entity* is an object about which we store data as *tables*, which are the same as a *relation* and constitute the basis for a relational DB. We specify *attributes* in the table columns, whereas the rows represent the contents of a table, a *tuple* or *instance* of a relation. We can link two tables using a *key*.

As a simple example, consider two entities or tables (Figure 10.4), one named "Location" contains a list of names and locations by station IDs. These three attributes are the columns, whereas the instances

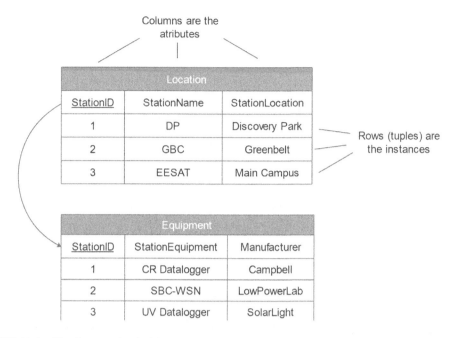

FIGURE 10.4 Simple example of tables and a key to relate two tables.

are the rows; in this case, only three and each one is a tuple, for example [1, DP, Discovery Park]. The other entity or table is named "Equipment" and it contains three attributes Station ID, Station Equipment, and Manufacturer of equipment used at the station. One example of tuple in this case is [1, CR datalogger, Campbell]. The station ID is a key and links both tables; in this example, StationID is underlined to signal that it is a key. This is an extremely simple example, typically a DB consists of many relations representing objects and the relationships among them.

Each attribute of a relation has a name and a *domain*, i.e., a set of allowed values for the attribute; for example, StationName has domain DP, GBC, EESAT. Attribute values are required to be atomic, that is, indivisible. The special value NULL is a member of every domain. Denote $A_1, A_2, ..., A_n$ as attributes of a table, then $R=(A_1, A_2, ..., A_n)$ is a relation *schema*. For example, for the table Location given above Location-schema = (StationID, StationName, StationLocation). Further, $r(R)$ is a relation on the relation schema R. Listing the attributes is the barebones of the schema, which should also specify the type of data for each attribute; for example, StationID would be integer, StationName would be character, and StationLocation would be character. Depending on the attribute, the type could be timedate, e.g., a timestamp of a datalog file, or numeric, e.g., temperature which would be a float and we would specify the number of decimal places.

K is a superkey of a schema R if values for K are sufficient to identify a unique tuple of each possible relation r(R). For example: (StationID, StationName) and StationName are both superkeys of table Location, assuming no two stations have the same name. K is a candidate key if K is minimal. Example: StationName is a candidate key for Location, since it is a superkey (assuming no two stations can have the same name), and no subset of it is a superkey (Huang et al. 2008).

A *functional dependency* is a generalization of the notion of a key requiring that the value for a certain set of attributes determines uniquely the value for another set of attributes. Example: Consider the schema:

```
Station-Sensor-schema = (StationName, StationLocation, SensorID,
SensorManuf, Phenomena).
```

We expect this set of functional dependencies to hold

```
StationName → StationLocation
SensorID → SensorManuf, Phenomena, StationName
```

but would not expect that

```
StationName → Phenomena
```

Other integrity constraints are given by using primary keys (PKs) and foreign keys (FKs). An FK is the PK of one table that is placed into another table to represent a relationship among those tables. For example

```
R1 = (StationName, StationLocation)
R2 = (SensorID, sensorManuf, Phenomena, StationName)
```

StationName in R2 is a FK referencing the PK of R1.

Sometimes selection of a PK is subjective, even if two designers have the same set of superkeys and candidate keys, it is possible for these two people to choose different PKs and FKs for the relationships. Therefore, an important process is *normalization*, which consists of expressing relations in "good form". First, decide whether a particular relation R is in good form, if not decompose it into a set of relations (R1, R2, ..., Rn) such that each one of these relations is in good form. When is a relation in good form? Relations preferably should be in either Boyce-Codd Normal Form (BCNF) or Third Normal Form (3NF). BCNF and 3NF eliminate redundancy (Huang et al. 2008). As an example of BCNF decomposition, consider:

```
R = (StationName, StationLocation, SensorID, SensorManuf, Phenomena).
```

With functional dependencies

```
StationName → StationLocation
SensorID → SensorManuf, Phenomena, StationName
```

We can decompose to R1, R2

```
R1 = (StationName, StationLocation)
R2 = (SensorID, sensorManuf, Phenomena, StationName)
```

DATA MODELS AND ENTITY RELATION DIAGRAMS

Visualizing entities and their relationships in the form of diagrams is very helpful when designing a data model for a DB, as well as debugging a complicated DB. Figure 10.5 shows the previous example

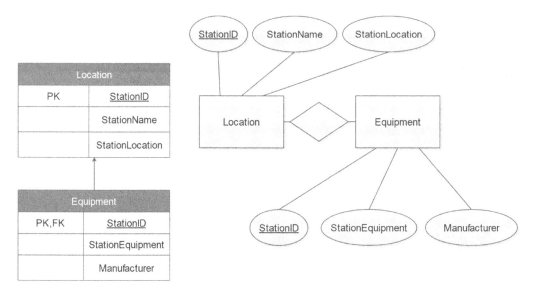

FIGURE 10.5   Data model as tables and as Chen ER diagram.

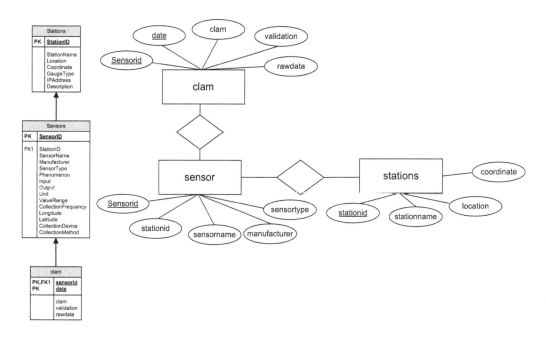

FIGURE 10.6   Normalized table example. From Huang et al. (2008).

of two relations, Location and Equipment, developed as two types of *Entity-Relation Diagram (ERD)*. The ERD shown on Figure 10.5 left uses a table for each entity and these tables are connected. This type of table should not be confused with table as equivalent to a relation, which uses columns for the attributes. The diagrammatic type of table has two columns, the right-hand side columns are fields and contain the attributes, Figure 10.5 (left) designates which ones of the attributes is the PK and FK.

Figure 10.5 right shows another type of ERD, called a Chen diagram, that consists of interconnected shapes. The boxes are entities, ovals linked to boxes are attributes of the relation, the diamond indicates that both tables are related. For a more complete example, consider Figure 10.6 that consists of three tables.

Figure 10.5 shows a low level of detail, limited to the entities, attributes, and keys. A more complete model using tables would add the data type for each attribute; for example, StationID would be integer, StationName would be character, and so on. In this case, the tables have three columns. In addition, the lines connecting tables can have a variety of symbols such as crow foot, dash, circle, or a combination of these. With these more complex lines, we can specify whether the relation is one-to-one or one-to-many.

There are a variety of types of ERD and ways of representing a data model as well as software that helps to develop these (Visual Paradigm 2022; Lucidchart 2022; Database Star 2022).

SQL

DB language commands are specified according to its purpose as data *definition* or data *manipulation*. A Data Definition Language (DDL) is used for changes in the DB schema. Example: create table, drop table, alter table, create index. A Data Manipulation Language (DML) is used to read or change the content of the database. For example, a DML would include commands such as insert, delete, select, and update.

DDL

DDL commands are used to change the schema for each relation, domain of values associated with each attribute, integrity constraints, and set of indices to be maintained for each relation, security and authorization information for each relation, and the physical storage structure of each relation.

For example, an SQL relation is defined using the `create table` command

```
create table r (A1 D1, A2 D2, ..., An Dn,
(integrity-constraint1),
...,
(integrity-constraintk))
```

where r is the name of the relation, each Ai is an attribute name in the schema of relation r, and Di is the data type of values in the domain of attribute Ai. Integrity constraints in `create table` include not null, PK (A1, ..., An), FK (A1, ..., An) references (B1, ...Bn), check (P), where P is a predicate. Figure 10.7 shows the process of using SQL to create the schema for tables in the simple example stations. In the lab guide companion to this textbook, we will practice developing examples using two software tools RSQLite and SQLiteStudio.

Let us look at another example creating two related tables. In the following, we create table `Sensors` with attributes `SensorID` as integer, `SensorManuf` as character, and `Phenomena` as character, and then as integrity constraint, we declare `SensorID` as the PK for table `Sensors`.

```
create table Sensors
      (SensorID      integer,
      SensorManuf   char(30)
      Phenomena     char(30),
      primary key (SensorID));
```

and now we create another related table `Wind` that has `SensorID` referencing the `sensorID` for Sensors (an FK), declares two additional attributes (`dateTime` and `windspeed`), two PKs, and lastly ensures that `windspeed` is non-negative.

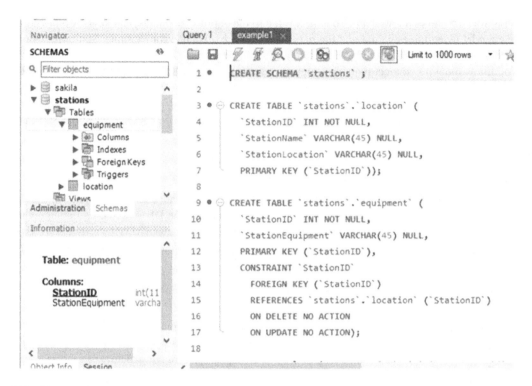

FIGURE 10.7   SQL creating the tables for the stations example.

```
create table Wind
(sensorID integer REFERENCES Sensors(SensorID),
        dateTime date,
        windspeed float,
        PRIMARY KEY(sensorID,dateTime),
        check (windspeed >=0));
```

After a table is created, we can use commands `drop table` and `alter table` to modify it. For example:

```
DROP TABLE Sensors;
```

deletes all information about `Sensors` from the database and

```
ALTER TABLE Wind add maxWindSpeed float;
```

adds an attribute called `maxWindSpeed` to the `Wind` table.

You can also drop attributes of a table, add integrity constraints, e.g., PKs, reference keys, to existing table.

DML

DML functions include read or change the content of the database. A typical SQL query has the form:

```
select A1, A2, ..., An
from r1, r2, ..., rm
where P
```

Here, the first line refers to attributes, the second line to the relations, and the third line to predicates. The result of an SQL query is a relation. The following are examples of queries.

Select the day and the maximum UV value from the table

```
SELECT date, suv
FROM uvdata
WHERE suv IN (select max(suv) from uvdata);
```

Select the daily UV intensity averages from January using only non-zero values

```
SELECT date_trunc('day', date), avg(suv)
FROM uvdata
WHERE date BETWEEN '2007-1-1' AND '2007-1-31'
AND suv>0
GROUP BY date_trunc('day',date);
```

Select the UV values from today

```
SELECT date, suv
FROM uvdata
WHERE date_trunc('day',date)=CURRENT_DATE;
```

Select the monthly average of rain

```
SELECT date_trunc('month',date),avg(value)
FROM rain
GROUP BY date_trunc ('month',date)
ORDER BY date_trunc('month',date);
```

Select the temperature values (corresponding to sensorid=6002) from March 4, 2002

```
SELECT date, value
FROM sonde
WHERE sensorid=6002 AND date between '2002-3-4 00:00:00' and '2002-
3-4 23:59:59';
```

Select all dates from January 2001 where percent saturation of dissolved oxygen (sensorid=6005) is between 0 and 5

```
SELECT date,value
FROM sonde
WHERE sensorid=6005 AND value BETWEEN 0 and 5
AND date BETWEEN '2001-1-1' and '2001-1-31';
```

Select all the distinct locations

```
SELECT location
FROM stations
WHERE location IS NOT NULL;
```

Select sensorid and sensorname from all the sensors whose names start with T

```
SELECT sensorid, sensorname
FROM sensors
WHERE sensorname like 'T%';
```

Count all the sensors

```
SELECT count(sensorid)
FROM sensors;
```

Select days where turbidity (sensorid=6016) average is between 0 and 2

```
SELECT date_trunc('day',date), avg(value)
FROM sonde
WHERE sensorid=6016
GROUP BY date_trunc('day',date)
HAVING avg(value) BETWEEN 0 and 2;
```

In addition, SQL also includes Insertion, Deletion, Update, Bulk upload, and Create index.

XML

After the advent of the web and HTML, recognizing the need for a more flexible platform to exchange data, XML was created to allow for a customizable definition of elements and tags beyond the set specified by HTML. XML has become popular for environmental monitoring data exchange motivated by several factors, which includes heterogeneity of sensors and sensor data and thus the need for a standard to exchange sensor configuration and data (Huang 2008). Furthermore, XML facilitates the process of making data from environmental sensor networks accessible through the Internet. Indeed, sensors and sensor data are typically heterogeneous and need some level of flexibility when organizing it for presentation via the web.

As in HTML, the structure of XML data includes tags and elements. A *tag* is a label for a section of data. An *element* is a section of data beginning with `<tagname>` and ending with a matching `</tagname>`. Elements must be properly nested, and every document must have a single top-level element. The following is an example of nested elements

```
<station>
        <Description>Doyle and Freeman</Description>
        < Phenomenon Name="Rain" />
        <Point>
                <pos>42, 24, 11</pos>
        </Point>
</station>
```

In the order station, Description, Phenomenon Name, Point, and pos (position). Elements can have attributes

```
< Phenomenon Name="Rain" />
```

An element may have several attributes, but each attribute name can only occur once. An element without any sub-element can be abbreviated

```
< Phenomenon Name="Rain"/>
```

instead of

```
< Phenomenon Name="Rain" > </ Phenomenon >
```

The same tag name may have different meaning in different organizations or domains, causing confusion. Therefore, use

```
unique-name:element-name
```

We can avoid using long unique names all over a document by using XML *Namespaces*. For example, in the following, we declare namespace `gml` in the `station` tag then `Point` and `pos` are unique names for that namespace

```
<station Xmlns:gml="http://www.opengis.net/gml">
<Description>Doyle and Freeman</Description>
        < Phenomenon Name="Rain" />
        <gml:Point>
                <gml:pos>42, 24, 11</gml:pos>
        </gml:Point>
</station>
```

To store string data use CDATA that stands for "character data". This is useful to store strings that may contain tags not meant to be interpreted as sub-elements,

```
<![CDATA[<station> … </station>]]>
```

contents between square braces [...] are treated as strings.

As discussed earlier in this chapter, DB schemas constrain the information that can be stored, and the data types of stored values. XML document schemas are not required but important for XML data exchange because parsers use schemas to automatically interpret data and users employ schemas to understand what the data are about. There are two ways to specify XML document schema: *Document Type Definition* (DTD), which is simple and popular, and *XML schema*, which is more complex and has data type constraints.

DTD specifies what elements can occur, what attributes an element can or must have, and what sub-elements can or must occur inside each element. However, DTD does not constrain data types, all values are represented as strings. The DTD syntax is

```
<!ELEMENT element (subelements-specification) >
<!ATTLIST   element (attributes)  >
```

for example,

```
<!DOCTYPE teo[
        <!ELEMENT teo( ( stations| databases)+)>
        <!ELEMENT stations(stationName stationLocation sensors*)>
        <! ELEMENT databases(databaseName databaseType )>
        <! ELEMENT stationName(#PCDATA)>
        <! ELEMENT stationLocation(#PCDATA)>
        <! ELEMENT sensors (sensorName phenomenon)+>
        <! ELEMENT sensorName(#PCDATA)>
        <! ELEMENT phenomenon(#PCDATA)>
        <! ELEMENT databaseName(#PCDATA)>
        <! ELEMENT databaseType(#PCDATA)>
        <!ATTLIST stations boundingBox CDATA  #required>
        ]>
```

Users define DTDs and schemas for their application domains specifying allowed tags and structures, such that the semantic meaning and information structure will have some level of consensus in the application domain. For example, in the environmental monitoring field: GML that specifies geographical features and geospatial datasets (European Commission 2022), SensorML that specifies sensor terms and semantics (SensorML 2022), KML (Keyhole Markup Language) for geographic annotation and visualization and used by Google Earth (Google 2022), and EML (Ecological Metadata Language) developed by the ecological discipline (KNB 2022).

For illustration, consider this GML example, where the tags are specified to follow their GML unique name

```
<gml:Polygon>
        <gml:outerBoundaryIs>
                <gml:LinearRing>
                        <gml:coordinates>0,0 100,0 100,100 0,100 0,0</
gml:coordinates>      </gml:LinearRing>
        </gml:outerBoundaryIs>
</gml:Polygon>
<gml:Point>
        <gml:coordinates>100,200</gml:coordinates>
</gml:Point>
<gml:LineString>
        <gml:coordinates>100,200 150,300</gml:coordinates>
</gml:LineString>
```

Consider also this KML example

```
<kml>
        <Placemark>
                <name>City of Denton: Doyle and Freeman</name>
        <description>Rain Gauge Reading: 0 inch
        </description>
                <Point>
                        <coordinates>-97.132,33.214
                        </coordinates>
                </Point>
        </Placemark>
</kml>
```

Google Earth provides simple application program interfaces based on KML that helps users to publish their geospatial data.

GIS

GISs are DB systems structured spatially, or geo-referenced DBs, such that we can collect, store, retrieve, and display data for positions on the Earth's surface. GIS facilitates geospatial analysis as well as making and editing maps. There are many analysis applications, just to mention a few consider demographic analysis, site selection, watershed analysis, resource inventories and monitoring, land management, and transportation modeling. Regarding environmental monitoring, GIS provides a way to organize data by station and sensor location, follow sensor measurements by spatial location, ingest remote sensing data, and merge with other geospatial data sources.

GIS Software

There are a variety of software tools to implement GIS, ranging from commercially licensed products as ArcGIS (ESRI 2022), academic-based emphasizing monitoring and modeling such as IDRISI/TerrSet (Clark Labs 2022), and open-source community-based (QGIS 2022; GRASS GIS 2022). In this chapter and in the lab guide companion to this book, we will develop some examples using QGIS.

GIS software includes tools to perform DB operations such as queries and DB management. Prominent operations are vector files management such as joining attributes by location and creating spatial indices. In addition, GIS software typically includes linkages to DB systems such as SQL PostGIS, and SpatiaLite. Very importantly, for monitoring and modeling, GIS software includes linkages to programming languages such as C++ and Python. Due to its closeness to spatial processes, GIS software also includes analysis of remote sensing images and tools to import raster files and processing them by themselves or in conjunction with other raster and vector files. Moreover, these tools include export of spatial analysis results as remote sensing formatted files.

GIS Layers

A GIS is organized by data *layers*, which are files containing data about positions on the spatial domain of interest and that can be superimposed or *overlaid* to query data from all layers for a target position. Consider these examples: a topographic elevation layer consisting of either the elevation of a set of coordinates for positions on the area or a set of lines representing the contour lines, a soil layer consisting of soil-type polygons covering the area, and a vegetation layer derived from remote sensing and having values on a grid of cells defined by the raster image. We can overlay these layers to find, for example the soil type for vegetated areas located in land of low elevation.

Raster Layers

From the material on remote sensing presented in Chapter 8, we understand the concept of raster images; this is also a type of GIS layer that divides the area of interest in a grid of cells of given size. Each cell has a unique value in the domain of values and data type of the layer. These can be logical, integer, and floating values. For example, a vegetation raster pixel could take a Boolean or logical 1 if the cell is vegetated or 0 if not or take an integer from 1 to 4, for three vegetation types (forest, grass, crop, bare), or float values between −1 and 1 for NDVI values (Figure 10.8). To further illustrate the raster concept, the NDVI image presented in Chapter 8 with values from about −0.2 to 0.5 and pixel size 30 m×30m is imported into QGIS as a raster layer and displayed in Figure 10.9 using gray scale from dark (lower values) to light (highest values) at a scale of 1:25,000. Note that at this scale, the size of the side of a pixel is 30 m/25,000=0.0012 m or 1.2mm.

Using a GIS, we gain insight about a raster layer by computing its histogram, as illustrated in Figure 10.10 for the NDVI layer discussed above. We can see a sharp peak around −0.02 that corresponds to the open water pixels and a broader peak around 0.3 that corresponds to vegetation. A useful tool typical of GIS is an *identifier* that can be used to query information about a pixel directly on the image; for example, Figure 10.11 displays the value and coordinates of a pixel that was clicked on the image using QGIS.

Importantly, GIS software allows querying information about the layer, including its Coordinate Reference system (CRS), extent, data type, range of values, number of rows and columns, and pixel

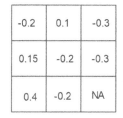

0	1	0
1	0	1
1	0	NA

Logical: Presence

4	3	4
2	4	4
1	4	NA

Integer: Types

-0.2	0.1	-0.3
0.15	-0.2	-0.3
0.4	-0.2	NA

Float: NDVI

FIGURE 10.8   Examples of raster layer for the same theme (vegetation) but with different data types.

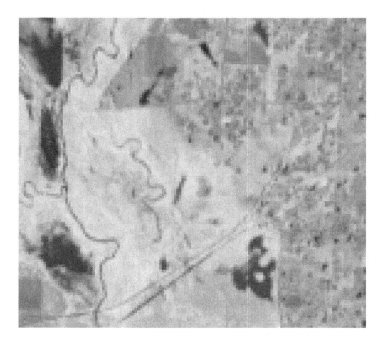

FIGURE 10.9   Example of a raster layer in GIS showing pixels of an NDVI image.

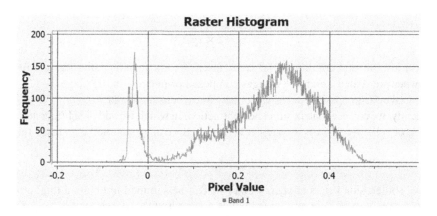

FIGURE 10.10   Histogram of the NDVI raster layer.

size. As an example, for the NDVI image that we have been discussing, querying the properties of the layer in QGIS we obtain that the CRS is EPSG:32614-WGS84, UTM Zone 14N, the datatype is 32-bit floating point, extent 664995,3671805:700005,3706815, rows 1167, columns 1167, pixel size 30×30 m, and values ranging from -0.2065 to 0.595.

In this chapter and the lab guide, we cover examples of just a few of the many raster layer manipulations available in GIS software, ranging from basic tools such as reprojecting a layer to a different CRS, georeferencing, and resampling to more complex operations such as statistics, terrain analysis, and modeling.

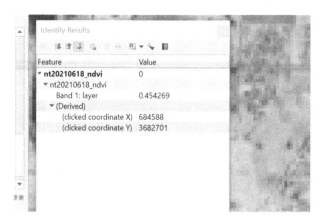

FIGURE 10.11 Example of identifier result.

RASTER ANALYSIS: ENTRY-WISE CALCULATIONS

The raster format is very convenient for analysis because the contents of the raster can be thought as a mathematical matrix with cells and rows, and therefore raster operations can be implemented as mathematical operations on matrices. This assumes that the raster layers have compatible data types, e.g., integers, real numbers, or Boolean variables.

It is often useful to scale a numeric layer multiplying all entries by the same number (integer or float). For example, a raster layer x of integers could be multiplied by a scalar a (a float) to obtain a new layer z with float type of cells

$$z_{ij} = a \times x_{ij} \tag{10.1}$$

where i and j are row and column of the raster layer. We encountered this situation in Chapter 8 and Lab 8 when we scaled Landsat 8 images composed of integers in the range $0 - (2^{16} - 1)$ by the maximum of all cells to obtain a raster with float values between 0 and 1.

Conveniently, we can use matrix sum and subtraction, in which we add or subtract numeric layers in a cell-wise or cell-by-cell basis; two raster layers x and y can be summed into a new raster layer z,

$$z_{ij} = x_{ij} + y_{ij} \tag{10.2}$$

For example, soil erosion layers at years t and $t+1$ can be summed to obtain a total erosion layer. The *Hadamard* or *Schur* matrix product, in which you multiply entry-wise the values of matrices is the basis for raster multiplication of the two raster layers x and y, to obtain a third z

$$z_{ij} = x_{ij} \times y_{ij} \tag{10.3}$$

Note that this operation is different from the standard matrix multiplication of conforming matrices. For example, we can calculate an erosion rate layer as the product of rainfall raster layer and a coefficient depending on soil type for the cell.

Furthermore, we can construct linear and non-linear functions that operate on a cell-by-cell basis, to build one raster from another or a new raster from two others. Consider, for example layers x and y and their normalized difference z

$$z_{ij} = \frac{x_{ij} - y_{ij}}{x_{ij} + y_{ij}} \tag{10.4}$$

We already encountered this operation in Chapter 8 when using Landsat 8 images, we calculated NDVI from the red and NIR bands and NDWI from the green and NIR bands.

Lastly, let us mention Boolean algebra of raster images consisting of Boolean valued cells. We can use the NOT, AND, OR, and XOR Boolean operators on a cell-wise manner. Take two raster layers x and y and perform the AND operation

$$z_{ij} = x_{ij} \text{ AND } y_{ij} \tag{10.5}$$

which would be useful when determining whether two required factors, e.g., soil nutrient availability and water infiltration, enable a third, say crop growth.

RASTER ANALYSIS: NEIGHBORHOOD AND ZONAL CALCULATIONS

The raster framework also allows to do calculations using a moving *neighborhood* or *window filter* of cells around a target cell or even extending this calculation to a larger neighborhood or *zone*. There are many of these methods, some prevalent ones are: calculating an average around the target cell, calculating filters based on weighted averages or on convolution, resampling to change raster resolution, and calculating slope and aspect of the terrain at a target cell. In the following, we will go over some examples.

Suppose we have a raster of $30\,\text{m} \times 30\,\text{m}$ and want to derive a lower resolution raster with $90\,\text{m} \times 90\,\text{m}$ cells, this means using a moving neighborhood or window of $3 \times 3 = 9$ cells around (and including) the target cell. A simple grid of $6 \times 6 = 36$ cells is shown in Figure 10.12 for illustration; we consider four target cells shown at positions (2, 2), (2, 5), (5, 2), (5, 5) by (row, column) and marked with circles. The raster is composed of integers indicating categorical values or classes, e.g., vegetation type, and thus we can take the *mode* of the values in the neighborhood 3×3 cells surrounding the target; for instance, the nine values for target cell at (2, 2) are 2,1,2,1,1,1,2,1 and the mode is 1 which will be the value assigned to the upper corner cell of the new raster of 2×2 cells of $90\,\text{m} \times 90\,\text{m}$. You can verify that the other cells have mode 4,4,1. This algorithm based on the mode

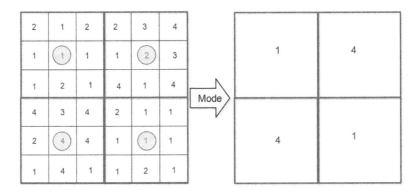

FIGURE 10.12 Resampling by using the mode or majority.

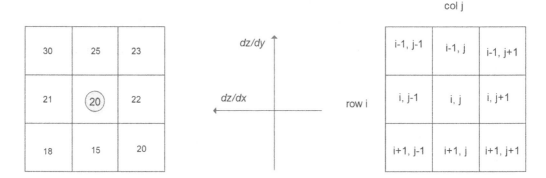

FIGURE 10.13   A nine-cell window to calculate slope angle and aspect for a DEM.

is also called *majority* resampling. Other algorithms for resampling include bilinear interpolation and cubic convolution.

One more example can be drawn from *terrain* calculations. Suppose a raster is a *digital elevation model* (DEM) with 30 m×30 m cells that has integer or real numbers giving elevation above sea level for each cell, and we want to calculate slope angle (steepness) and aspect (direction) for each cell on the raster. Scanning over the entire raster with a 3×3 window, we can calculate the slope and aspect for each cell based on the elevation of the cells in the window. Figure 10.13 (left) shows one of these windows located at a target cell, the numbers are elevation values as integers. Denote by z the elevation value, and by x and y the horizontal (E–W) and vertical (S–N) directions on the DEM oriented toward the left (W) and toward the top (N) (Figure 10.13 center). We will present the calculation following the algorithm proposed by Horn (1981), which is implemented in many GIS software tools, consisting of calculating the elevation gradient dz/dx in the horizontal x direction (E–W) at the target cell and dz/dy in the vertical y direction (S–N) at the target cell and composing a slope angle based on

$$\alpha = \tan^{-1}\sqrt{\left(\frac{dz}{dx}\right)^2+\left(\frac{dz}{dy}\right)^2}\qquad(10.6)$$

Denote rows and columns by i and j and assume that the center cell is at i, j. You can look at Figure 10.13 (right) to help understand how we use the indices i and j in the following equations. The horizontal gradient is computed by all elevation differences along the columns from right to left and dividing by the number of weights used $2\times(1+1+2)=8$ times the cell side length d

$$\frac{dz}{dx}=\frac{\left(\left(z_{i-1,j+1}+2z_{i,j+1}+z_{i+1,j+1}\right)-\left(z_{i-1,j-1}+2z_{i,j-1}+z_{i+1,j-1}\right)\right)}{(8\times d)}\qquad(10.7)$$

and the vertical gradient by all elevation differences along the rows from bottom to top

$$\frac{dz}{dy}=\frac{\left(\left(z_{i+1,j-1}+2z_{i+1,j}+z_{i+1,j+1}\right)-\left(z_{i-1,j-1}+2z_{i-1,j}+z_{i-1,j+1}\right)\right)}{(8\times d)}\qquad(10.8)$$

A few lines of code in R can help make this computation easier for the values shown in Figure 10.13 (left)

```
z <- matrix(c(30,25,23,21,20,22,18,15,20), byrow=T, ncol=3)
d <- 30; i=2;j=2
dz.dx <- ((z[i-1,j+1]+2*z[i,j+1]+z[i+1,j+1]) -
          (z[i-1,j-1]+2*z[i,j-1]+z[i+1,j-1]))/(8*d)
dz.dy <- ((z[i+1,j-1]+2*z[i+1,j]+z[i+1,j+1]) -
          (z[i-1,j-1]+2*z[i-1,j]+z[i-1,j+1]))/(8*d)
ang <- atan(sqrt(dz.dx^2+dz.dy^2))*180/pi
> ang
[1] 8.327143
```

Resulting in a slope angle of 8.32°. Note that the atan function returns the angle in radian, so we converted from radian to degrees using $180/\pi$. When writing the slope angle raster layer at the target cell i, j the GIS would write 8.32 if treated as real number or rounding 8 as an integer raster layer.

The aspect (or compass direction) of the slope is calculated in two steps. First use the two-argument arctangent to obtain the angle in Cartesian x, y coordinates (Figure 10.14 left) as

$$\theta_2 = \text{atan } 2\left(\frac{dz}{dy}, \frac{-dz}{dx}\right) \tag{10.9}$$

For a brief review, the two-argument arctangent is the same as the arctan calculated as

$$\theta = \tan^{-1}\left(\frac{dy}{dx}\right) \tag{10.10}$$

for positive x values but takes a different value when the x values are negative. Note that two points $(-a, b)$ and (a, b) with the same b value for the y coordinate would have the same angle as calculated by equation (10.10) that is to say $\theta = \tan^{-1}(b/a)$. For instance, a point of coordinates $(-1, 1)$ would have an angle of 135° calculated by equation (10.9) instead of 45° calculated by Equation (10.9) and a point with coordinates $(-1, -1)$ would have an angle of $-135°$ instead of $-45°$. The atan2 function was developed for this purpose, so that result can be distinct for negative values of x.

Using the atan2 function of R on the $-dz/dx$ and dz/dy calculated previously

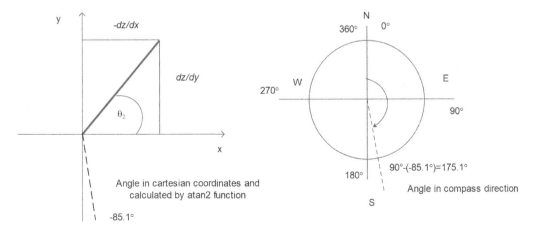

FIGURE 10.14   Left: calculating the two-argument arctangent. Right: converting to compass direction.

```
asp2 <- atan2(dz.dy,-dz.dx)*180/pi
> asp2
[1] -85.10091
>
```

Resulting in an angle θ_2 of $-85.1°$. However, we need to express this as angle in a compass direction; that is to say, increasing angle clockwise in degrees from $0°$ (north) to $180°°$ (south), to $270°$ (west), and to $360°$ (again due north) coming full circle. By looking at Figure 10.14 (right), we can see that we would use $\theta = 90 - \theta_2 = 90 - (-85.1°) = 175.1°$ resulting an angle which is pointing SE but almost due south, indeed just $180 - 175.1 = 4.9°$ to the east. When writing the slope aspect raster layer at the target cell i, j, the GIS would write 175.1 if treated as real number or rounding 175 as an integer raster layer.

VECTOR LAYERS

Vector layers can be of various types depending on the geometry of their elements: *points*, *lines*, or *polygons* (Figure 10.15). A point layer is a collection of pair of coordinates, latitude, and longitude or UTM, such that each point has a value for a set of *attributes*. Take for example, groundwater monitoring wells, such that the layer would have coordinates of the wells, and values of a parameter of water quality and other relevant information such as depth and geological formation.

More complex, a line layer would be a set of connected series of points, that is a set of connected pairs of coordinates, and each line would have values or a set of attributes. Take, for example streams and rivers, which would be represented by a series of pair of coordinates; with attributes such as name (e.g., Clear Creek) or stream order (e.g., 1).

A polygon layer would be a series of points with coinciding start- and end-points. Each polygon also would have values or a set of attributes. For example, consider lakes, where the series of coordinates represent shoreline, and the attributes may be name (e.g., Lake Lewisville) or water quality (e.g., average Total Dissolved Solids, TDS). Depending on scale, the same theme, e.g., cities, would be displayed as points (e.g., City of Denton in a state map) or polygon (e.g., City of Denton boundary map).

VECTOR ANALYSIS

As an example of points layer, Figure 10.16 shows 100 points with coordinates that were generated at random within the extent of the NDVI layer. Using an identifier tool, a click on each point will

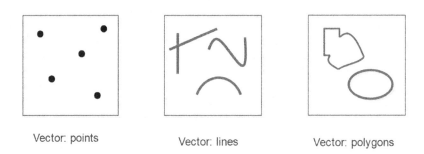

Vector: points   Vector: lines   Vector: polygons

FIGURE 10.15 Examples of vector layers with different geometric types.

FIGURE 10.16 Example of points layer with points coordinates generated at random within the extent of the NDVI layer.

display the coordinates. We can conduct calculations on a point layer that lead to global numbers for the entire layer, or for parts of a layer selected by extent, or by a polygon, or by random sample of all the points. Another set of interesting operations pertains to generate one type of vector layer from another, or an interplay of raster and vector layers.

For example, point coordinates can be used to extract the values of a raster layer into a new points layer that has the raster layer values, for example using the 100 points layer of Figure 10.16, one could extract the NDVI values for those points from the NDVI raster layer (Figure 10.17). The resulting vector layer now has, in addition to the coordinates, the NDVI values for the points.

One common operation for point layers is the calculation of distance between points, which in GIS work is typically the Euclidian distance in 2D

$$d_{pq} = \sqrt{(p_1 - q_1)^2 + (p_2 - q_2)^2} \tag{10.11}$$

where (p_1, p_2) and (q_1, q_2) are the coordinates of points p and q, respectively. Once distance is calculated for all pairs of points, the result is a square symmetrical matrix of dimension $n \times n$ for a layer with n points. A distance layer can be created as a point layer itself where the attributes of each point contain the distance to all the other points.

FIGURE 10.17   Points shown together with raster layer.

Based on distance calculations, one could conduct a *nearest neighbor analysis*, which could lead to results of whether the points are clustered or distributed. This type of calculation could be useful, for example the patterns of air quality monitoring stations in each area or the distribution of WSN nodes. For instance, using the points of Figure 10.16 analysis of nearest neighbor yields an expected mean distance of 1723 and observed mean distance of 1811, meaning that the points are not clustered. We will discuss nearest neighbor analysis in Chapter 14 in the context of animal distributions.

As mentioned above, an interesting set of operations pertains to generate one type of vector layer from another type. For instance, using Voronoi polygons we can generate a polygon layer from a points layer (Figure 10.18).

When making an overlay of various layers, we can analyze the interaction between the attributes of the layers, e.g., we can operate on two layers to obtain a third, or on three layers to generate new information. Consider the cartoon example in Figure 10.19 that illustrates an overlay of raster land cover layer, with three vector layers, wells, streams, and a lake. We could use this to understand the effect of land cover on water quality measured at wells, the lake, and rivers upstream and downstream from the lake.

One commonly needed operation is to count for each polygon of layer *x*, the points of another layer *y* contained in that polygon. A new polygon layer *z* has the original polygons plus a new attribute giving points count for each polygon. For example, we may want to count the number of

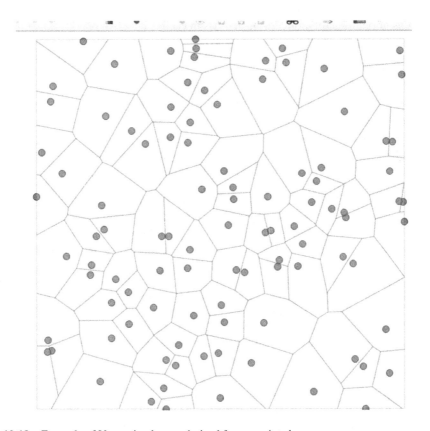

FIGURE 10.18 Example of Voronoi polygons derived from a points layer.

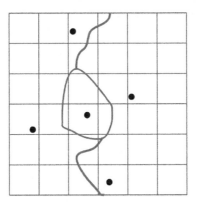

FIGURE 10.19 Example of an overlay of raster land cover layer, with three vector layers wells, streams, and a lake.

sampling wells or of air quality monitoring stations in each county. Another common operation on vector layers is to sum the length of lines in a polygon, or the number of intersections of lines with polygon boundaries. In this case, a new polygon layer would have new attributes with the sum of lengths of lines or the number of intersections. This would be useful, for example to calculate miles of road in each county or number of roads that cross the county lines.

BACKUP

Digital information is perishable and consequently it is important to back up the data that are created with a lot of effort and investment. The easiest example consists of files that are trivially portable between computers (Gustavus 2006). Copying every bit of data accumulated on the computer in perpetuity, although ideal, cannot reasonably be expected in practice. Implementable backup strategies are always compromises. Some guidelines include determining which files to backup and assessing potential risks and vulnerability. Backups can be full, differential, and incremental. When restoring files from backup, we need to consider many details such as types of backups.

There are many security issues associated with backups such as how to provide secure access to remote hosts, network file encryption, how to secure the backup system, and how to secure the backup media. Some free and open-source software tools available are Bacula (Bacula 2022) and BackupPC (BackupPC 2022). Bacula is an enterprise model designed for removable backup media, employs dedicated daemons on both server and client. BackupPC is optimized for ease of installation using hard disk for backup media and employs network sharing technology supported by the operating system (Gustavus 2006).

WEB SERVICES

Once monitoring data are on a DB server, it can be shared via the WWW using web services. These are based on Client-Server and Service-consumer/Service-provider concepts and require protocols and languages. SOAP (Simple Object Access Protocol) specifies Request/Response, WSDL (Web Service Description Language) describes how to connect and query information, and UDDI (Universal Description Discovery and Integration) specifies service registration and discovery (Figure 10.20).

The term *sensor web* is used to denote a system of sensors that uses the web intensively to provide data. The OGC SWE framework defines web service interfaces and protocols allowing for high heterogeneity of sensors and data (OGC 2022a). Sensor location is usually a key piece

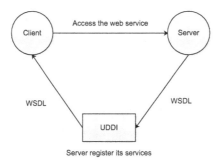

FIGURE 10.20 Web services. From Huang et al. (2008).

of sensor data information, and SWE standards make it easy to integrate this information into thousands of geospatial applications that implement the OGC's other standards (OGC 2022b). There are currently some sensor webs in the environmental monitoring field, some in a particular area such as NASA Volcano Sensorweb (NASA 2022), and others more broader scope such as OpenSensorWeb (2022).

METADATA, STANDARDS, INTEROPERABILITY, PRESERVATION

Metadata are data about data, or in other words data or information that describe other data, thus, allowing for easier data discovery and retrieval. A good example is a library catalog that provides information such as subject and author, and location of the publication, thus allows finding a publication as an item on a shelf or its digital version. Besides facilitation of searching for data, metadata can be used to analyze a network of monitoring stations, for example by analyzing spatial coverage and frequency of various environmental variables and avoiding temporal and spatial biases (Schröder et al. 2006; Desaules 2012).

One important application of metadata standards is providing for *interoperability* of monitoring networks, i.e., to operate together by exchanging information. One example of standards is *Dublin Core* that is a core metadata framework for simple and generic resource descriptions (DCMI 2022). There are also metadata standards specific to disciplines; for example, relevant to environmental monitoring, EML already mentioned above applies to the ecological discipline, Darwin Core to geographical occurrence of species.

As database management acquires more and more relevance in environmental monitoring, the field of *ecoinformatics* has emerged to encompass among other themes, many aspects of ecological and environmental data management. Similarly, the field of *hydroinformatics* has emerged playing the same role is data related to water resources.

Environmental monitoring data have a long-term value and therefore we need to understand principles of data preservation. A digital *repository* is a useful tool in managing data for the long term and supports interoperability (Moen 2008). The Open Archival Information System reference model provides a conceptual design for an archive, including its primary components and their associated functions and relationships (CCSDS 2012). Retrieving information from a repository is facilitated by metadata, thus the need for a protocol to request metadata from a repository. An example is the Open Archives Initiative Protocol for Metadata Harvesting (OAI 2015), used to gather metadata from separate repositories for discovery (Moen 2008). Open-source software for repositories includes DSpace and Fedora.

EXAMPLE: DATA COLLECTED FROM DISTRIBUTED SENSOR SYSTEMS

As an example, of moving from sensors to DB, we present the TEO's station for distributed soil moisture monitoring that was located in a hardwood bottomland forest in north central Texas. As explained in Chapter 6, this system includes a WSN, an SBC acting as remote field gateway (RFG), and a datalogger-based wired sensor system. The SBC itself provides a DB server, and therefore, sensor data can be retrieved through a uniform interface and securely managed in the field, even in the event of network failures between the RFG remote station and the central data collection (CDC) server located on campus. After acquiring all data, the RFG server notifies the CDC server that new data are ready for retrieval. The CDC server then synchronizes its database to the RFG database and hides the heterogeneity of different physical layer devices. Finally, the CDC server passes data to a PostgreSQL DB server that will provide data to be served on the web for the public. Daemon programs running on the CDC server pre-process the data before insertion into the DB and periodically perform synchronization tasks (Yang et al. 2010).

EXERCISES

Exercise 10.1

At Location EESAT Building in the UNT Main Campus, we have two atmospheric data stations: one measures UV and the other measures ozone by two methods: CAS and DOAS. The UV station reports UV every 30 minutes, the ozone sensor reports CAS and DOAS ozone every day. Define tables and specify keys. Draw an ERD of relations between the tables.

Exercise 10.2

The City of Denton operated eight rain gage stations, and in one of them (Ponder) simultaneous stream water level is measured for street closure upon impending flash flood. Every ten minutes, rain gage reports look like this

```
RG    LOCATION            RAINFALL     RATE
 1    Pecan Creek WRC     .00 INCHES   .00 IN/HR
 2    Doyle & Freeman     .00 INCHES   .00 IN/HR
 3    Ponder              .26 INCHES   .00 IN/HR
 4    Masch Branch        .00 INCHES   .00 IN/HR
 5    South Fork          .00 INCHES   .00 IN/HR
 6    Wimbleton           .00 INCHES   .00 IN/HR
 7    Grissom             .00 INCHES   .00 IN/HR
 8    Hobson              .00 INCHES   .00 IN/HR
```

and gate closure reports from station Ponder are like this

```
RG    LOCATION            Gate
 1    Ponder      .UP
```

where the gate can be UP or DOWN. Define tables and keys. Draw an ERD.

Exercise 10.3

Write SQL code to create a table for solar radiation data collected by several sensors that include a timestamp and is recorded as floating-point numbers, with integrity constraint such that only non-negative values are entered in the DB. We want to be able to relate the data to the location where it was measured, based on an existing table that contains sensor location and has PK SensorID.

Exercise 10.4

Write an example of XML code to specify data from nodes of a WSN with tags that would correspond to node, location name, variable measured, distance to gateway, and RSS.

Exercise 10.5

Consider a GIS raster calculation of slope angle and aspect that yields 10.2° for angle and 20.3° for atan2 calculation at a target cell. Assume we want to create two raster layers with degrees as integer values, one for angle and one for aspect as compass direction. What would be values to write for the angle and aspect layers?

Exercise 10.6

Consider a GIS vector layer for 30 water quality monitoring wells specifying location by UTM eastings, and northings, and average EC for year 2005. How would you calculate distance between wells? What units would you use for these distances?

REFERENCES

Acevedo, M.F. 2024. *Real-Time Environmental Monitoring: Sensors and Systems, Second Edition – Lab Manual.* Boca Raton, FL: CRC Press, Taylor & Francis Group. 463 pp.

BackupPC. 2022. *BackupPC.* accessed December 2022. https://backuppc.github.io/backuppc/.

Bacula. 2022. *Bacula.* accessed December 2022. https://www.bacula.org/.

CCSDS. 2012. *Reference Model for an Open Archival Information System (OAIS).* accessed March 2015. http://public.ccsds.org/publications/archive/650x0m2.pdf.

Clark Labs. 2022. *TerrSet 2020 Geospatial Monitoring and Modeling Software.* accessed December 2022. https://clarklabs.org/terrset/.

Database Star. 2022. *A Guide to the Entity Relationship Diagram (ERD).* accessed December 2022. https://www.databasestar.com/entity-relationship-diagram/.

DCMI. 2022. *Dublin Core Metadata Initiative.* accessed December 2022. http://dublincore.org/.

Desaules, A. 2012. The Role of Metadata and Strategies to Detect and Control Temporal Data Bias in Environmental Monitoring of Soil Contamination. *Environmental Monitoring and Assessment* 184 (11):7023–7039.

ESRI. 2022. *ArcGIS.* accessed December 2022. https://www.esri.com/en-us/arcgis/about-arcgis/overview.

European Commission. 2022. *Basic Concepts of XML and GML.* accessed December 2022. https://inspire.ec.europa.eu/training/basic-concepts-xml-and-gml.

Google. 2022. *Keyhole Markup Language.* accessed December 2022. https://developers.google.com/kml/documentation/kml_tut.

GRASS GIS. 2022. *GRASS GIS.* accessed December 2022. https://grass.osgeo.org/.

Gustavus, D. 2006. *Writing in water.* UNT, accessed 2014. http://www.unt.edu/benchmarks/archives/2007/october07/backup.htm.

Horn, B. 1981. Hill Shading and the Reflectance Map. *Proceedings of the IEEE* 69 (1):14–47.

Huang, Y.. 2008. *Converting to XML.* accessed March 2015. http://www.teo.unt.edu/ci-team/tutorial.html.

Huang, Y., C. Zhang, and H. Cuellar. 2008. *CI-TEAM Mini-Course: Environmental Monitoring, Database and Beyond.* UNT, accessed March 2015. http://www.teo.unt.edu/ci-team/tutorial.html.

KNB. 2022. *Knowledge Network for Biocomplexity (KNB).* accessed December 2022. https://knb.ecoinformatics.org/.

Lucidchart. 2022. *Entity-Relationship Diagram Symbols and Notation.* accessed December 2022. https://www.lucidchart.com/pages/ER-diagram-symbols-and-meaning.

Moen, W. 2008. *Digital Repositories, Data Stewardship, and Preservation.* accessed March 2015. http://www.teo.unt.edu/ci-team/tutorial.html.

NASA. 2022. *Volcano Sensorweb.* accessed December 2022. https://ai.jpl.nasa.gov/public/projects/sensorweb/.

OAI. 2015. *The Open Archives Initiative Protocol for Metadata Harvesting.* accessed March 2015. http://www.openarchives.org/OAI/openarchivesprotocol.html.

OGC. 2022a. *Sensor Web Enablement DWG.* accessed December 2022. http://www.opengeospatial.org/projects/groups/sensorwebdwg.

OGC. 2022b. *Why is the OGC involved in Sensor Webs?* accessed December 2022. http://www.opengeospatial.org/domain/swe.

OpenSensorWeb. 2022. *Environmental Data at Your Fingertips.* accessed December 2022. https://www.opensensorweb.de/.

QGIS. 2022. *QGIS A Free and Open Source Geographic Information System.* accessed December 2022. https://www.qgis.org/en/site/.

Schröder, W., R. Pesch, and G. Schmidt. 2006. Identifying and Closing Gaps in Environmental Monitoring by Means of Metadata, Ecological Regionalization and Geostatistics Using the UNESCO Biosphere Reserve Rhoen (Germany) as an Example. *Environmental Monitoring and Assessment* 114 (1–3):461–488.

SensorML. 2022. *Examples, SensorML 2.0.* accessed December 2022. http://www.sensorml.com/sensorML-2.0/examples/index.html.

Silberschatz, A., H. F. Korth, and S. Sudarshan. 2010. *Database System Concepts.* New York, NY: McGraw Hill Higher Education.

Visual Paradigm. 2022. *What is Entity Relationship Diagram (ERD)?* accessed December 2022. https://www.visual-paradigm.com/guide/data-modeling/what-is-entity-relationship-diagram/;WWWSESSIONID=7FA448A91A0DA266076AEEADD9B9A4F1.www1.

Yang, J, C. Zhang, X. Li, Y. Huang, S. Fu, and M. Acevedo. 2010. Integration of Wireless Sensor Networks in Environmental Monitoring Cyber Infrastructure. *Wireless Networks* 16 (4):1091–1108.

11 Atmospheric Monitoring

INTRODUCTION

This chapter is the first of several where we present applications of monitoring to areas of environmental systems, such as atmospheric processes, hydrology and water quality, terrestrial ecosystems, and wildlife monitoring. In this chapter, we focus on atmospheric processes; this includes concentration of various atmospheric gases, as well as concentration of particles and aerosols. To supplement learning statistical methodologies, this chapter also introduces nonlinear regression to model the dynamics of increasing atmospheric carbon dioxide (CO_2) concentration and global temperature. We emphasize the spectral characteristics of incoming solar radiation particularly UV as it relates to total column concentration of important atmospheric gases such as ozone. In terms of air quality, this chapter provides an overview of standards, monitoring stations, and devices to measure important gases. We cover in detail the process of measuring total column of atmospheric gases from the ground and more briefly from space. We end the chapter with a discussion of weather, including common weather instruments, weather radar, and weather satellites.

EARTH'S ATMOSPHERE

The Earth's atmosphere is a gaseous envelope, held by gravitation, extending for ~10,000 km from the surface of the planet, but denser near the surface of the Earth. In addition to *gases*, it contains small particles in suspension termed *aerosols*.

COMPOSITION AND VERTICAL STRUCTURE

Almost all, ~99%, of the gases are nitrogen (~78%) and oxygen (~21%) with concentrations that are relatively constant. The rest includes *water vapor* and *trace gases* such as methane CH_4, carbon dioxide CO_2, and ozone O_3; these are present in small quantities and are more variable. Trace gases, however, can have important effects; for example, CO_2 and CH_4 affect weather and climate, and O_3 affects human health since it serves as filter for ultraviolet (UV) radiation as well as air pollutant. Particles come from volcanic eruptions, salt spray, fires, and dust storms; they serve as condensation nuclei for clouds and can absorb or reflect sunlight. Water vapor stays mostly near the planet's surface, it is spatially variable, and can form clouds upon condensation.

Temperature changes with increasing altitude with varying rates and signs depending on vertical layers, which alternate between warm and cold layers. At the lowest elevation from the planet surface, temperature decreases with altitude in a layer called the *troposphere*; its depth is spatially and seasonally variable, being deepest in tropical areas and higher in the summer season. This negative temperature gradient extends until the top of the troposphere named the *tropopause*, at which point it reverses to increasing temperature with altitude in the *stratosphere* layer (Figure 11.1). This temperature gradient continues positive until the top of the stratosphere called the *stratopause*.

Atmospheric composition changes according to this vertical structure; for example, water vapor is present mostly in the troposphere, CO_2 is well mixed, while ozone is more abundant in the stratosphere, at altitudes that varies from 10 km to 50 km, acting as a filter of UV. There is also a ~2% variation in an 11-year period due to the solar cycle that has been monitored as sunspots since the 1700s. *Total column* atmospheric concentration of a gas refers to the integrated concentration over the entire atmosphere above a position on Earth's surface. Profiles of atmospheric concentration of various gases are measured by sondes launched from the ground and that also measure meteorological variables at increasing altitude (km) or decreasing air pressure (hPa) intervals. Figure 11.2

DOI: 10.1201/9781003425496-11

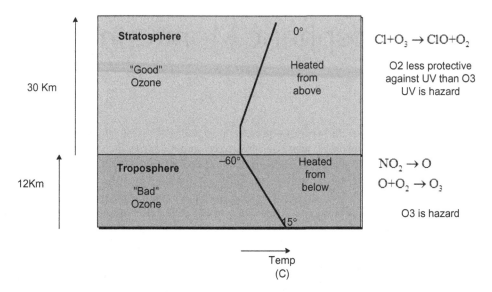

FIGURE 11.1 Atmosphere layers nearest to the surface: troposphere and stratosphere. Good and bad ozone: processes in upper and lower atmosphere.

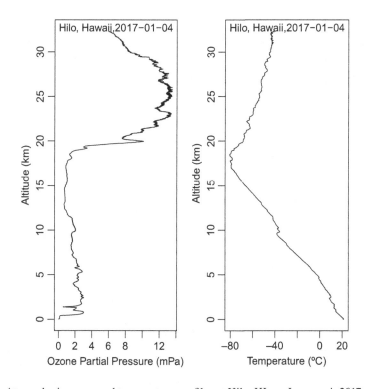

FIGURE 11.2 Atmospheric ozone and temperature profiles at Hilo, HI, on January 4, 2017.

illustrates ozone and temperature profiles using data from WOUDC (2023) obtained by a sonde launched at Hilo, Hawaii, on January 4, 2017. The integrated ozone concentration to the lowest atmospheric pressure (8.09 hPa) reached by the sonde is 205.47 DU or Dobson units. DU are defined in terms of what the thickness of a layer of pure ozone would be at the ground if the total column of ozone was reduced to this layer at standard conditions of temperature and pressure; 100 DU correspond to an ozone layer of 1 mm at 1 atm and 0°C (Graedel and Crutzen 1993).

Depletion of stratospheric ozone, which was caused by CFCs (chlorofluorocarbons), allowed more UV to reach the Earth's surface which became an important health problem since excess UV produces skin cancer, cataracts, and other issues. In opposite manner, an increase of ozone in the lower troposphere or ground level leads to a local air pollution problem since excess ozone in the air causes respiratory health issues (Figure 11.1).

Ozone's high abundance in the stratosphere is explained by photochemical principles. Oxygen is photo-dissociated into atomic oxygen O by UV, then collision of O and molecular oxygen O_2 produces ozone O_3. There is natural variability of ozone in the stratosphere consisting of day-to-day variation, related to the movement of surface pressure systems, latitudinal and seasonal variation, related to a meridional circulation. Nitrogen dioxide NO_2 is a highly reactive gas and has high concentration near the ground in the troposphere and in the stratosphere, playing an important role in the production of tropospheric ozone being the main sink of stratospheric ozone (Crutzen 1970). Natural variability of NO_2 shows a diurnal variation, related to several photochemical processes; its abundance is greater at sunset than at sunrise. In addition, NO_2 has a latitudinal and seasonal variation.

DIRECT AND DIFFUSE SOLAR RADIATION

As we discussed in Chapter 7, the extraterrestrial solar radiation I_0 or solar radiation received by Earth outside the atmosphere, given as power density or power per unit area, varies with the day of the year and its average is ~1.377 kW/m². As the solar radiation flux goes through the atmosphere, a good part of it is absorbed by atmospheric gases and scattered by particles. Direct radiation reaching the surface of the Earth can be as high as ~70% of I_0, or ~1 kW/m². There are two major components of the radiation reaching the Earth's surface: direct and diffuse radiation, and can be modeled by the Bougher-Lambert-Beer exponential attenuation law

$$I = I_0 \exp(-\tau \times m^a) \tag{11.1}$$

where I can be applied to the direct beam (normal) I_{bn} or to the diffuse I_d horizontal portion of clear-sky radiation reaching the Earth's surface, m is the *air mass*, τ is an atmosphere pseudo-optical *depth*, and a is a coefficient. The optical depth can be applied to either direct or diffuse components τ_b, τ_d, these values are location-specific and vary through the year (Gueymard and Thevenard 2013). Likewise, the power coefficient a also applies to either direct or diffuse a_b and a_d, relating to the optical depths by empirically derived equations (Acevedo 2018).

GREENHOUSE EFFECT

One of the trace gases, carbon dioxide CO_2 is important in photosynthesis, it is an emission product of fossil fuel combustion and other compounds containing carbon. Atmospheric CO_2 absorbs infrared (IR) radiation thus preventing heat to escape to space, which is the essence of the *greenhouse* (GH) effect. A good absorber at a certain wavelength is also a good emitter at that wavelength, therefore heat absorbed by the GH gases (IR absorbers) is emitted back as heat and can be re-radiated back to Earth's surface, warms the surface, producing more heat release from the surface and this leads to warming.

Methane, CH_4, is less abundant than CO_2, but it can absorb and emit long-wave radiation ~30 times more effectively than CO_2 and therefore is a much more powerful GH gas and has potentially greater impact on warming (US EPA 2017). Methane is released to the atmosphere by a variety of natural processes occurring on land (e.g., termites), in the oceans (e.g., microorganisms in the seafloor), and in inundated ecosystems (e.g., decomposition in wetlands). These emissions are mostly offset by natural uptake processes. However, CH_4 atmospheric emissions due to human activities have increased. These correspond, for instance, to cultivating rice under inundated conditions and decay of solid waste in municipal landfills.

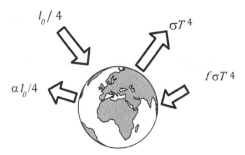

FIGURE 11.3   A very simple model of energy balance for Earth.

It is important to realize that the GH effect is a natural process that occurs in other planets as well. Mars has very little and thus its average temperature is low, but Venus has too much and therefore its average temperature is high. On Earth, human activities enhance the GH effect; these activities are primarily emission of CO_2 from fossil fuel burning and deforestation, as well as methane from landfills and agricultural activities.

As we explained in Chapters 7 and 8, *incoming* solar radiation reaching Earth is distributed by wavelength, increasing for short wavelengths from UV to visible, reaching a peak at in the visible range, and then decreasing as wavelength increases. As Earth's surface warms, it emits *outgoing* radiation in longer IR waves. Earth's average temperature results as a balance of incoming and outgoing radiation. A fraction of the incoming short-wave radiation is reflected by surfaces like clouds, snow, and particles. The coefficient representing this fraction is termed *albedo*. A fraction of the outgoing long-wave radiation is re-radiated to Earth due to the GH effect. Thus, we can reduce warming by lowering of GH atmospheric concentrations by energy conservation, using non-carbon sources of energy, reforestation, and enhancing carbon sequestration in the soil (Acevedo 2018).

An extremely simple model of energy balance for planet Earth (Figure 11.3) is given by equating incoming and outgoing radiation (Graedel and Crutzen 1993)

$$\frac{I_0}{4}(1-\alpha) = \sigma T^4(1-f) \qquad (11.2)$$

where I_0 was defined in the previous section, the coefficient α is albedo or reflectivity (incoming reflected loss), f is factor due to GH (fraction of outgoing radiation trapped by GH effect), σ is the Stefan-Boltzmann constant 5.6704×10^{-8} Wm^{-2}K^{-4}, and T is the temperature of the Earth in K. Outgoing radiation is calculated here as σT^4 using blackbody radiation.

For example, assume albedo is $\alpha=0.28$ and $f=0.39$. What would be the Earth's temperature at equilibrium in °C? We can answer this question by solving for T in (11.2)

$$T = \left(\frac{\frac{I_0}{4}(1-\alpha)}{\sigma(1-f)} \right)^{1/4} = \left(\frac{\frac{1380}{4}\frac{W}{m^2}(1-0.28)}{5.67\times10^{-8}\frac{W}{m^2}K\times(1-0.39)} \right)^{1/4} = \left(\frac{248.4}{3.45\times10^{-8}} \right)^{1/4} K = 291.3 \text{ K}$$

and converting K to °C yielding $T = 291.3 - 273 = 18.3\,°C$. This is an extremely simple calculation and cannot be used to predict changes; also, it is not intended to prove or disprove the GH effect on planetary temperature. In the next section, we will examine what are the trends in CO_2 and global temperature from existing data.

INCREASING ATMOSPHERIC CO_2 CONCENTRATION

An important piece of our knowledge of planetary carbon dynamics comes from the measurement of atmospheric CO_2 concentrations recorded in Mauna Loa, Hawaii (Vaughan et al. 2001;

Lovett et al. 2007). A visit to the web site of the National Oceanic and Atmospheric Administration (NOAA) Global Monitoring Division (NOAA 2020) will inform us of recent values of monthly average of CO_2 concentration in parts per million (ppm). For example, in July 2020, it was 414.38, ~3 ppm up from 411.74 ppm for the same month of the previous year (July 2019).

Concentration in ppm expresses dry air mole fraction defined as the number of molecules of CO_2 divided by the number of all molecules in air, including CO_2 itself, after water vapor has been removed (NOAA 2020). The July 2020 value of 414.38 ppm represents a mole fraction of 0.000414. On the web site, we can see a graph of CO_2 in ppm as monthly average and its trend (seasonal correction) for the last five years of record. The trend is calculated by a moving average of seven (an odd number) adjacent seasonal cycles centered on the month to be corrected (NOAA 2020). The trend changes from 395 to 406 ppm in 5 years, which is an average increase of ~2 ppm/year.

Besides the graph, the web site offers the data for download. Figure 11.4 illustrates the CO_2 trajectory for the measurement record (since March 1958) using the data downloaded from this web site. From the dataset, we plot the same two lines shown at the web site. The dashed line represents the monthly average values (centered on the middle of each month) and which fluctuates up and down during the year according to the seasons. Removing the average of this seasonal cycle yields the solid line that shows a clear accelerating increase during the entire record. Figure 11.5 uses a time window of the most recent ten years (2007–2017) so that we can visualize the graphs more clearly.

An interesting feature of Figure 11.5 is a nonlinear trend, e.g., it displays an increase of the rate of change over time. A first thought may be that the rate is itself proportional to the concentration $X(t)$, so that as concentration increases, so does the rate. This is modeled as a linear ordinary differential equation. Using the derivative of X with respect to time t for the rate of change of X, we write $\dfrac{dX(t)}{dt} = kX(t)$ where the coefficient k is a *per unit* rate of change or *rate coefficient*. The solution is well known, and it is exponential function $X(t) = X(0)\exp(kt)$ that can be calculated once we know the initial condition $X(0)$. This is a commonly used model for many processes, and we will refer to it simply as the *exponential model*. We have seen an instance of this model in Equation (11.1) when referring to attenuation of solar radiation by the atmosphere. Divide the exponential by the initial condition to obtain $X(t)/X(0) = \exp(kt)$, take natural logarithm of both sides, $\ln(X(t)/X(0)) = kt$ which says that the log of the ratio of current values to the initial value is proportional to the time interval. In other words, the log of the ratio should plot as a straight line with respect to time, with

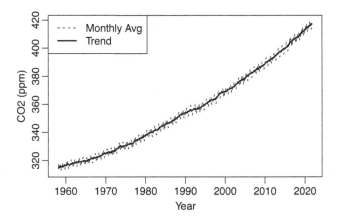

FIGURE 11.4 Monthly mean CO_2 at Mauna Loa – record from 1958 to 2019. Data downloaded from NOAA (NOAA 2020).

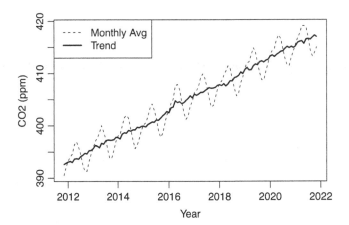

FIGURE 11.5 Monthly mean CO_2 at Mauna Loa, zooming in the period 2010–2020. Data downloaded from NOAA (NOAA 2020).

slope equal to the coefficient. The latter can be calculated solving for k to obtain $k = \dfrac{\ln(X(t)/X(0))}{t}$.

A derived concept from an exponential model is the doubling time t_d. By making $X(t_d) = 2X(0)$ we can write $k = \dfrac{\ln(2)}{t_d} = \dfrac{0.693}{t_d}$.

The exponential model applies to several other environmental variables since they have dynamics displaying a changing rate of change of the variable Y with respect to other independent variable X; this can be represented by an exponential model $Y = Y_0 \exp(-kX)$ which can be calculated once we have the condition Y_0 and the rate coefficient k which is a per unit rate of change. For example, this is a good explanatory model for light attenuation through the atmosphere at certain wavelengths as we will see in the upcoming section on optical absorption spectroscopy (OAS) and differential OAS (DOAS). In addition, it can be used to model light attenuation through a water column (Beer–Lambert law, Chapter 12) and vegetation canopy (Chapter 13). For water, k is extinction coefficient and Y_0 is the downwelling light just below the surface which is the measurement reported at $X = 0$.

The process of parameter estimation, referred in some cases as model calibration, consists of finding the values of model parameters. In this chapter, we explain the method using as an example, the parameter k in the exponential model. To estimate the value of k, using n data points: $X(t_i), i = 1,\cdots,n$, we can perform simple linear regression of $X(t_i)$ vs. t_i yielding k. We covered basics of simple regression in Chapter 1, a simple way of performing regression in this case is to apply linear regression after logarithmic transformation of $X(t_i)$

$$\ln\left(\frac{X(t_i)}{X_0}\right) = kt_i \tag{11.3}$$

We can see that a linear regression of $\ln(X(t_i)/X_0)$ vs. t_i yields a slope that should correspond to k. In this case, we want regression with *zero intercept* ($b_0 = 0$) since the first position of $\ln(X(t_i)/X_0)$ is zero because $\ln(1) = 0$. We apply this method in Lab 11 of the lab manual (Acevedo 2024). The estimated rate coefficient has a value of 0.399% per year with $R^2 = 0.994$ which is close to 1, and the p-value is 2.2×10^{-16} which is negligible. This value of k translates to a doubling time of $0.693/0.00399 = 173.7 \approx 174$ years. Doubling CO_2 concentration with respect to a reference year is often used as a scenario for climate change modeling.

However, plotting the log of ratios vs. time, we realize that it cannot be approximated to a straight line (Figure 11.6). Therefore, our first guess of a linear rate of change or exponential model is not a good approximation, and the rate may be a nonlinear function of the concentration. This implies

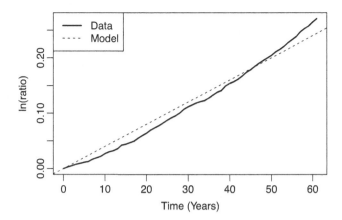

FIGURE 11.6 Determining possible exponential increasing trend of CO_2.

that the prediction of doubling time would be off, and the doubling time could be shorter. The trajectory before $t \sim 46$ years ($\sim$2005) indicates under prediction while thereafter it is overpredicted. We will now explore a more complicated model that could account for these differences.

DOUBLY EXPONENTIAL

We just saw that simple exponential growth $X(t) = X(0)\exp(kt)$ cannot account for data on CO_2 increase. There are various ways of modeling faster increase than the one predicted by an exponential. One of these is by a doubly exponential

$$X(t) = X(0)\exp(\exp(kt) - 1) \tag{11.4}$$

where you take the exponential of an exponential, meaning the rate coefficient is itself increasing exponentially. More flexibility in fitting data is possible by using two rate coefficients

$$X(t) = X(0)\left(\exp(k_1 \exp(k_2 t)) - \exp(k_1) + 1\right) \tag{11.5}$$

where we add the term $-\exp(k_1) + 1$ to force the function $X(t)$ through $X(0)$ for $t=0$.

How do we calculate doubling time in a doubly exponential model? Make $X(t_d) = 2X(0)$ in Equation (11.5) and group terms to obtain $2 + \exp(k_1) - 1 = \exp(k_1 \exp(k_2 t_d))$, now take natural log of both sides $\ln(2 + \exp(k_1) - 1) = k_1 \exp(k_2 t_d)$, divide by k_1 and take logarithm again to get $\ln\left(\dfrac{\ln(2 + \exp(k_1) - 1)}{k_1}\right) = k_2 t_d$, and solve for t_d to get

$$t_d = \ln\frac{\dfrac{\ln(2 + \exp(k_1) - 1)}{k_1}}{k_2} \tag{11.6}$$

We use this opportunity to introduce *nonlinear regression*, which consists of a numerical procedure to minimize the error of the fit of a function to the data.

NONLINEAR REGRESSION

In many cases, we need to estimate parameters of a nonlinear equation relating Y to X of the general form

$$Y = f(X, p) \tag{11.7}$$

where $f(\ldots)$ is a function and p is a vector of parameters. For example, the exponential model with parameters k and Y_0 is nonlinear function. As we learn in the previous section, sometimes we can transform Equation (11.7) into a linear regression problem. However, this is not always possible and then we should apply the process of nonlinear regression. This consists of postulating the function that may fit the data, e.g., an exponential curve, and then use an optimization algorithm to minimize the square error with respect to the coefficients. In other words, find the value of the coefficients that would yield a minimum square error.

The *error* (*residual*) for data point i is

$$e_i = y_i - \widehat{y}_i = y_i - f(x_i, p) \tag{11.8}$$

take the square and sum over all observations to obtain the *total squared error*

$$q = \sum_{i=1}^{n} e_i^2 = \sum_{i=1}^{n} \left(y_i - f(xi, p) \right)^2 \tag{11.9}$$

We want to find the values of the coefficients p that minimize the sum of squared errors (over all $i = 1, \ldots, n$) that is to say, find p such that

$$\min_p q = \min_p \sum_{i=1}^{n} e_i^2 = \min_p \sum_{i=1}^{n} (y_i - f(x_i, p))^2 \tag{11.10}$$

An optimization algorithm works in the following manner. It reads an initial guess of the values of the coefficients then recursively changes the parameter values in a small amount and moving down gradient (derivative) in the q surface until changes in parameter values no longer produces a decrease in q. Sometimes, we can obtain the initial guess of the parameter values by means of a linear regression performed on some approximation or transform of the nonlinear function.

It is often difficult to determine the functional relationship or model, and sometimes this is possible by knowing or postulating how the system works. For example, if we are trying to find a coefficient of light attenuation in the water column of a lake, we know that light attenuation follows an exponential law, because the rate of decay is linear with depth.

A convenient way of applying nonlinear regression is to use function `nls` of R. We learn how to do this in Lab 11 of the lab manual, where we obtain values $k_1 \approx 0.172$ and $k_2 \approx 0.014$ for the rate coefficients. With these values, we can predict CO_2 concentration and plot for comparison to the data points, illustrating a good fit to the data (Figure 11.7).

Calculating doubling time using Equation (11.6), we get an estimate of ~107 years instead of 177 years as obtained by the simple exponential. Looking back in time, we may guess that 107 years ago (first decade of the last century), the concentration would have been half of current values or about 200 ppm. Looking ahead in time, double CO_2 concentration with respect to 1959 ($\sim 2 \times 315 = 730$ ppm) would be obtained in year 2068. These are predictions based only on the trend and do not consider modifications due to changing emission rates or other climate dynamics conditions.

GLOBAL TEMPERATURE: INCREASING TREND

Now let us visit NASA's Global Climate Change web site (NASA 2020). You can see a graph of the change in global surface temperature since 1880. Global surface temperature refers to average over land and ocean. The record is expressed as an anomaly or difference relative to the 1951–1980 average temperature. It can be visualized as an annual average and as a 5-year average (Figure 11.8). We see a clear increasing trend and positive anomaly after 1980, that is in the last 40 years, raising to 0.99°C above the 1951–1980 average. The 10 warmest years in the examined record have occurred in the last 40 years or since 2000 (except 1998).

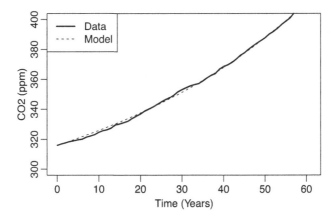

FIGURE 11.7   Modeling CO_2 increase using the doubly exponential.

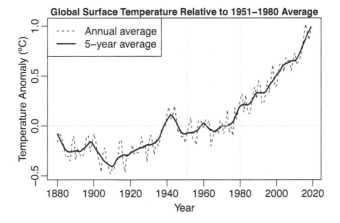

FIGURE 11.8   Global average temperature of Earth during 1880–2020 as anomaly with respect to the 1951–1980 average.

Do the global temperature data show an exponential increase? We can proceed as we did for CO_2 in previous sections. First, we add an arbitrary positive value (+1.0) to the anomaly to make all anomaly values positive. Applying a log transform regression as we did for CO_2, the results indicate an estimate of $k = 0.0026$ per year, with poor R^2 (0.4344) but significant (negligible p-value). This means approximately 0.26% per year rate coefficient, which translates to a doubling time of $0.693/0.0026 \approx 266$ years. We notice that R^2 is 0.4385, which is not very good. Moreover, by plotting this log of ratios, we realize that a straight line is not a good estimate (Figure 11.9). Therefore, the rate may be a nonlinear function of temperature. This implies that the prediction of doubling time would be off, and the doubling time could be shorter. The trajectory after $t \sim 100$ years (~1980) indicates a faster increase. Therefore, let us apply the doubly exponential to the global temperature data, to see if we get a better fit. In the lab 11 guide, using nls of R, we get coefficient values of $k_1 = 0.0068$, $k_2 = 0.0345$. Using these estimates, we can predict temperature and plot together with

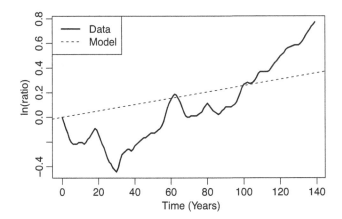

FIGURE 11.9   Determining possible exponential increase of global temperature.

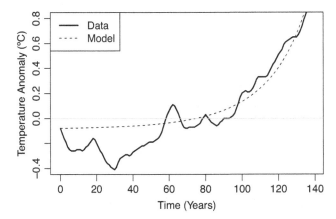

FIGURE 11.10   Modeling temperature increase using the doubly exponential.

the data to get Figure 11.10 which indicates a better fit. The doubling time calculates to ~134 years instead of ~231 years that we obtained when using the simple exponential. To put this in perspective, we would double the anomaly (~+0.9°C) of 2019 entering the next century.

ATMOSPHERE – NEAR-SURFACE AIR QUALITY

Ambient or ground-level atmospheric processes are of paramount importance for several reasons, including that the resultant air quality relates to human health. Two major sets of variables require monitoring: pollutant concentrations and meteorological conditions; the latter affects movement and dynamics of the former. More specialized monitoring relates to the pollutant sources, such as stack emissions, and relates to issues of permitting and controls. Similarly, other specialized monitoring efforts focus on indoor air quality, particularly in schools and industrial environments. Ambient air quality monitoring helps to determine areas and time periods of high levels of pollution, emergency and warning systems, health-based impact assessment, and control and mitigation measures. Results of monitoring efforts inform air quality models and support public education and outreach.

STANDARDS

In the USA, the Clean Air Act (U.S. EPA 2014a, c) is the major legislation on air quality and requires the U.S. EPA to set National Emissions Standards and National Ambient Air Quality Standards (NAAQS) (U.S. EPA 2014b) which establishes *primary* standards to provide protection of public health and *secondary* standards to protect public welfare, including visibility, vegetation, crops, animals, and buildings. The standards apply to six principal pollutants (criteria pollutants): carbon monoxide (CO_2), lead (Pb), nitrogen dioxide (NO_2), ozone, particulate matter (PM, of two different sizes PM_{10} and $PM_{2.5}$), and sulfur dioxide (SO_2). For example, ozone is subject to both primary and secondary standards, by eight-hour averaged values, the annual fourth-highest daily maximum eight-hour concentration, averaged over three years should not exceed 0.075 ppm.

To clarify why PM is classified by size, consider that it includes a large variety of particulates ranging from coarse such as dust to fine such as smoke and haze. Coarse particles (PM_{10}) are found near roadways and dusty industries, range in size from 2.5 to 10 μm in diameter, whereas fine particles ($PM_{2.5}$) have diameters smaller than 2.5 μm; these are either directly emitted from sources (e.g., forest fires) or formed when gases emitted from power plants, industries, and automobiles react in the air. Even smaller, there are ultrafine particles of less than 100 nm in diameter. PM has human health and environmental effects. Health effects include decreased lung function, aggravated asthma, and the development of chronic bronchitis. Environmental effects include visibility reduction and esthetic damage, atmospheric radiation balance, and global climate change. For $PM_{2.5}$, the NAAQS annual standard is that the annual mean, averaged over three years, should not exceed 12 μgm^{-3} as primary standard and 15 μgm^{-3} as secondary standard. The 24-hour standard is that 3-year average of 98th percentile of 24-hour average should not exceed 35.5 μg m^{-3}.

There are emission standards for type of pollutants and specific type of sources. Some pollutants are emitted directly, whereas others are formed by reactions involving directly emitted pollutants. For example, of the six criteria pollutants CO, Pb, NO_2, PM_{10} and $PM_{2.5}$, SO_2 are emitted directly, whereas ozone is formed by precursors.

AIR MONITORING STATIONS

There are four types of air monitoring stations (AMS) used by the U.S. EPA (Matthias 2004): State and Local AMS (SLAMS), National AMS (NAMS), Photochemical AMS (PAMS), and Special Purpose AMS (SPMS).

SLAMS conform a large (~4000) network of stations operated with the purpose of helping the local and state agencies meet the requirements of their implementation plans. A subset of these (numbering ~1000) is designated NAMS to monitor air pollution in areas of high concentration or high population density. PAMS are used in areas that fail to attain ozone standards; there are about 90. Finally, SPMS are non-permanent and used to meet specific needs of implementation plans and other purposes (Matthias 2004).

Monitoring networks have a variety of purposes such as ozone transport, modeling, public information, research, and compliance. An example is the network of stations monitoring ozone levels in Texas (TCEQ 2015). Depending on purpose, there is a variety of measurement frequencies employed, from direct continuous measurements to periodic sampling, coverage from localized analysis (e.g., in situ) to wide area. Typically, it is required to measure low concentration (ppm, ppb, or even ppt) that demands sensitive instruments.

OPTICAL DEVICES

Intensity and spectrum of light transmitted through a medium can be used to determine characteristics of that medium. Therefore, optical devices measuring intensity and spectrum of light are very useful to measure atmospheric gases and PM concentrations. In this section, we study a variety of optical devices used for this purpose.

LINEAR PHOTODIODE ARRAY (PDA) AND CHARGED COUPLED DEVICES (CCD)

A *photodiode* (PD) converts light into current and it is based on the same principle we studied for PV (Chapter 7), but instead of using it as a power generation device is designed to allow for fast response and production of a current that is proportional to light intensity (Pallás-Areny 2000). PDs can be arranged in an array (PDA) such that each diode would respond to light received at its position in the array, providing a way, for example of measuring dispersed light by wavelength. In essence, a PDA becomes a multichannel detector for which each element is a pixel. The number of PDs in the array relates to how many different points can be measured, for example how many wavelengths can be detected, and corresponds to powers of two; for example, 1024, 2048, or 4096.

Linear CCD are also arrays but based on the principle of storing charge in each element or pixel as a light arrives by means of a capacitor; the charge of the element is "read" and considered proportional to light intensity received, then reset or discharged for the next reading. CCDs can detect low intensity of light. An example is the Sony ILX511 that has 2048 pixels.

Dispersive Spectrometers

A *diffraction grating* is an optical device used to separate light into its component wavelengths by means of a series of grooves engraved or etched into its surface. As light reflects from the grating, the grooves cause the light to diffract, dispersing the light into its component wavelengths. As the name suggests, *dispersive spectrometers* are based on dispersing light according to wavelength and detecting these different components separately. Modern spectrometers disperse light using a grating and a multichannel detector such as a CCD or PDA reads the resultant dispersed beam. The light from the source fist goes through a slit and then it is sent to the grating (Figure 11.11). Mirrors can take the light from the slit to the grating or from the grating to the detector. The spectral resolution and wavelength range is determined by the number of lines of the grating, size of the slit, and number of diodes of the detector.

Photomultiplier Tubes

Photomultiplier tubes (PMTs) can detect light of very low intensity by amplifying current and are available for UV, visible, and near-IR ranges of light. To understand PMTs, we need to review the *photoelectric effect* and the *secondary emission* effect. The photoelectric effect consists of electron emission by metals when stroke by light of sufficiently short wavelength such that the photons have enough energy to overcome the *work function* or energy binding the electrons. Recall from Chapter 7 that the energy of a photon is inversely related to wavelength $E = h\nu = hc/\lambda$. Albert Einstein explained the photoelectric effect in 1904, receiving the Nobel prize in 1923 for this discovery. Secondary emission consists of the emission of particles from a material when stroke with particles

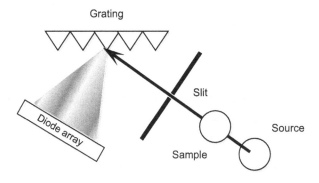

FIGURE 11.11 Dispersive spectrometer: general principle.

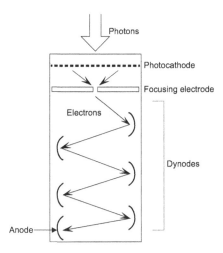

FIGURE 11.12 Photomultiplier.

of sufficient energy. For example, when electrons gain sufficient energy by acceleration by a high voltage can elicit emission of secondary electrons from materials.

These principles are employed in a PMT using a photocathode, a series of intermediate anodes (called dynodes), and an anode operating in a vacuum created inside a glass envelope (Figure 11.12). The photocathode is manufactured to have low work function using a combination of cesium, rubidium, and antimony. Incoming photons strike the photocathode, which by the photoelectric effect emits electrons when the light exceeds the work function. These electrons successively hit the dynodes set at increasing voltage levels (~100 V each stage), and by the secondary emission effect, provoke a cascade of electrons that are collected at the anode where the current is measurable and depends on the incident photons. Common configurations are end-on (transmission mode) where light enters the top of the tube, and side-on (reflection mode) where the light enters at the side of the tube.

Beam Splitter

A beam splitter is an optical device that divides a beam in two parts; one is reflected, and the other is transmitted. A beam splitter has an important application to conduct light in two different paths as required by an interferometer (next section) or when a reference light beam needs to be carried to a measuring device as well as a beam running through a sample (as required in fluorescence as we will see later).

Fourier Transform Interferometer

A Fourier transform *interferometer* is a non-dispersive instrument capable of producing the spectrum of light based on interference patterns. A movable mirror and a beam splitter are combined to generate an *interferogram* that is converted to absorption spectra using inverse Fourier transform techniques. The basics of an interferometer are depicted in Figure 11.13 that shows a stationary mirror and a movable mirror. The difference in length between two paths is the *retardation δ*, which is expressed in cm. One path is from the source to the detector via the stationary mirror and the other path is from the source to the detector via the movable mirror. Retardation varies as the movable mirror changes position and affects the constructive and destructive peaks of the light received at the detector; the resulting signal as a function of retardation is the interferogram, which is the Fourier transform of the spectrum of incoming light. Then we can apply a numerical inverse Fourier

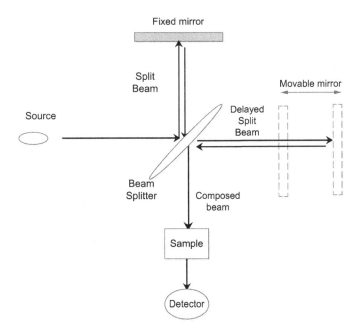

FIGURE 11.13   Interferometer.

transform to convert the interferogram to the spectrum. Instead of wavelength λ, the spectrum is given by its inverse wavenumber $\nu = 1/\lambda$ in cm or reciprocal cm.

FIBER OPTICS

An *optical fiber* is a flexible transparent fiber that can function as a waveguide for light or the equivalent of a light pipe and is made from glass or plastic with a diameter from tens to hundreds of μm. At this point, it is useful to recall the concept of refraction index or *refractive index* as the ratio of the speed of light in a medium to that of a vacuum. The higher the refractive index of a medium, the slower the light wave propagates. When the light travels slower in a medium and bounces on a medium of lower refractive index at steep angles, the light experiences total reflection. This is the principle employed by optical fibers. A fiber consists of a *core* surrounded by a *cladding*, both made from dielectric materials. The refractive index of the core is greater than that of the cladding, therefore the light transmitted along the axis remains contained in the core experiencing total internal reflection.

Optical fibers are well known for their use in communication, but they can also be used to transmit or pipe light from one location to another, where they can be analyzed. We can also use fibers to build sensors since imposing stress, pressure, or temperature changes can modify the intensity, phase, or wavelength of light transmitted in the fiber. Thus, measuring changes in the received light properties can be correlated with the changes of the intended measurand. Later in this chapter, we will use an optical fiber as a means of piping light from one location to a dispersive spectrometer.

MEASUREMENT METHODS USING SAMPLES IN CLOSED PATH

In this section, we cover those optical methods that are easier to implement using spectrometers and PMT, together with electronic signal conditioning and processing, making them amenable to real-time monitoring. For this reason, we do not discuss other methods such as gas chromatography and flame ionization detectors that are used to measure volatile organic compounds (VOCs), inertial and filter methods used for PM, and absorption atomic spectrophotometry used to analyze Pb (Matthias 2004). Neither we discuss the use of chemical sensing electrodes in this chapter. To use these sensors for air samples, the gas to be analyzed is dissolved in a solution at a known pH, which is passed through an ion-selective electrode. The ion concentration (proportional to the concentration of the pollutant) is absorbed and measured electronically. This method is used to measure SO_2, NO, and NO_2. The principles of chemical sensing electrodes are covered in Chapter 3.

OAS

Many optical methods are based on the Beer-Lambert Law stating that a light beam passing through a medium of thickness L (cm) is attenuated exponentially at each wavelength according to

$$I(\lambda) = I_0(\lambda) \exp(-L\sigma(\lambda)c) \tag{11.11}$$

where: λ is the wavelength (nm); $I(\lambda)$ is the light intensity at λ, after passing through the column (W m^{-2}nm^{-1}); $I_0(\lambda)$ is the intensity of incident light at λ emitted by a light source, c is the concentration of a substance (molecules cm^{-3}); and $\sigma(\lambda)$ is the absorption cross-section of the substance (molecules^{-1}cm^2). The medium can be liquid solution or a gas chamber (Figure 11.14). This is the basis of OAS.

Therefore, by knowing L, $I_0(\lambda)$, and $\sigma(\lambda)$, we can determine c after measuring light transmitted $I(\lambda)$ taking logarithm of the ratio and solving for c

$$c = \frac{1}{-L\sigma(\lambda)} \ln \frac{I(\lambda)}{I_0(\lambda)} \tag{11.12}$$

Measurements are conducted at specific wavelengths using a spectrometer. In its simplest form, the process is to shine the light of only one wavelength (*monochromatic*) and measure the transmitted light at that wavelength. For example, ozone can be measured by using wavelengths in the UV range at which ozone is absorbed; CO and CO_2 can be measured using wavelengths in the IR range radiation of specific wavelengths that are absorbed by CO and by CO_2. Spectroscopy is also employed to measure SO_2 using the *wet chemical method*. In this method, SO_2 is mixed in an aqueous solution and the transmitted light is related to SO_2.

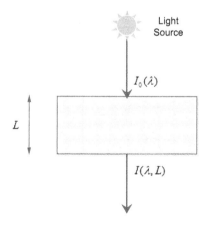

FIGURE 11.14   Beer-Lambert law.

Absorption cross-sections for many trace gases can either be found in published literature or existing databases. The HITRAN2012 molecular spectroscopy database (Rothman et al. 2013) keeps a collection of high-resolution absorption cross-sections of many compounds including atmospheric gases such as ozone, nitrogen dioxide, sulfur dioxide, bromine oxide, and formaldehyde. These datasets can be downloaded as indicated on the web site (HITRAN 2014).

CHEMILUMINESCENT ANALYZER

A chemiluminescent analyzer uses the light emitted when molecules of an excited chemical species fall back to their ground state, converting this light to a voltage by a PMT. In this chapter, we use asterisk to denote the excited state of a molecule; for instance, NO_2^* is the excited form of NO_2. Chemiluminescent analyzers can be used to measure SO_2, NO, NO_2, and O_3. For example, a chemiluminescent analyzer is designed to measure ozone based on the reaction of ozone with ethylene, which produces light that is detected by a PMT (Figure 11.15).

Another example is to measure NO by a reaction with ozone to form excited NO_2^* molecules and release oxygen

$$NO + O_3 \rightarrow NO_2^* + O_2$$

and when NO_2^* molecules return to the ground state, they emit IR radiation

$$NO_2^* \rightarrow NO_2 + h\nu$$

In this method, we can measure NO_2 in two stages (Figure 11.16). Total NO_x is first measured by reducing NO_2 to NO by a heated catalyst and drawing this stream into the sample chamber together with an ozone stream (Matthias 2004). In this stage, the PMT reading represents the total NO_x. Then $NO_2 + NO$ from the air bypass the catalysis and run through the chamber where it is detected by the PMT; in this stage, the PMT reading represents the NO component of the NO_x. Finally, NO_2

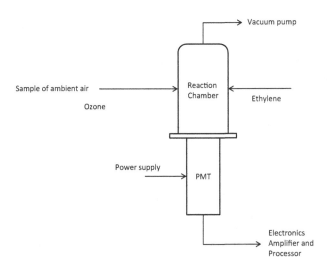

FIGURE 11.15 Chemiluminescent detector measuring ozone.

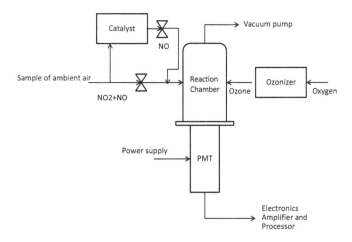

FIGURE 11.16 Chemiluminescent detector measuring NO$_2$.

concentration is estimated as the difference in measurement between the NO (bypassing catalysis) and the NO$_x$ (NO$_2$ reduced by catalyst). This type of instrument can detect a minute concentration (~500 ppt) of NO$_2$. Switching between these two streams uses an automatically controlled valve (Matthias 2004). The reaction is more efficient when the chamber is maintained at vacuum pressures (5–25 mbar).

Microprocessor technology has enabled modern gas analyzers to evolve into sophisticated data collection nodes. Different measurement technologies are incorporated as modules into a complex platform that may include serial communications, user interfaces, and standardized components. Features include automatic self-monitoring, diagnostics, and programmable calibration.

FLUORESCENCE INSTRUMENTS

Fluorescence occurs when a gas absorbs light of a given wavelength and forms excited molecules that release energy when falling back to the ground state. A fluorescence instrument is based on measuring this energy by optical means. For example, SO$_2$ excited by UV radiation forms an excited molecule SO$_2^*$ that releases energy in the IR which is measured by a PMT (Figure 11.17). Fluorescence is different from chemiluminescence because in the latter, the excited state is formed by a chemical reaction, whereas in fluorescence, the excited state is formed by the absorption of light. A beam splitter transmits reference light beam to a detector (e.g., a PD) that is processed together with the reflected light from the splitter that goes through the sample chamber.

NON-DISPERSIVE INFRARED

The non-dispersive infrared (NDIR) method also uses the Beer-Lambert law; however, the incident monochromatic light is at a specific wavelength in the IR range, and transmitted light is detected by a PMT instead of a spectrometer (Figure 11.18). In contrast with a dispersive spectrometer that breakdowns light into many wavelengths, an NDIR instrument is based on selective absorption of that wavelength by the substance of interest; for example, to measure CO a wavelength of 46 μm is used. NDIR can be employed to measure SO$_2$, NO, NO$_2$, VOC, and CO.

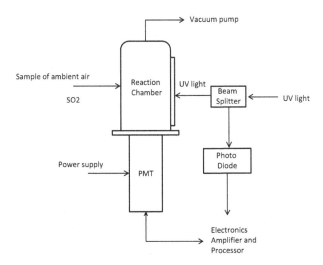

FIGURE 11.17   Fluorescence detector for SO_2 measurement.

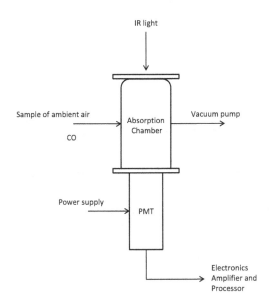

FIGURE 11.18   NDIR to measure CO concentration.

MEASUREMENT METHODS USING OPEN PATH

As an alternative to drawing a sample into a chamber, pollutant concentrations can be measured directly by transmitting light of a specific wavelength across an air column, detecting intensity at the wavelength of interest at the receiving side, and applying the Beer-Lambert law (Figure 11.19). Transmitted light is then processed by a spectrometer and applying Differential Optical Absorption Spectroscopy (DOAS) methods or by an interferometer and applying Fourier Transform IR (FTIR) methods. DOAS employs numerical methods to separate the narrow or fast-changing absorption components from the broad or slowly varying components, while FTIR calculates the spectrum from the interferogram.

This open-path approach is used to measure emitted pollutants in a cross-section of a power plant stack, across a roadway, airport, and urban areas (Matthias 2004). A variation of this method consists of deploying the light source and instrumentation together on one side and reflecting light from a mirror on the other side (Figure 11.20) (Matthias 2004). An important aspect is the influence of dust, particles, and other gases, which may affect the measurement.

TOTAL COLUMN ESTIMATION FROM THE GROUND

In this section, we explain how to accomplish total column estimation from the ground using OAS. The classical OAS principle can be applied to total column measurements by assuming incident light I_0 to be the extraterrestrial radiation or incoming light at the top of the atmosphere and absorbed light I to be the light received at the ground measured by a detector (Figure 11.21). Extending the Beer-Lambert law of Equation (11.11) to discount scattered light and considering a slanted sun path, we can write

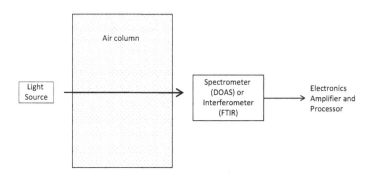

FIGURE 11.19 Open-path OAS or FTIR. Bistatic mode. Requires power at the light source end but the light path is single.

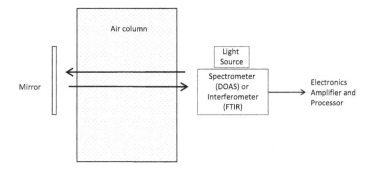

FIGURE 11.20 Open-path OAS or FTIR. Monostatic mode. Does not require power at the light source end but the light path is twice as long.

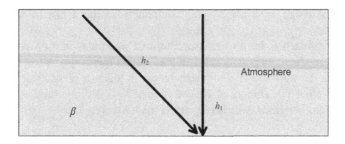

FIGURE 11.21 Principle of measuring total column by OAS.

$$I(\lambda) = I_0(\lambda)\exp\left(-L_s\left[\sigma(\lambda)c + \varepsilon_R(\lambda) + \varepsilon_M(\lambda)\right]\right) \tag{11.13}$$

where the slanted column is of length L_s, the Rayleigh scattering coefficient $\varepsilon_R(\lambda)$ accounts for light reflected by particles of diameter less than the wavelength of the incident light, and Mie scattering coefficient $\varepsilon_M(\lambda)$ accounts for light reflected by particles of a diameter similar to the wavelength of incident light (Moran 2006). These scattering processes discount the amount of light not reaching the optical detector (Platt and Stutz 2008).

Using the ratio of light intensity measured at two different wavelengths λ_1 and λ_2 and if $\varepsilon_M(\lambda_1) = \varepsilon_M(\lambda_2)$, we can write

$$\frac{I(\lambda_1)}{I(\lambda_2)} = \frac{I_0(\lambda_1)}{I_0(\lambda_2)}\exp\left(-L_s\left[\left(\sigma(\lambda_1)-\sigma(\lambda_2)\right)c + \varepsilon_R(\lambda_1) - \varepsilon_R(\lambda_2)\right]\right) \tag{11.14}$$

For example, to measure the total column of ozone, we can use cross-section at two of the wavelengths $\lambda = 305\,\text{nm}$, $\lambda = 311\,\text{nm}$, and $\lambda = 316\,\text{nm}$ at which ozone absorbs light differentially.

Equation (11.14) can be solved for c to obtain

$$c = \frac{\ln\dfrac{I_0(\lambda_1)}{I_0(\lambda_2)} - \ln\dfrac{I(\lambda_1)}{I(\lambda_2)} - \left(\varepsilon_R(\lambda_1) - \varepsilon_R(\lambda_2)\right)L_s}{\left(\sigma(\lambda_1) - \sigma(\lambda_2)\right)L_s} \tag{11.15}$$

This means that we can determine concentration if we measure I at both wavelengths calculate $\ln\dfrac{I(\lambda_1)}{I(\lambda_2)}$ and estimate the remainder quantities, which we tackle next one by one.

To determine the slanted length L_z, we need air mass that is described in Chapter 7; for easy reference, Figure 11.22 shows how the slanted distance h_2 and vertical distance h_1 relate as a function of sun elevation angle β, namely $h_2 = h_1/\sin\beta$ by simple trigonometry. In addition, correcting by pressure, since density diminishes with altitude, we obtain the mass ratio as $m = \dfrac{1}{\sin(\beta)}\dfrac{P}{P_r}$ where P is the measured atmospheric pressure at the station and P_r is the standard atmospheric pressure value (1 atm). When the sun is overhead $\beta = 90°$ and $\sin(\beta) = 1$, at standard pressure conditions, i.e., $P = P_r$, we have $m = 1$ or the air mass ratio is 1 (referred to AM1). Note that at the top of the atmosphere $P/P_r = 0$ and so $m = 0$ (referred to AM0) regardless of sun elevation angle. Therefore, we conclude that the slanted path length is $L_s = L_z m$ given the length of the vertical column above the observing station.

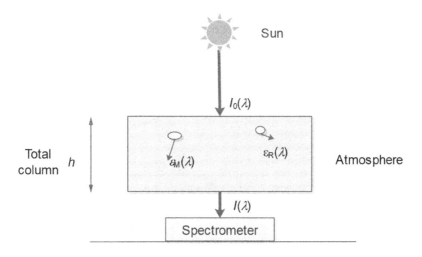

FIGURE 11.22 Geometry for slanted path used for the calculation of air mass ratio.

A polynomial approximation of Rayleigh scattering gives values for $\varepsilon_R(\lambda_1)$ and $\varepsilon_R(\lambda_2)$, and similarly a polynomial approximation of the absorption cross-section of the trace gas gives values for $\sigma(\lambda_1)$ and $\sigma(\lambda_2)$.

Lastly, we need $\ln\dfrac{I_0(\lambda_1)}{I_0(\lambda_2)}$ and you may be wondering how to obtain incoming radiation at the top of the atmosphere from the ground. The answer is provided by the *Langley extrapolation* method that is based on the Beer-Lambert law applied as a function of air mass $I = I_0 \exp(-m\tau)$ which is Equation (11.1) for $a = 1$. Take the natural log to obtain $\ln(I/I_0) = -m\tau$ and rewrite

$$\ln(I) = \ln(I_0) - m\tau \tag{11.16}$$

which is the equation for a straight line. Next, we measure I at time intervals at known sun elevation β and thus m, in the morning or afternoon, assuming that the atmospheric attenuation factor contained in parameter τ and pressure P do not change substantially during the measurement time. Doing this, we have a dataset of $\ln(I)$ vs. m values, which plot as a straight string of points with a negative slope starting at a minimum $m = 1$ at $\beta = 90°$ (noon) assuming standard pressure (Figure 11.23). The Langley extrapolation consists of projecting these points back to $m = 0$ or the intercept with the vertical axis which corresponds to the top of the atmosphere. This extrapolation is facilitated by a linear regression that estimates the intercept parameter as shown in Figure 11.23. Examples of the Langley extrapolation method giving approximate measured values of I_0 for three values of wavelength in the UV relevant to ozone. Were measured on March 3, 2005, at 305, 312, and 320 nm (Nebgen 2006) and measured on June 28, 2011, at 305 nm (Jerez 2011).

DOAS is a generalization of OAS. Light intensity is measured in a wide range of wavelengths. Absorption structures of several trace gases are separated from each other as well as from other extinction processes. DOAS has some advantages over the classic OAS approach: it relies on the measurement of absorption spectra instead of the intensity of monochromatic light and there is no need to estimate a value for $I_0(\lambda)$.

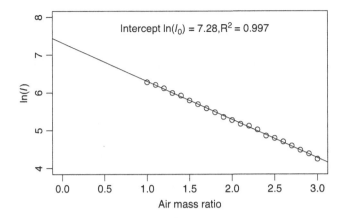

FIGURE 11.23 Langley extrapolation method explained. ratio

EXAMPLE: MEASURING UV AND TOTAL COLUMN OZONE CONCENTRATION BY OAS AND DOAS

At UNT in 1999 we started measuring UV in Denton, Texas, as part of what was then the U.S. EPA's Environmental Monitoring for Public Access and Community Tracking program. We sustained UV monitoring for several years by deploying a UV detector (Solar Light 501) on the roof of the EESAT building of the UNT main campus in the summer of 1999. Data was logged every 30 minutes and transferred to a PC using a serial port upon command from proprietary Solar Light Co. software, which was then transmitted to a database server, and subsequently to a web server. Using a browser, a user can request select periods to view and download historical data. With the advent of low-cost SBC as the Raspberry Pi, we converted the system by interfacing the 501 Biometer Recorder to a Raspberry Pi and replacing the PC. The Pi SBC then transfers the data to the database and web server.

Subsequently, at UNT we developed a low-cost ozone automated monitor using classical OAS as described earlier in this chapter (Nebgen 2006; Acevedo et al. 2009). To aim at the sun, we used an optical fiber with a narrow field-of-view collimated lens mounted on a pan-and-tilt device that tracked the sun's path during the day following the equations for azimuth and sun elevation given in Chapter 7. The fiber piped the light to a fiber optic-based spectrometer housed indoors. We used an Ocean Optics S2000-TR2 configured with grating #10: 1800 lines (200–350 nm), with a slit of 10 μm, yielding a spectral resolution of 0.1 nm, and optical resolution of 0.234 FWHM (full width at half maximum). The resulting instrument provided ozone concentrations comparable to those obtained with MICROTOPS and those reported by satellite. Error budgets and comparisons can be consulted in (Nebgen 2006). Jerez (2011) developed a routine based on DOAS adapting this DOAS routine to Nebgen's instrument. This work included estimating the error of DOAS measurements, comparing the total column ozone derived from DOAS and Nebgen's routine (based on OAS), and published data online using a web site (TEO 2015).

Later developments included integrating the software and hardware into a stand-alone, outdoor-ready instrument by Faschingbauer et al. (2014). This integration provides an example of using a small form factor computer but with high processing power. It employs the Intel D2500HN – Mini-ITX Motherboard; it has an embedded 1.86-GHz processor, low power consumption and heat generation, and a solid-state drive for primary storage. In this application, the computer handles all the automation scripts and log data to run the whole system without user interaction.

ATMOSPHERIC GASES AND AIR QUALITY FROM REMOTE SENSING

As discussed previously in this chapter, the concentration of atmospheric gases varies with altitude, with vertical profiles characteristic of each gas, and the total column atmospheric concentration of a gas refers to the integrated concentration over the entire atmosphere above a position on Earth's surface. Total column and air quality variables (such as ozone, SO_2, NO_2, and aerosols) can be monitored from spaceborne platforms using backscattered light in UV wavelengths. The NASA (2023b) Ozone and Air quality site offers a variety of data related to various methods of measurement. For ozone, these include the OMI (Ozone Monitoring Instrument), OMPS (Ozone Mapping and Profiler Suite), and TOMS (Total Ozone Mapping Spectrometer). OMI data on this site include measurements of ozone columns and profiles, aerosols, clouds, surface UV irradiance, and the trace gases NO_2, SO_2, HCHO, BrO, and OClO.

Using spaceborne platforms allows to compose images for large areas of Earth as opposed to a limited set of stations on the ground.

ATMOSPHERE – WEATHER

AIR TEMPERATURE

One of the most often measured meteorological variables is air temperature. We already presented in Chapter 3 an example of a thermistor-based temperature transducer, the "107 Temperature Probe" manufactured by Campbell Scientific. It measures the temperature of the air in a weather station. Its range of measurement is from –35°C to 50°C. Please refer to Chapter 3 to refresh concepts of self-heating and radiation shield.

PRECIPITATION

Rain is easily measured by a rain gage consisting of a funnel and a tipping cup, calibrated to a given volume of water. Water flows down the funnel into the cup. Then each time the cup dumps the water, its movement is detected by a magnetic reed switch and this pulse is counted by a datalogger (Figure 11.24). Consider a simple example, the TE525 of Campbell Scientific manufactured by Texas Electronics. It has a funnel input area of 15.4 cm in diameter and measures rainfall in 0.254-mm increments.

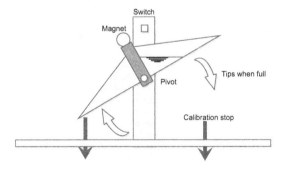

FIGURE 11.24 Rain gage.

Relative Humidity

Relative humidity (RH) is the moisture content in the air expressed as the percentage of moisture of the saturation value or water vapor capacity of the air. RH changes by either changing water vapor in the air (air gets damper if air moisture increases) or by changing the air temperature; RH increases as temperature decreases (air gets damper when cold). RH can be sensed by a capacitor that changes capacitance as the dielectric has more water. Recall from Chapter 3 the concept of permittivity. Many times, an RH probe is housed together with an air temperature probe, forming a single unit that provides two signals, one for temperature and one for RH. For example, the Vaisala HMP155 senses RH in the 0%–100% range and air temperature in –80°C to 60°C.

Solar Radiation

There are several ways of measuring solar radiation, and instruments reflect these various methods. In this section, we will focus on *pyranometers* and net *radiometers*, and later in Chapter 13, we will discuss *quantum sensors* and their relationship to primary productivity. In addition, there are several values of the field of view for collecting solar radiation for transmission to the electronic detector area.

A pyranometer measures broadband solar irradiance or solar flux density or insolation (in W/m^2) on a planar surface from a hemispherical field of view (180°). A uniform cosine response for varying sun angles is provided by a glass dome or by a plastic diffuser. The detector itself can be based on a *thermopile*, converting the thermal energy to electricity, or on a *silicon cell* similar to a PV cell. A thermopile consists of thermocouples connected in series generating a signal that would depend on temperature, which in turn depends on radiation. The detector must have a *cosine response* that represents a directional response to the incident beam; the response would follow a cosine function, corresponding to full response when the sun is at zenith, and zero response when the sun is at the horizon or 90° with respect to the zenith (Figure 11.25).

A net radiometer measures *net radiation* or balance between incoming and outgoing solar radiation. A net radiometer provides four components of net radiation: incoming and reflected (which allows calculation of albedo), and downwelling and upwelling IR, to account for long-wave radiation balance. The simplest type of sensor provides the sum of all four components.

Wind Velocity and Direction: Sonic Anemometers

In Chapter 3, we discussed the fundamentals of a wind vane and a simple cup anemometer to measure wind speed and wind direction. In this chapter, we introduce a sonic anemometer. Ultrasound is a pressure wave at a frequency higher than the human hearing range or approximately 20 kHz. Ultrasonic technology offers a no-moving parts alternative wind sensor to the cup and vane anemometer. Sonic sensors are based on measuring ultrasound across a perpendicular axis (Figure 11.26). For example, a two-axis ultrasonic wind sensor provides wind speed and direction data as two analog outputs or if converted to digital at the level of the sensor, it can be provided as digital via a serial port. Sonic anemometers perform well in harsh environmental conditions without fear of damage

FIGURE 11.25 Pyranometer and cosine response.

FIGURE 11.26 Two axis sonic anemometer.

often associated with cup and vane sensors. They can operate up to high wind speed conditions and wind direction has no dead area near 0 or 360 degrees.

A two-axis sonic anemometer measures the difference in time taken for a sound pulse to travel in the N-S direction vs. the S-N direction as well as the time difference between W-E and E-W directions. This is possible because sound time of flight is longer against the wind, and therefore wind speed and direction are calculated from the differences in the times of flight for each axis. Denote by L the distance between sensors; then upwind T_u and downwind T_d travel times are

$$T_u = \frac{L}{Cs - W} \quad T_d = \frac{L}{Cs + W} \tag{11.17}$$

Where C_s is sound speed and W is wind speed. Invert and subtract

$$\frac{1}{T_d} - \frac{1}{T_u} = \frac{C_s + W}{L} - \frac{C_s - W}{L} = \frac{2W}{L} \tag{11.18}$$

and therefore, we can solve for W to obtain wind speed

$$W = \frac{L}{2}\left(\frac{1}{T_d} - \frac{1}{T_u}\right) \tag{11.19}$$

When wind speed is very low, the difference between T_d and T_u is very small and W cannot be detected.

A three-axis anemometer has sensors facing vertically in addition to the perpendicular horizontal axis sensors providing a 3D measurement of the wind field. Either a pair of three sensor arrays or three pairs of sensors achieve this. In three-pair sensor arrangements, each pair represents the x, y, and z axes (Figure 11.27). In a pair of three-sensor arrays model, each array has three sensors separated horizontally by 130 degrees. The two arrays face each other vertically, one in a low position and the other in the upper position (Figure 11.27). In this case, we want the difference in time taken for the sonic pulse to travel from the upper to the opposite lower sensors vs. travel from the lower to the upper sensors. Wind speed along the axis between each pair of transducers is then calculated from the times of flight on each axis. The frame holding the two arrays can have one to three poles (spars) that set a reference (North). The axis definition is as follows: +x is the direction in line with the reference spar, y is perpendicular (90 degrees from N), and +z is the vertical pointing up. From the three axis velocities, the wind speed is calculated, as either signed x, y, and z.

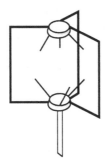

FIGURE 11.27   Three axis sonic anemometer.

A 3D sonic anemometer provides measurements of air turbulence which is relevant around structures (such as bridges, buildings, wind turbine sites), as well as meteorological and gas flux measurement sites which we discuss in Chapter 13.

NEXT-GENERATION WEATHER RADAR (NEXRAD)

Precipitation and wind are measured by weather radar, using the reflectivity of precipitation as well as of moving air mass. RF energy emitted in pulses from the radar antenna strikes objects (e.g., rain drops and snowflakes) in the atmosphere scattering in many directions, and with some energy reflected to the radar; the reflected energy increases with the size of the objects, while its time to return relates to the distance to the object. *Doppler* radar systems can provide information regarding the movement of targets as well as their position. The US National Weather Service (NWS) operates a radar network of Weather Surveillance Radar-1988 Doppler, or WSR-88D for short; a WSR-88D transmits ~450 kW of an RF signal for ~1.5 μs every ~1 ms, listening in between pulses to keep track of the phase shift between transmitted and received pulses (National Weather Service 2023).

This phase shift is analogous to the Doppler shift of sound waves, with the sound pitch (frequency) increasing for an object moving toward you or decreasing for an object moving away from you. For a Doppler radar, the direction (toward or away) and velocity of movement of the object depend on the phase shift between a transmitted pulse and a received pulse. The WSR-88D antenna increases to preset elevation angles or slices as it rotates, comprising a volume coverage pattern, with complete volume covered upon completion of all elevation slices, providing a 3D image of the atmosphere around the radar site. In precipitation mode, the volume scan is completed every four to six minutes. In addition, the pulse can be polarized in two directions (vertical and horizontal) allowing to discern between rain, hail, snow, and ice. This *dual polarization* also detects rotation, allowing to forecast tornadoes, and airborne tornado debris allowing to forecast tornado touch down.

Returned energy represents the reflectivity of the precipitation in the atmosphere around the radar and is modeled by a reflectivity factor Z of precipitation, which is the integral of the number of reflector objects (called hydrometeors), multiplied by their diameter distribution $f(D)$ and the 6th power of the diameter

$$Z = \int_0^{D\max} n f(D) D^6 \, dD \qquad (11.20)$$

Using a Z_0 defined as the Z for 1 mm drop in 1 m^3, we calculate the dBZ values as

$$L_Z = 10 \log_{10} \frac{Z}{Z_0} \qquad (11.21)$$

The Z value is related to the rainfall rate by formulas such as the Marshall-Palmer $Z = 200R^{8/5}$ (Marshall and Palmer 1948) which can be solved for R to obtain $R = \left(\dfrac{Z}{200} \right)^{5/8}$.

Base reflectivity corresponds to a low elevation slice (at 0.5 degrees) and represents a precipitation survey of rainfall in the region of the radar, with levels representing the strength of returned energy in dBZ. These values are color-coded in the radar images you typically see in weather reports or online images. Light rain begins at ~20 dBZ, and 1-inch hail may start ~60 dBZ, but this reflectivity value may not necessarily indicate severe weather. As the name indicates, *composite reflectivity* is aggregated from returns from all elevation scans taking the highest dBZ value from all elevations; thus, it is available at the end of the scan cycle, whereas base reflectivity is available after the first scan. Composite reflectivity allows to study the structure of a storm.

Precipitation can be estimated from reflectivity up to a maximum range, e.g., 230 km, from the radar location. Accumulated precipitation amounts are given by *one-hour precipitation* and *storm total precipitation* images. Besides estimating rainfall, both the static and looping one-hour precipitation images can provide other useful information such as the motion of the storms. Storm total is of course the estimated accumulation since the precipitation began and until there is no more precipitation for one hour. This accumulation can exceed several days during rainy periods.

As discussed above, a Doppler radar can detect motion toward and away from the radar, i.e., *radial* velocity, which means that it can detect only the velocity of that part of the wind field in the radial direction. Decomposing, the wind field in two parts: perpendicular (to the radar beam) and radial, and considering that the radar is scanning the area, there will be times when the beam is more aligned with the overall wind flow and times when it does not detect the wind motion. The Doppler radar calculates a velocity based on the perpendicular and radial vectors. The system can provide two velocity images: *base velocity* and *storm relative motion*. Base velocity, like base reflectivity, provides a picture of the basic wind field from the lowest (0.5°) elevation scan and it is useful for determining areas of strong wind or detecting the speed of cold fronts. The storm's relative motion is an image of the wind circulation around a storm after subtracting the overall motion of the storm.

In the United States, the NEXRAD is a network of 160 high-resolution S-band Doppler weather radars jointly operated by the NWS, the Federal Aviation Administration, and the U.S. Air Force. The National Centers for Environmental Information provides access to archived NEXRAD Level-II data and Level-III products (NOAA NCEI 2023). At this web site, the public has access to NEXRAD inventory that can be accessed by location on a map. For example, drilling at the location of Dallas Fort Worth, Texas pulls up the station KFWS – DALLAS/FTW, TX and an inventory of data available for the period 1994–2023. NEXRAD files have many variables; in terms of precipitation, for instance, there are files for instantaneous rate, one-hour, and storm total. Historical NEXRAD data are useful to estimate rainfall over a watershed and provide input to hydrological models based on rainfall-runoff relationships.

WEATHER SATELLITES

We discussed satellite orbits in Chapter 8 as geostationary, equatorial and sun-synchronous (Zhu et al. 2018). In this chapter, we will mostly be concerned with *geostationary* orbits that rotate at a period equal to Earth's rotation period (24 hours), therefore the satellite always stays over the same location on Earth; this orbit type is useful to monitor weather.

We will focus on U.S. coverage by GOES that stands for *Geostationary Operational Environmental Satellite* Program, a joint effort of NASA and NOAA (NASA 2023a; NOAA 2023a, 2023b), it currently consists of GOES-13, or GOES-East, at 75°W longitude and GOES-15, or GOES-West, at 135°W longitude. Latest series of GOES satellites include GOES-R, S, T, and U. GOES platforms help observe and predict weather, including thunderstorms, tornadoes, and hurricanes, as well as monitoring dust storms, volcanic eruptions, and forest fires. GOES-12 was the first satellite to carry a Solar X-Ray Imager type instrument, that helps monitor solar activity. The GOES-R series covers Earth's western hemisphere and provides advanced imagery and atmospheric measurements, real-time mapping of lightning activity, and monitoring of space weather (NASA 2023a).

Instruments on GOES, Advanced Baseline Imager, sense EM energy at several wavelength bands, of which IR and visible are commonly shown on weather broadcasts. IR bands show heat or long wave re-emitted by the Earth's surface, atmosphere, and clouds with the advantage of showing this energy during nighttime as well, thus providing 24-hour coverage. Visible band images consist of reflected solar radiation by clouds, the atmosphere, and the planet's surface, with the advantage of higher resolution (1 km) than IR images (4 km). Images are often colorized to enhance cloud patterns (NOAA 2023b).

Details on the multiple bands available from GOES can be consulted at GOES-R 2023. Two visible bands are blue (470 nm), excellent to monitor aerosols, and red (640 nm) with a finer spatial resolution (0.5 km) and able to detect finer detail like top of cumulus clouds. NIR bands are of 0.86, 1.37, 1.6, and 2.2 μm. The latter is useful to detect cloud particle size. Ten IR bands span from 3.9 to 13.3 μm, with some of these able to detect water vapor (3.9, 6.2, 6.9 μm), and others centered on ozone and CO_2.

EXERCISES

Exercise 11.1

Assume the simple energy budget of planet Earth as in Figure 11.3. Assume the albedo is $\alpha = 0.28$ (incoming reflected loss) and $f = 0.39$ (outgoing trapped loss by GH effect). What is the Earth's temperature in °C?

Exercise 11.2

Consider the following six annual values of CO_2 concentration in Earth's atmosphere

```
Year,   CO₂ (ppm)
1960,   316.91
1970,   325.68
1980,   338.75
1990,   354.39
2000,   369.55
2010,   389.90
```

Assume that the increase of CO_2 is exponential and estimate the rate coefficient for these 50 years using a simple calculation of log of ratio divided by the time interval.

Exercise 11.3

Assume the doubly exponential model for CO_2 with values for the coefficients $k_1 \approx 0.172$ and $k_2 \approx 0.014$. Calculate the doubling time for CO_2 increase.

Exercise 11.4

Assume the doubly exponential model for Earth's global temperature with values for the coefficients $k_1 = 0.013$, $k_2 = 0.030$. Calculate the doubling time for global temperature.

Exercise 11.5

Assume you have the following values of ln(I) corresponding to values of air mass from 1 to 3 in steps of 0.1.

```
6.29, 6.23, 6.11, 6.01, 5.87, 5.79, 5.73, 5.57, 5.45, 5.36, 5.29,
5.22, 5.06, 4.95, 4.87, 4.75, 4.65, 4.56, 4.48, 4.35, 4.33
```

Use Langley extrapolation to determine ln(I_0). You can use a plot and draw a line or run a linear regression.

Exercise 11.6

Visit the NEXRAD web site and explore precipitation datasets available for the station KFWS – DALLAS/FTW, TX.

Exercise 11.7

Visit the NOAA GOES Image viewer web site (NOAA 2023a) and explore images on various bands, available for Texas.

REFERENCES

Acevedo, M. F. 2018. *Introduction to Renewable Electric Power Systems and the Environment with R*. Boca Raton, FL: CRC Press. 439 pp.

Acevedo, M. F. 2024. *Real-Time Environmental Monitoring: Sensors and Systems, Second Edition – Lab Manual*. Boca Raton, FL: CRC Press, Taylor & Francis Group. 463 pp.

Acevedo, M. F., W. T. Waller, and G. B. Nebgen. 2009. *Instrument, System and Method for Automated Low Cost Atmospheric Measurements*. US Patent Number 7,489,397. US Patent Office.

Crutzen, P. J. 1970. The influence of nitrogen oxides on the atmospheric ozone content. *Quarterly Journal of Royal Meteorological Society* 96:320–325.

Faschingbauer, A., J. Stumberg, and T. Eminger. 2014. *Instrumentation Panel for Total Column Ozone Monitoring System. Senior Design Project*: Electrical Engineering Department Denton, TX: University of North Texas.

GOES-R. 2023. *ABI Technical Summary Chart*. accessed January 2023. https://www.goes-r.gov/spacesegment/ABI-tech-summary.html.

Graedel, T. E., and P .J. Crutzen. 1993. *Atmopsheric Change: An Earth System Perspective*. New York: W.H. Freeman. 446 pp.

Gueymard, C. A., and D. Thevenard. 2013. Revising ASHRAE climatic data for design and standards-part 2: Clear-sky solar radiation model. *ASHRAE Transactions* 119:194–209.

HITRAN. 2014. *The HITRAN Database*. CFA, accessed 2014. http://www.cfa.harvard.edu/hitran/.

Jerez, C. 2011. *Measuring Atmospheric Ozone and Nitrogen Dioxide Concentration by Differential Optical Absorption Spectroscopy*. PhD dissertation, Environmental Sciences, Denton, TX: University of North Texas.

Lovett, G. M., D. A. Burns, C. T. Driscoll, J. C. Jenkins, M. J. Mitchells, L. Rustad, J. B. Shanley, G. E. Likens, and R. Haeuber. 2007. Who needs environmental monitoring? *Frontiers in Ecology and the Environment* 5 (5):253–260.

Marshall, J. S., and W. M. Palmer. 1948. The distribution of raindrops with size. *Journal of Meteorology* 5: 165–166.

Matthias, A.D. 2004. "Monitoring Near-Surface Air Quality." In *Environmental Monitoring and Characterization*, edited by J. F. Artiola, I. L. Pepper and M. L. Brusseau, 163–181. Burlington: Academic Press.

Moran, J. M., ed. 2006. *Weather Studies. Introduction to Atmospheric Science*. Boston: American Metereological Society.

NASA. 2020. *Global Climate Change. Vital Signs of the Planet*. accessed August 2020. http://climate.nasa.gov/vital-signs/global-temperature/.

NASA. 2023a. *GOES Satellite Network*. accessed January 2023. https://www.nasa.gov/content/goes.

NASA. 2023b. *Ozone and Air Quality*. accessed January 2023. https://ozoneaq.gsfc.nasa.gov/.

National Weather Service. 2023. *Doppler Radar*. accessed January 2023. https://www.weather.gov/jetstream/doppler_intro.

Nebgen, G. 2006. *Automated Low Cost Instrument for Measuring Total Column Ozone*. PhD dissertation, Environmental Sciences, Denton, TX: University of North Texas.

NOAA. 2020. *Trends in Atmospheric Carbon Dioxide*. accessed August 2020. http://www.esrl.noaa.gov/gmd/ccgg/trends/.

NOAA. 2023a. *GOES Image Viewer*. accessed Jan 2023. https://www.star.nesdis.noaa.gov/GOES/index.php.

NOAA. 2023b. *NOAA Geostationary Satellite Server*. accessed January 2022. https://www.goes.noaa.gov/.

NOAA NCEI. 2023. *Next Generation Weather Radar (NEXRAD)*. accessed January 2023. https://www.ncei.noaa.gov/products/radar/next-generation-weather-radar.

Pallás-Areny, R.. 2000. *Amplifiers and Signal Conditioners*. Boca Raton, FL: CRC Press.

Platt, U., and J. Stutz, eds. 2008. *Differential Optical Absorption Spectroscopy: Principles and Applications*. Berlin: Springerpp.

Rothman, L. S., I. E. Gordon, Y. Babikov, A. Barbe, D. Chris Benner, P. F. Bernath, M. Birk, L. Bizzocchi, V. Boudon, L. R. Brown, A. Campargue, K. Chance, E. A. Cohen, L. H. Coudert, V. M. Devi, B. J. Drouin, A. Fayt, J. M. Flaud, R. R. Gamache, J. J. Harrison, J. M. Hartmann, C. Hill, J. T. Hodges, D. Jacquemart, A. Jolly, J. Lamouroux, R. J. Le Roy, G. Li, D. A. Long, O. M. Lyulin, C. J. Mackie, S. T. Massie, S. Mikhailenko, H. S. P. Müller, O. V. Naumenko, A. V. Nikitin, J. Orphal, V. Perevalov, A. Perrin, E. R. Polovtseva, C. Richard, M. A. H. Smith, E. Starikova, K. Sung, S. Tashkun, J. Tennyson, G. C. Toon, Vl G. Tyuterev, and G. Wagner. 2013. The HITRAN2012 molecular spectroscopic database. *Journal of Quantitative Spectroscopy and Radiative Transfer* 130:4–50.

TCEQ. 2015. *Current Ozone Levels*. accessed March 2015. http://www.tceq.state.tx.us/cgi-bin/compliance/monops/select_curlev.pl.

TEO. 2015. *Texas Environmental Observatory*. accessed March 2015. www.teo.unt.edu.

U.S. EPA. 2014a. *Clean Air Act*. accessed 2014. http://www.epa.gov/air/caa/.

U.S. EPA. 2014b. *National Ambient Air Quality Standards (NAAQS)*. accessed 2014. http://www.epa.gov/air/criteria.html.

U.S. EPA. 2014c. *Summary of the Clean Water Act*. accessed 2014. http://www2.epa.gov/laws-regulations/summary-clean-water-act.

US EPA. 2017. *Landfill Methane Outreach Program (LMOP)*. accessed August 2017. https://www.epa.gov/lmop.

Vaughan, H., T. Brydges, A. Fenech, and A. Lumb. 2001. Monitoring long-term ecological changes through the Ecological Monitoring and Assessment Network: Science-based and policy relevant. *Environmental Monitoring and Assessment* 67:3–28.

WOUDC. 2023. *World Ozone and Ultraviolet Radiation Data Centre, Data Products*. accessed January 2023. https://woudc.org/data/products/.

Zhu, L., J. Suomalainen, J. Liu, J. Hyyppä, H. Kaartinen, and Haggren. H. 2018. "A Review: Remote Sensing Sensors." In *Multi-Purposeful Application of Geospatial Data*, edited by R. B. Rustamov, S. Hasanova and M. H Zeynalova, 19–42. London: Intech Open.

12 Water Monitoring

INTRODUCTION

This chapter covers a variety of topics related to monitoring of Earth's hydrosphere, describing common measurements of water *quantity*, such as level, velocity, and flow, as well as *quality* in terms of natural constituents and pollutants. Various water quality parameters are temperature, electrical conductivity (EC), total dissolved solids (TDS), pH, dissolved oxygen (DO), and turbidity. The light environment in water is very important since it impacts the production of aquatic ecosystems. We will discuss *hydrodynamics*, *water quality*, and *hydrological* models in terms of their linkage to monitoring. After covering remote sensing of water quality, this chapter describes ocean monitoring, emphasizing temperature, and groundwater monitoring, focusing on the dynamics of water level and water quality. We end this chapter presenting methods for predicting time series using autoregressive models. Although water in the *vadose* zone, i.e., the terrestrial subsurface from the surface to the groundwater table, including the soil, is an important component of hydrology, we leave soil water measurements for Chapter 13 in conjunction with other terrestrial ecosystem measurements. The concepts of water monitoring are further elaborated by computer exercises in Lab 12 of the companion Lab Manual (Acevedo 2024).

WATER

Water covers 70% of Earth's surface, and almost all (97%) of the water is in the oceans which is saline. The remainder 3% is freshwater of which 2% is iced in glaciers, only the rest 1% is surface (lakes, rivers), and in the soil and groundwater, and water vapor (Artiola 2004).

Water can exist as gas, liquid, and solid phases. Changes from one phase to another imply changes in energy, either gain or loss. Processes involved in changing between gas and liquid are condensation and evaporation, and in between solid and liquid are freezing and melting. When liquid water becomes water vapor, energy is required as latent heat of vaporization. Increasing temperature promotes evaporation and increases vapor pressure of the air.

These processes are involved in water cycles on Earth. The *hydrologic cycle* consists of evaporation from surface water and soil, condensation of atmospheric water vapor into liquid, precipitation from atmosphere, runoff to surface water and infiltration into soil moisture, and back to the atmosphere by evaporation.

Human beings and aquatic organisms depend on the quality of freshwater. Dissolved solids, pH, and DO are some important parameters that affect aquatic organisms. Sustainability of water resources requires monitoring of water quality and quantity (Artiola 2004).

WATER LEVEL AND DEPTH

We can estimate depth of a column of water by a submersed sensor or from ultrasonic sensors placed above the water surface. Submersed sensors estimate depth from the pressure exerted by the water column after subtracting atmospheric pressure. To accomplish this, we can use a strain gage transducer (Chapter 4) housed in a waterproof case and placed underwater; the transducer measures pressure difference with one side of the transducer exposed to the water and the other side exposed to the air. This measurement is affected by variations in barometric pressure, water density, and temperature and therefore it is important to calibrate the sensor for desired conditions and perform regular re-calibrations. Level loggers are small probes enclosed in a waterproof case made of stainless steel

DOI: 10.1201/9781003425496-12

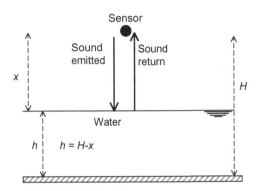

FIGURE 12.1 Principle of the operation of ultrasonic water level sensor.

or titanium that be submersed to measure pressure and therefore depth; we saw a picture of a probe like this in Figure 5.14. This device is used to measure changing water level in streams, lakes, estuaries, and groundwater; for the latter purpose, they can be deployed inside a pipe or a well.

Ultrasonic sensors of water level are deployed above the water surface and operate under the principle of measuring the time for the return of sound emitted from the sensor and reflected from the surface (Figure 12.1). The sensor continuously transmits pulses of sound (~42 kHz) which strike the surface of the liquid and return to the sensor. The travel time of sound is related to the distance between the sensor and the water surface x, which translates into water level h by knowing a reference or datum H. The relation is $h = H - x$. However, air temperature affects the speed of sound, and therefore ultrasonic level sensors include a built-in temperature sensor to compensate for temperature changes.

Hydrometric stations monitor stream or river water levels, usually termed "stage". In the United States, the USGS maintains monitored data of the network of stations across the country (USGS 2023e). For illustration, USGS station 08051500 of Clear Creek near Sanger, Texas, shows a stage averaging 6.22 ft (~1.9 m) during January 2023, and the flood stage at this location is 25 ft (~7.6 m). Also important are lake and reservoir levels, typically reported as elevation above sea level; for example, for the same North Texas area, USGS 08052800 gage at Lewisville Lake near Lewisville, Texas, showed 520 ft (150.5 m) elevation, which is only 2 ft below the conservation pool elevation of 522 ft (159.1 m). The conservation pool level is defined as the level for which the lake has a specified amount of water storage.

Another illustration of water level comes from *tides* that shows a periodic variation of water level. The lunar tide occurs twice daily or *semidiurnal*, since as the Earth rotates, any point on the sea will experience a bulge due to gravitational pull of the Moon every 12 hour and 24 minutes, which is half of a lunar day. The gravitational effect of the Sun also produces a tidal bulge but of lower height of that of the Moon because it is further away, and occurring every 12 h, which is half of a solar day. When the Moon and Sun are aligned (full or new moon phases) the lunar and solar bulges superimpose, creating a larger bulge and producing the *spring tides* or the highest tides. In contrast, when the Moon and the Sun alignment with the Earth is perpendicular, the lunar and solar bulges do not superimpose, leading to lower height, producing neap tides, or the lowest tides. This occurs in first and third quarter phases of the Moon. The time in between neap and spring tides is one quarter of the 29.5 days of the lunar cycle; or about 7 days. *Tidal range* for a particular tide cycle is the total excursion from high to low tide, and thus increases during spring tides and decreases during neap tides.

Tidal data for the USA are available from NOAA (2023b). Take, for example Port Aransas, Texas on the Gulf of Mexico. Data are reported with respect to a datum ($z=0$) set at Mean Lower-Low Water. In this case, mean sea level is 0.83 ft (0.253 m) above this datum. For the same day, January 15, 2023, the tide level at this station showed −0.01 m at 3 am, 0.19 m at 1:23 pm, and a predicted level of 0.61 m for 6:36 pm. Data are available for download from this web site with timestamps separated by six-minute intervals. Long-term tide data allow calibration of *harmonic* modeling of tidal dynamics by using *Fourier series*, or a summation of sine and cosine functions of several periods related to the astronomical process we just described. In the United States, this approach consists of adding *harmonic*

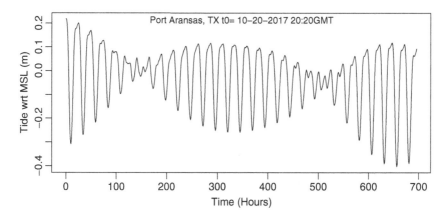

FIGURE 12.2 Tide levels at Port Aransas, Texas, derived from harmonics modeling.

constituents, which represents the periodic variation of the relative positions of the Earth, Moon, and Sun (NOAA 2017). Using this method, we can produce plots such as Figure 12.2 for Port Aransas; these plots let you appreciate the variation of neap and spring tides within the modeling period.

WATER VELOCITY AND FLOW

There are many instances when we need to measure water velocity. One example is when estimating runoff water flow after rain events, subsumed in the stream water flow. Once we measure water velocity and know stream's cross-sectional area, we can calculate flow q as the product of cross-sectional area A and velocity v, that is $q = vA$. When v is given in m/s and A in m^2, we obtain flow in m^3/s. To calculate cross-sectional area, we can determine the stream profile, monitor the water level using a water level sensor, and calculate the area based on the profile and depth of the channel (Figure 12.3). Irregular cross-sections require integrating the velocity field along the profile of the cross-section and thus we need to measure several velocity values. In small streams, a flume or weir may be used to simplify the cross-sectional area calculation.

A simple typical water velocity probe includes a propeller, which rotates on its shaft and would turn faster as water velocity increases. This is the same principle as a cup anemometer to measure wind speed. For small streams, an approach is to use positive displacement sensing; magnetic material in the propeller tip passes a pickup point producing electrical impulses that are passed to the electronics by an internal cable (Figure 12.4). For larger streams, a bigger propeller is needed

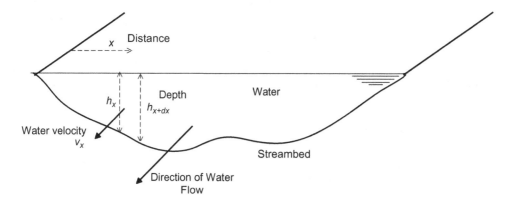

FIGURE 12.3 Estimating stream flow from velocity and cross-section.

FIGURE 12.4   Positive displacement water velocity sensor for small streams.

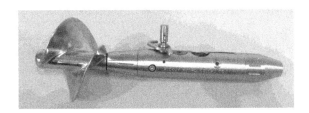

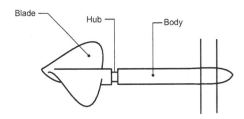

FIGURE 12.5   Water velocity propeller sensor for large streams.

(Figure 12.5). An electronic circuit, based on a microcontroller, would then count the pulses, and convert the rate to water velocity. The microcontroller or a datalogger would record minimum, maximum, and average velocity in given periods.

The *Doppler* effect, which we discussed in Chapter 11 in the context of wind, also plays a role in water flow measurements by transmitting sound through the flowing liquid which reflects the sensor from solids or bubbles in the fluid, in a manner equivalent to a Doppler radar receiving a return from rain drops and other hydrometeors. The echoes return at a different frequency proportional to fluid velocity, and the instrument measures this frequency shift to calculate flow. For better performance, Doppler flow meters are deployed away from features that create turbulence in the flowing fluid. Acoustic Doppler current profilers allow to measure water flow or discharge in rivers and other waterways, which can be used to predict flooding and low-water conditions. At the same time, the data can also be used to create a detailed map of the water velocity distribution in a river (Fricker 2014).

Along with the stage, hydrometric stations monitor stream flow termed *discharge*. We can find these data on the same web site as river stage (USGS 2023e). For illustration, take the same USGS station 08051500 of Clear Creek near Sanger, Texas, which shows a discharge of 15.6 cfs (ft³/s) or 0.44 m³/s on January 14, 2023. As discussed above, real-time flow is not measured directly but

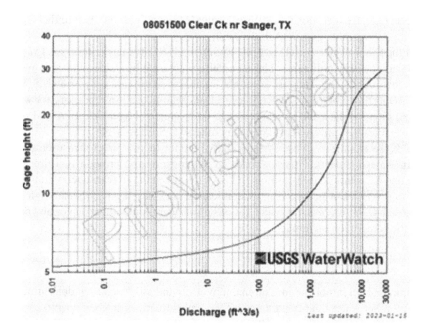

FIGURE 12.6 USGS rating curve for Clear Creek, near Sanger, TX.

rather estimated from a relationship of flow and level, this is called the rating curve or stage-discharge curve. For each hydrometric station, the USGS site includes the *rating curve* employed to predict flow from stage, often on a log-log scale (base 10). Figure 12.6 shows the rating curve for the example we have been discussing, and you can see how it is non-linear because of the geometry of the streambed controlling how cross-sectional area depends on level. Rating curves may undergo changes, i.e., *shifts*, due to changes in the streambed (e.g., erosion or deposition) or the growth of riparian vegetation.

Water velocity is also very important for tidal and ocean currents. For the sake of illustration, looking at the same web site NOAA (2023b) take the same location, Port Aransas, Texas on the Gulf of Mexico, and the same day January 14, 2023. We find that the tidal current was −80.4 cm/s at ebb (tide current going offshore) and 88.9 cm/s at flood tide (current flowing onshore) measured at a depth of 4.9 m. Information is also available on the direction of the current.

WATER QUALITY PARAMETERS

In regard to water quality monitoring, we can consider three types of variables: physical, chemical, and biological (Artiola 2004). In this field, it is common to refer to "variables" as "parameters", which may be confusing when first exposed to this literature. For consistency with the usage of these terms in this field, we will use variable and parameter indistinctly in this chapter. Major physical water quality variables are temperature, total suspended solids (TSS), EC, and turbidity, whereas major chemical variables are pH, oxidation-reduction potential (ORP), DO, TDS, major cations (Ca, Mg, Na, K, NH_4), major anions (Cl, SO_4, HCO_3, CO_3, PO_4, H_2S, NO_3), total organic carbon, and biological oxygen demand (BOD). Some of the chemical parameters are inferred from measurements of temperature, turbidity, and EC (indication of TSS and ionic strength). Other variables of interest include chlorophyll and volatile organic hydrocarbons. Major biological parameters include bacteria (e.g., fecal coliforms), viruses, protozoans (e.g., giardia), helminthes (e.g., parasitic worms), and algae (e.g., blue-green) (Artiola 2004).

As part of the Clean Water Act (U.S. EPA 2014c), the U.S. EPA has criteria for 157 priority pollutants and 45 non-priority pollutants. These guidelines are used by states and tribes to develop regulatory programs such as the National Pollutant Discharge Elimination System (U.S. EPA 2014a). The STORET (short for STOrage and RETrieval) Data Warehouse is a repository for water quality, biological, and physical data and is used by state environmental agencies, EPA and other federal agencies, universities, private citizens, and many others (U.S. EPA 2014b).

In this book, we emphasize those systems that allow for in situ direct measurements by inserting probes in water connecting these probes to dataloggers and provide real-time monitoring. Temperature and EC sensors are common and easy to implement in transducer circuits, and many probes such as for DO and pH are electrochemical devices based on glass electrodes as we discussed in Chapter 4. We discuss the principles of sensor operation in this chapter and provide some examples in the companion lab guide.

TEMPERATURE

Water temperature is an important measurement by itself and also because it determines the conditions of several other water quality parameters. We have already described how to use thermistors and thermocouples, and therefore we will focus on a different sensor named *Resistive Temperature Detector* (RTD) made from metals; therefore, resistance increases with temperature, because increased atom vibration increases resistance to electron displacement. Like a thermistor, and different from a thermocouple, an RTD is a passive sensor, since it needs power to force a current that produces a voltage drop that is measured as the transducer output, and the current must be limited to avoid self-heating (Pallás-Areny and Webster 2001).

A common model for the sensor response is a series approximation of the resistance R vs. temperature T curve

$$R(T) = R_0 \left(1 + \alpha_1 (T - T_0) + \alpha_2 (T - T_0)^2 + \cdots + \alpha_n (T - T_0)^n \right) \tag{12.1}$$

Where T_0 is a reference temperature, typically 0°C, R_0 is the resistance at a reference temperature, a common value is 100 Ω, α_i $i = 1, \ldots, n$ are coefficients, and n is the number of terms included in the series expansion, which for many sensors is $n = 1$ because $\alpha_1 \gg \alpha_2, \cdots, \alpha_n$ whenever the temperature does not affect the sensor dimensions. Exceptions to the latter are thin-film RTDs that may require including up to the third term. For sensors with dominance of the first term, the response is very linear compared to thermistors.

A commonly used metal is Platinum for which $\alpha_1 \approx 3.85 \times 10^{-3}$ K^{-1} dominates other coefficients by a factor much larger than 10 within temperature range below 650°C. In this case, Equation (12.1) reduces to

$$R(T) \simeq R_0 \left(1 + \alpha_1 (T - T_0) \right) \tag{12.2}$$

Note that since T_0 is subtracted from T we can calculate this equation in °C or K. In this case, RTD sensitivity is nearly constant with T and given by the slope of the R vs. T curve which calculated from equation (12.2) is

$$\frac{dR}{dT} \simeq R_0 \times \alpha_1 \tag{12.3}$$

However, it actually decreases slightly with increasing temperature due to the higher order terms in (12.1). For example, a platinum RTD with 100 Ω at 0°C and $\alpha_1 \approx 3.85 \times 10^{-3}$ / °C has a sensitivity of ~0.385 Ω/°C. Standards provide RTD specifications, for instance the DIN standard specifies a base resistance of 100 Ω at 0°C, a temperature coefficient of 0.00385/°C, and tolerance in various classes.

To avoid self-heating, one must consider the measuring environment because the sensor will dissipate heat differently if immersed in air, still water, or moving water. In this respect, the *heat dissipation factor* δ in mW/K depends on conditions; following Pallás-Areny and Webster (2001), we can calculate the maximum current to limit self-heating below a threshold ΔT by solving for I in

$$\Delta T = \frac{I^2 R(T)}{\delta}, \text{ that is } I = \sqrt{\frac{\Delta T \times \delta}{R(T)}}.$$ For instance, suppose a 100 Ω at 0°C sensor has $\delta = 100$ mW/K when immersed in still water at about 25°C, and we want to limit ΔT to 0.1°C. Using Equation (12.2),

we have $R(25) = 100 \times (1 + 0.00385 \times (25 - 0)) = 109.625\,\Omega$ and then $I = \sqrt{\frac{0.1 \times 100 \times 10^{-3}}{109.625}} = 9.5$ mA.

Transducer RTDs can be configured as two-wire, three-wire, and four-wire. The two-wire configuration is the simplest but the least accurate due to the resistance of the transducer leads, which causes an offset in the resistance measurement. Three-wire sensors compensate for the resistance of the leads, by letting the controller make two measurements; one for the total resistance and another for the compensation resistance, which is subtracted from the total resistance to obtain the net resistance. The four-wire sensor configuration allows to remove the influence of lead wires and is more complicated to implement. In practice, two-wire sensors are simple to use and sufficient for many applications, and three-wire sensors are common and provide a good trade-off of accuracy and complexity of measurement technique. RTD sensors can be connected in voltage divider circuits as well as bridge circuits.

ELECTRICAL CONDUCTIVITY

Water conducts electric current as a function of ions present in a solution, which is subsumed as the EC of the solution. EC is measured by injecting a current I, reading the voltage drop V produced by the solution (Figure 12.7), calculating the conductance from Ohm's law, and inferring conductivity from geometric considerations. In its simplest form, we can use the resistance between a simple pair of wires or electrodes as a sensor in a voltage divider circuit or bridge. As a simple experiment, one can learn how conductance measurements work by immersing foils and connecting them together with a 10 kΩ resistor to form a voltage divider sensor that is connected to an Arduino port (Gertz and Di Justo 2012). The microcontroller uses this port providing power, allowing for a delay, and performing a reading after the voltage is stable. A script converts voltage to resistance and resistance to conductivity. We simulated this experiment in lab guide 5 of the lab manual.

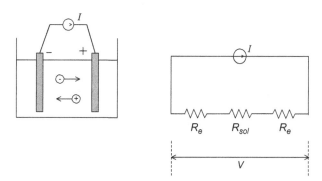

FIGURE 12.7 Simple EC probe schematic.

Recall from Chapter 3 that conductance G is the ratio of current I to voltage V, per Ohm's law $G = I/V$. Conductivity σ, or conductance per unit length, is then inferred from conductance based on a *cell constant* related to the probe geometry and construction described a little bit later (Decagon Devices 2014; Radiometer Analytical 2004). As stated in Chapter 3, conductance is measured in units of Siemens (S = 1 Amp/Volt), then EC has units of Siemens per unit length, S/m. To see this, note that if the current flow is through a cross-section of area A, and along length l, we can relate conductance and conductivity σ by

$$G = \sigma \frac{A}{l} \text{ or } \sigma = G \frac{l}{A} \tag{12.4}$$

Water EC is also denoted by the Greek symbol kappa κ. The cell constant k (1/cm) accommodates this geometric component $k = l/A$, where l is the distance (cm) between electrodes and A is the effective area (cm²) of the electrodes.

Electrode *polarization* occurs when ions accumulate near the surface of the electrodes, thus creating an additional parasitic resistance to the solution resistance. Applying alternating current (AC) of the proper frequency (<100 Hz) and increasing electrode area A reduces this effect. When the electric field lines stray outside the cross-section facing the electrodes, there is an additional *field-effect* resistance (Radiometer Analytical 2004).

A problem with a simple two-electrode water conductivity sensor is that we measure the conductance of the solution (G_{sol}) along with additional conductance due to electrode polarization and field effects (G_{pfe}) as shown in (Figure 12.7). Besides using AC and optimizing area, a more elaborate and practical probe uses four wires or poles (or electrodes) arranged in two separate pairs; an external pair serves to inject a current, and the voltage between the inner pair is measured with very small current draw avoiding the effect of extra conductance (Figure 12.8). When this voltage equilibrates to a reference voltage, current applied is proportional to conductance (Radiometer Analytical 2004).

Many water quality measurements are reported in µS/cm because typical values would be low when expressed in S/m. Note that 1 µS/cm = 0.0001 S/m. EC of distilled water is in the range of 0.5–3 µS/cm. Inland fresh waters with EC between 150 and 500 µS/cm are considered normal. Values outside this range could indicate that the water is not suitable for certain species of fish or macroinvertebrates. The conductivity of rivers in the United States generally ranges from 50 to 1500 µS/cm. However, industrial waters can range as high as 10,000 µS/cm (U.S. EPA 2014e). At high values of conductivity, expressing it in units of S/m or dS/m avoids using a high numerical value. For example, 10,000 µS/cm would be 100 dS/m or 10 S/m.

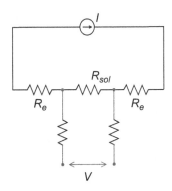

FIGURE 12.8   EC 4 pole schematic.

EC depends on temperature and all the above values refer to a standard temperature (25°C). Therefore, an EC reading $\kappa(T)$ needs adjustment to the standard 25°C using water temperature T of the sample by a factor $f_{25}(T)$ determined from a polynomial

$$\kappa(25) = f_{25}(T) \times \kappa(T) \tag{12.5}$$

For moderate EC values a linear approximation for this factor is

$$f_{25}(T) = \frac{1}{1 + \theta \times (T - 25)} \tag{12.6}$$

where θ is a temperature coefficient in 1/°C. A common approach is to include a temperature sensor near the EC electrode sensors to measure water temperature and adjust the EC measurements to their 25°C value (Radiometer Analytical 2004). Calibration of an EC probe requires comparing EC readings with known amounts of KCl in a calibrating solution.

TDS AND SALINITY

TDS relates to the total organic and inorganic dissolved solids in water, measured as the total weight of cations, anions, and the non-dissociated dissolved species in one liter of water. Therefore, it has units of concentration in mg/L or ppm. A practical estimation of TDS when performing real-time monitoring using probes is to calculate TDS from EC, multiplying by a scale factor $TDS = f \times EC$. The factor may be obtained by calibration using a standard of known TDS_a at a given temperature using $f = \dfrac{TDS_a}{\kappa_a(18)}$ TDS_a is expressed in mg/L and $\kappa_a(18)$=conductivity of the standard corrected to 18°C (in S/cm) as explained in the previous section. As a simplification, for practical purposes, one could use the 500 scale or 640 scale which consists of multiplying the value in mS/cm by 500 or by 640 to obtain TDS_{500} or TDS_{640}.

Salinity represents the weight of dissolved salts in water and is given in a variety of units that refer directly to weight and concentration or to a relationship with EC. In terms of weight, *practical salinity units* (psu) are defined as the concentration by weight of dissolved salts, 1 psu is 1 g of salt in 1 kg of water, for example 5 psu is the same as 5 g of salt dissolved in 1 kg of water. In oceanography it is common to use a per thousand unit denoted as ppt or $_o/_{oo}$, for example 35 ppt. However, this may lead to confusion because ppt is often used for parts per trillion. A practical estimation of salinity when performing real-time monitoring using probes is to relate salinity to EC and is determined from EC using a standard seawater solution that has a given salinity with known EC at a standard temperature. For example, a standard with conductivity $\kappa_a(15)$ at 15°C. Conductivity of the sample $\kappa_b(T)$ is measured and then we calculate salinity S using another polynomial $g(T)$ and the ratio of conductivities

$$S = g(t) \times \frac{\kappa_b(T)}{\kappa_a(T)} \tag{12.7}$$

pH AND ORP

Please refer to the description of electrochemical sensors given in Chapter 4. As a brief summary of that discussion, a pH sensor is a glass electrode that senses the concentration of hydronium ions, producing an electrical potential proportional to the logarithm of the concentration of free hydrogen

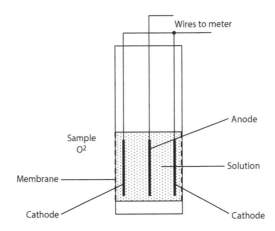

FIGURE 12.9 DO sensor schematic. A temperature sensor is typically added.

ions (Figure 16 of Chapter 4). ORP sensors produce a voltage proportional to the potential of the measured solution to act as oxidizing or reducing (loss or gain of electrons from other substances). ORP is related to pH, and the ORP sensor is similar to the pH sensor but made to produce a voltage proportional to ORP instead of pH. Redox values represent the overall potential of mixed oxidation-reduction processes.

DISSOLVED OXYGEN

These are glass electrodes that have a redox reaction in the presence of oxygen. DO diffuses through the porous membrane into the chamber; here it is reduced by electrons product of the oxidation of the lead electrode (anode)

$$2Pb+O_2+2H_2O+4e^- \rightarrow 2Pb^{++}+4OH^-+4e^- \qquad (12.8)$$

The current flow cathode (silver) to the anode is proportional to the concentration of molecular oxygen O_2 (Artiola 2004). Figure 12.9 illustrates schematically a DO probe. Besides providing DO that is valuable by itself, a DO probe can be applied to monitor DO dynamics and infer BOD and respiration rates from the depletion of DO.

TURBIDITY

Turbidity is a measure of the lack of clarity of water or its cloudiness due to individual solid particles that remain suspended in water and are not visible to the naked eye. These particles include silt, clay, algae, and organic matter (Sadar 2014; U.S. EPA 2014d). Turbidity is determined by the amount of light that is scattered by suspended solids in water. Turbidity is not a measure of the amount of suspended solids but, instead, an aggregate measure of the combined scattering effect of the suspended particles on an incident light source (Sadar 2014). However, turbidity is relatively easy to measure in real time on a continuous basis and then becomes a useful indicator of other water quality parameters that are difficult or costly to measure.

A turbidity probe includes a light source and a detector in various geometrical arrangements. The light source is one of three types: incandescent (monochromatic, including short wavelength), light emitting diode (LED) that emits in the low near-infrared (NIR) (830–890 nm), and laser at a

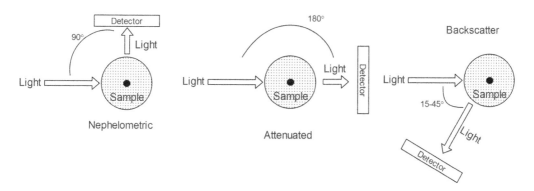

FIGURE 12.10 Turbidity probe schematic. Various detection angles.

single wavelength. Low wavelengths in the incandescent source are effective because it is scattered by small particles. The LED low NIR wavelengths just above visible is effective because it reduces interference by colored particles. Sensitive instruments at very low turbidity ranges use lasers as light sources.

A detector measures scattered light from the source at a selected detection angle. There are several options for the detection angle (Figure 12.10). The *nephelometric* angle is a 90° angle with respect to the light source and sensitive to a broad range of particle size, the *attenuated* angle is 180° from the source and includes absorption as well as scattering, and the *backscatter* angle is in between 315° and 360° with respect to the source and applicable to high turbidity measurements. The amount of light being scattered directly into the detector converts to volts and translates into turbidity units.

Other designs include multiple detectors set at a combination of the above angles and using a ratio approach integrated by software to obtain turbidity and two-detector dual light sources also integrated by software (Figure 12.11). In the latter, measurement occurs in two steps; first, one source turns on and two detectors are measured (at nephelometric and attenuated angles) then the other source turns on and the detectors are measured (but their angles are exchanged because of the geometric arrangement of the sources).

The most common turbidity probe is a *nephelometer*. It uses an incandescent source, a nephelo-metric angle, and it measures *nephelometric turbidity units* (NTU). These units are widely used. However, standards now include other units based on technology, methods, or detection angles used. For example, when attenuated or backscatter angles are used the units reflect this fact by using letters A or B in the units reported (AU or BU).

A Formazin standard is typically used to calibrate a turbidity meter. A concentration providing 20 NTU is a common value in the middle of the NTU range. Surface water is in the range of 0–50 NTU but of course can be higher due to runoff with suspended sediments occurring after rain events.

FLUOROMETER

As we discussed in Chapter 11 in relation to gases, fluorescence is the emission of energy by a sub-stance when excited by light of a given wavelength; upon absorption, the excited molecules release energy when falling back to the ground state. Then we detect and measure the intensity and spec-trum of this energy by optical means to infer the presence of chemicals in the substance. We can apply this principle to water and measure the fluorescence of algae, dyes for tracing, and pollutants.

In vitro measurements correspond to those conducted in the lab, whereas in vivo corresponds to those measurements conducted directly in the field with fluorometer probes and are amenable to

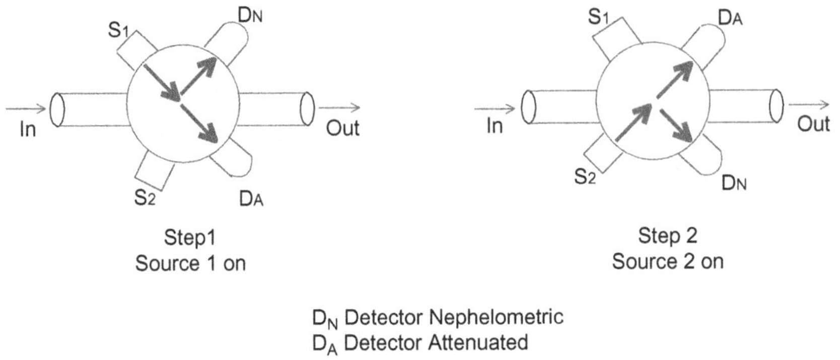

Step1
Source 1 on

Step 2
Source 2 on

D_N Detector Nephelometric
D_A Detector Attenuated

FIGURE 12.11 Turbidity probe schematics. Dual source.

real-time continuous monitoring when connected to a datalogger. However, submersible probes provide only a measure of in vivo relative concentration of chlorophyll or phytoplankton biomass, but not an absolute concentration (say in mg/L). When needing the latter, the relative in vivo measures are calibrated versus quantitative in vitro measurements (Sanders 2014).

Fluorometer probes measure a variety of water quality parameters by using a variety of excitation source wavelengths. For example, chlorophyll (blue range), blue-green algae (BGA) (orange range), dissolved organic matter (DOM), Rhodamine (green range), Fluorescein (blue range), Phycocyanin (yellow-orange range), Phycoerythrin (green range), Optical Brighteners (UV range), Crude Oil (UV range), Refined Fuels, PTSA Dye (UV range), Tryptophan (UV range), and turbidity (IR range) (Fondriest Environmental 2014; Turner Designs 2014).

In fluorometer probes, small windows allow the excitation light to shine into the water subject to measurement, and to detect the emitted energy from the substance (Figure 12.12).

MULTIPLE PARAMETER PROBES

Since it is desirable to measure multiple parameters of water quality instead of only one, it is now common to install multiple sensors in one device, i.e., a multiple-parameter sonde or multiparameter water quality sonde (Figure 12.13). Typical sensors installed in a multiparameter sonde are EC, temperature, light, DO, DOM, pH/ORP, depth, Chlorophyll, Rhodamine WT, BGA, Chloride, Nitrate, Ammonia, and turbidity.

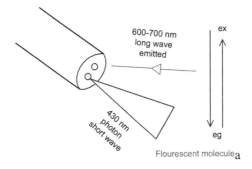

FIGURE 12.12 Principle of the operation of fluorometer probes.

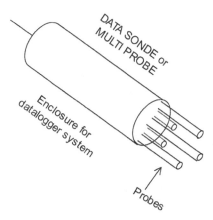

FIGURE 12.13   Typical multiparameter probe.

Multiparameter probes are configurable by varying the sensors installed. It can be powered by internal batteries or by an external battery. The probes are arranged so that they are in a container or cup where we can put a calibration solution. A datalogger collects data from all sensors. A multiparameter sonde is deployed in the field for real-time monitoring using buoys or floating platforms (Figure 12.14). We can also use these probes to profile a water body by lowering the probe at various depths.

IMPORTANCE OF IONIC PROFILE OF WATER QUALITY

The variables discussed above provide valuable information on water quality for many applications; however, they may not suffice for other applications when the concentration of the various ions is of importance. While many sensors amenable to be monitored by dataloggers, give an aggregate view of water quality, e.g., EC, there are fewer that give a concentration of many ions of interest. In many cases, then it is necessary to supplement real-time monitoring of water quality with in situ measurements using a photometer or a spectrophotometer and laboratory measurements with more elaborate instrumentation.

Just as an example, EC is important to determine if water is suitable as irrigation water for crop production (Artiola 2004). Some crops are tolerant of high EC water; for example, barley and

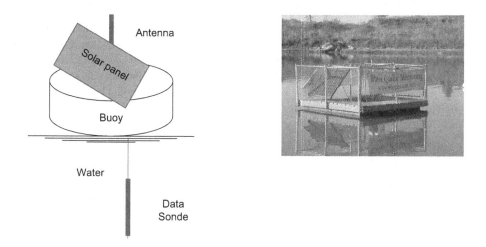

FIGURE 12.14   Examples of multiparameter probe deployment. Buoy and platform.

cotton, but others cannot tolerate irrigation with water with high EC. However, EC by itself may not provide enough information about the suitability of water, and other criteria based on ions are important. A water quality concept related to dissolved solids is the *sodium adsorption ratio* (SAR), utilized in agriculture and defined as $SAR = \dfrac{[Na^+]}{\sqrt{\dfrac{1}{2}\left([Ca^{2+}]+[Mg^{2+}]\right)}}$ where concentrations of sodium, calcium, and magnesium are in milliequivalent/liter (meq/l) (Glover 1996). The equivalent units consider valence; for univalent ions, the mmol amount is the same as meq, while a concentration of divalent ions provides twice the amount of meq. For instance, 1 mmol of Na^+ is equal to 1 meq, but 1 mmol of Ca^{++} is equal to 2 meq.

To convert mass concentration to mmol/l we can divide mg/l by the molecular weight; for example for sodium $\dfrac{[Na^+]\,mg/l}{22.98g/mol}$, calcium $\dfrac{[Ca^{++}]\,mg/l}{40.078g/mol}$ and magnesium $\dfrac{[Mg^{++}]\,mg/l}{24.31g/mol}$.

SAR is also a measure of soil *sodicity* when the analysis pertains to soil water. High values of SAR indicate less quality of the water for irrigation. Irrigation with high SAR leads to the substitution of calcium and magnesium by sodium (Glover 1996). Long-term effects of irrigation with high SAR water are soil degradation in terms of decreases in infiltration rate and structure.

LIGHT AS A FUNCTION OF DEPTH

Many aspects of water quality and processes in aquatic ecosystems depend on the amount of solar radiation reaching various depths in the water column. Solar radiation decreases when it goes through water and can be measured by submersed sensors or also estimated from surface measurements if we have models of light reflectance and attenuation. As with other media, the Beer-Lambert law (Chapter 11) applies to water and can be stated based on light intensity L which is attenuated exponentially according to

$$L(z) = L_s \exp(-kz) \tag{12.9}$$

where z is the depth, L_s is the water subsurface level, and k is the attenuation or extinction coefficient in [1/m]. The coefficient k includes absorption by organic and inorganic compounds as well as by photosynthesis. The attenuation coefficient is estimated by regression from values of light intensity going downwards (i.e., downwelling) at various values of depth. To measure total irradiance at each depth, a photosensor is submersed while connected via cable to a datalogger or hand-held meter above the surface. A logarithmic transformation can be used to perform regression on the light vs. depth data using the process explained in Chapter 11 (Figure 12.15) or we can alternatively use polynomial regression (introduced in Chapter 3), in which the predictor is a linear combination of increasing powers of X. In this case, we formulate the non-linear relation as a polynomial instead of a functional relationship, which is useful when you do not know what model to apply.

Although a solution is always found, we may not know the meaning or interpretation attached to coefficients. As mentioned in Chapter 3, we write the predictor as

$$\hat{Y} = b_0 + b_1 X + b_2 X^2 + \cdots + b_m X^m \tag{12.10}$$

where the polynomial is mth order. Figure 12.15 shows examples of second- and third-order polynomial regression applied to the exponential decay problem. In the case of $m = 3$, we estimated four coefficients

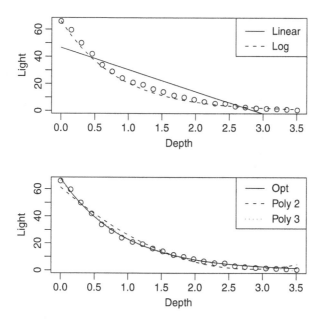

FIGURE 12.15 Comparing linear regression to polynomial regression for exponential attenuation.

$$\hat{Y} = b_0 + b_1 X + b_2 X^2 + b_3 X^3 \tag{12.11}$$

As we see in the figure, we have achieved a good fit. However, because we do not know the meaning of the coefficients, we cannot claim that we have a better understanding of a generic response of light to depth than the exponential model. One practical application of polynomial regression is for the calibration of sensors and instruments. By transforming each power of X into a variable, the polynomial regression is analogous to a multivariable linear regression problem (which we explained in Chapter 11).

The Beer-Lambert equation applies as well if we consider light intensity $L(\lambda)$ at a given wavelength λ, and in this case, the attenuation coefficient is specific to that wavelength. Measurements of light intensity spectra can be conducted by a spectrometer above the surface and piping light from underwater measurements depth to the spectrometer via an optical fiber.

For biological processes occurring in the water column, such as photosynthesis discussed in the next section, light intensity is measured as *photosynthetically active radiation* (PAR) or radiation in the 400–700 nm range. A *quantum sensor* provides PAR by measuring photosynthetic photon flux density as quanta or photons per unit time per unit area. Typical units are $\mu mols^{-1}m^{-2}$. Quantum sensors are based on silicon photovoltaic cells and require *cosine correction* to follow the ideal cosine function (Chapter 11). Quantum sensors for use underwater require watertight enclosures.

PRODUCTIVITY AND RESPIRATION

Primary producers convert solar energy to chemical energy stored in the form of carbohydrates, by photosynthesis, which is wavelength dependent and therefore we specify energy to that in the PAR range. Photosynthesis requires carbon dioxide CO_2 and water H_2O. The chemical reaction is

$$6CO_2 + 6H_2O \xrightarrow{\text{light}} C_6H_{12}O_6 + 6O_2 \tag{12.12}$$

Converting one mole of CO_2 to chemical energy requires eight E of solar radiation. E is a unit called Einstein. In terms of Joules, it takes 0.472 MJ to convert 1 mole of CO_2 or 0.00004 MJ to fix 1 mg of C.

The efficiency of photosynthesis as a percent of light energy converted to chemical energy is about 18%. The rate of photosynthesis can be expressed as a function of photosynthetic mass; for example, CO_2 fixed per unit time per chlorophyll concentration, [mmole CO_2 sec^{-1}/mg Chl m^{-3}].

Photosynthesis rate varies with light intensity; at low light levels, light reaction rates increase linearly with light. At higher levels, there is a saturation of Calvin cycle enzymes, and the photosynthesis rate drops at high light levels by photo inhibition. The net primary productivity rate is the gross productivity minus respiration. To obtain data for these rates in aquatic systems, we can use DO as an indicator and automatic dataloggers (data-sonde) to infer DO dynamics.

AUTOMATED REAL-TIME BIOMONITORING

Traditionally, built chemical and physical sensors are used to monitor water quality. In this section, we examine the use of organisms themselves to monitor the environment (Gruber et al. 1994). This approach introduces a *sentinel* organism that would be responsive to the variables we intend to monitor and then use transducers and sensors to measure the response of the sentinel organism (Figure 12.16). In other words, we build a real-time automated biological monitor or automated *biomonitor*, which should be able to integrate the array of environmental quality indicators we intend to measure. An example using clams was discussed in Chapter 9 when we illustrated the application of Bayes' rule to infer presence of an stressor.

Biomonitors employed for ecotoxicological monitoring respond to a great number of toxic conditions. Biological sensors employ sentinels that include bacteria, algae, invertebrates, and fish. Let us consider some examples. First, an algae-type biomonitor would use a fluorescence detector to measure the response of the algae to the stressor (a toxic compound) we intend to measure. Note that the fluorescent detector is not directly measuring the toxic concentration but rather the organism's response to that toxic. Going up in the food chain, consider a biosensor made with zooplankton, such as *Ceriodaphnia* typically used for ecotoxicology (Acevedo et al. 1995), and IR video recording of the movement and behavior of individual animals. Then, we can relate the video tracking data to the toxic concentration (Korver and Sprague 1988). We discuss video technology in more detail in Chapter 14. Sessile animals, such as bivalves, offer an opportunity to keep the biosensor fixed; in this example, we can attach a metal target to a valve of the animal and detect the valve movement by an electromagnetic proximity sensor (Figure 12.17). The valve opening and closing or feeding behavior relates to stressors in the water (Allen et al. 1996; Waller et al. 1994). As a last example, and going up in complexity, we can measure the ventilatory behavior of fish by non-contact submerged electrodes when subject to toxic stressors (Gruber et al. 1994).

Of interest are techniques allowing in vivo remote query measurement of bacteria cultures. For example, measuring the complex permittivity spectra of a biological culture solution by means of

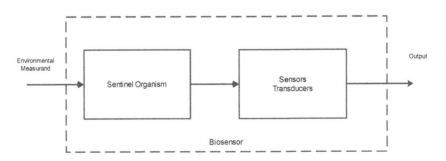

FIGURE 12.16 Basic concept of biosensor.

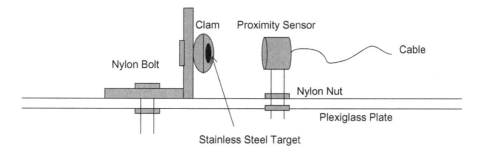

FIGURE 12.17   Clam biosensor. Adapted from Allen et al. (1996).

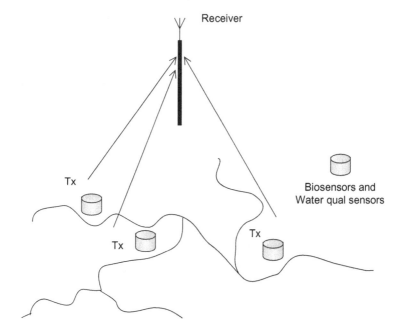

FIGURE 12.18   Biosensors as part of environmental observatories. Adapted from Allen et al. (1996).

a printed inductor-capacitor resonant circuit (Ong et al. 2001). The circuit is placed within the culture solution of interest, and by measuring the impedance spectrum of the sensor using a remotely located loop antenna; one can estimate permittivity, and then be able to monitor bacteria growth.

Biomonitors are effective early warning sensors and thus a combination of biomonitors with traditional environmental monitors can be part of observatories and early warning networks (Figure 12.18); for example, the Upper Mississippi River Early Warning Network (Allen et al. 2014). We will read in Chapter 14 how environmental sensors can be integrated with wildlife trackers thus providing data on the environmental parameters affecting the animal.

MODELING AND MONITORING OF SURFACE WATER

Hydrodynamic Models

Hydrodynamic models are computer models composed of differential equations of fluid dynamics that are used to predict water movement, i.e., water levels and water velocity (including direction) at all points of a water body. Long-term real-time monitoring of water levels and the velocity field plays an important role in providing these models with data for calibration and evaluation. Spatially, these models can be 1D, 2D, or 3D. For small streams and rivers, these models are 1D, with position along the course of the river as the only spatial variable of interest. Larger rivers may require 2D models when the velocity has significant variations across the river or significant variations with depth. Lakes, bays, and estuaries require 2D or 3D models, and these can become very complicated for large water bodies, particularly for interactions with much larger water bodies such as the seas. While some models may have a full fluid dynamics equation to be solved at high resolution, many models are simplified by dividing the water body into compartments or segments with a value for level and velocity representative of the segment. An example, of this type of model was developed for Lake Texoma, a reservoir created by impounding the Red River, which is located between the states of Texas and Oklahoma (García-Iturbe 2005). GIS software (Chapter 10) is useful to display the results of hydrodynamics models as well as ingesting spatial data to parameterize the models

WATER QUALITY MODELS

Water quality models are computer models composed of differential equations of chemical transport and fate of compounds in water that are used to predict concentrations of these compounds at all points of a water body. A hydrodynamic model is coupled to a water quality model to be able to provide the driver for the advection or transport of the compounds, while chemical reactions and biological processes drive the fate component of the model. Long-term real-time monitoring of water quality plays an important role in providing these models with data for calibration and evaluation. As with hydrodynamic models, water quality models can be 1D, 2D, or 3D, and the same considerations we made apply to these models. In addition, a benthic or bottom layer must be included to predict the processes occurring at the bottom of the water body. As mentioned for hydrodynamic models, many water quality models are simplified by dividing the water body into compartments or segments with concentrations representative of the segment. An example of this type of model was developed for Lake Texoma, was coupled to the hydrodynamics model of the previous section and predicts chlorides concentration change under hypothetical scenarios (García-Iturbe 2005).

HYDROLOGICAL MODELS

Hydrological models predict runoff and drainage from a watershed based on precipitation; their output consists of simulated *hydrographs*, i.e., time series plots of flow at important points of the watershed, for example points where hydrometric stations can provide measured hydrographs for calibration and evaluation. This type of model requires parameterization based on terrain and land use, which is accomplished using remote sensing and GIS to capture land cover and perform terrain analysis. Hydrological models can be conducted at various resolutions and approaches, ranging from a full raster at high-spatial resolution to a watershed divided in polygons representing homogeneous areas in terms of land use and terrain. Precipitation data from networks of weather stations around the watershed or NEXRAD data can be used to input rainfall events. In addition to water discharge, hydrological models may include the transport and fate of compounds that can be used to predict the contribution of the watershed to water quality of the receiving water body. For example, consider the watershed of Clear Creek that we mentioned previously as an example of a hydrometric station monitoring level and flow real time. A hydrological model based on curve numbers was developed for a set of sub-basins and hydrographs measured at the hydrometric station used for calibration and comparison (Redfearn 2005).

REMOTE SENSING OF WATER QUALITY

Remote sensing technologies are useful to monitor water quality data for all types of water bodies, increasing temporal and spatial coverage. You may have noticed when we studied the remote sensing image of Chapter 8, how the lakes showed some differences in water in certain areas. In general, solar radiation reaching water bodies are subject to direct reflection from the surface and absorption as discussed in the earlier section on light as a function of depth. A remote sensor will receive reflected light from the water surface and the emergent part of the upwelling light within the water. Both components are affected by the water quality of water, and therefore it is possible to analyze an image to extract water quality characteristics. Hyperspectral data offer many possibilities to extract water quality from spectral features, and with the advent of unmanned aerial vehicles, the capabilities of monitoring the water quality of smaller water bodies at higher resolution have improved. Many algorithms have been derived to infer TSS, chlorophyll-a, colored DOM, chemical oxygen demand, total nitrogen, and total phosphorus (Yang et al. 2022).

OCEAN MONITORING

Ocean monitoring includes temperature, salinity, sea level, and currents. Given the magnitude of the ocean, monitoring programs rely on multiple instruments and platforms deployed from boats, buoys, and floats, as well as remote sensing. In this section, we focus on monitoring ocean temperature and salinity. Ocean temperature includes sea surface temperature (SST) and temperature profiles as a function of depth. Ocean water absorbs incident solar energy incoming from the sun in the shortwave ranges as well as heat or IR reradiated from the atmosphere and changes its temperature according to its heat capacity. Absorbed heat is redistributed and not all oceans increase the temperature in the same manner, nor is temperature evenly distributed within oceans. Differences in SST are important factors in climate variability such as the El Niño Southern Oscillation that produces either droughts or excessive rainfall in different parts of the world. Ocean temperature monitoring encompasses large networks of observing stations on boats, buoys, and floats, as well as remote sensing from satellites.

Conductivity, Temperature, and Depth (CTD) instruments provide distribution and variation of water temperature, salinity, and depth of the oceans, down to thousands of meters, and can be deployed from boats, buoys, or floats. A shipboard CTD is composed of a set of probes attached to a large frame lowered on a cable down to the seafloor and collecting data real-time transmitted via wire to a computer on the ship, which also remotely operates water sample bottles as the instrument ascends. Small, low-power CTDs are used on autonomous instruments like the moored profiler, gliders, profiling floats, and autonomous underwater vehicles (AUVs). Additional instruments associated with the basic CTD can provide more information; for example, DO sensors, and Doppler current profilers that measure horizontal velocity.

Argo is a global array of thousands of free-drifting profiling floats that measure the temperature and salinity, at a function of depth of the upper 2 km of the ocean, allowing continuous monitoring of the upper ocean. Data are transmitted from the floats wirelessly and made available near real-time. Deployments began in 2000 and continue today at the rate of about 800 per year (NASA 2023a). Argo floats work on a 10-day mission cycle, spending most of the time drifting with deep ocean currents, taking measurements of EC, temperature, and pressure, and returning to the ocean surface for a brief interval where it gets location via GPS, communicates with a satellite to send its data. Currently, more than half of floats use GPS for position and transmit data to Iridium satellites which can provide data services anywhere on the Earth (Iridium 2023), thereby shortening the time that the float stays at the surface (Argo 2023).

An Argo float has (1) an antenna to send data, obtain the position, and receive new mission instructions; (2) a CTD instrument with an accuracy of 0.001°C for temperature, 0.1 dbar for pressure, and 0.001 psu for salinity. The latter calculated from conductivity, temperature, and pressure;

(3) a controller with a program to run the float; (4) a hydraulic pump that moves oil between an internal tank and an external bladder to control the buoyancy of the float; (5) batteries to power the pumps, sensors, controller, and communication system, and which limits the operational lifetime of the float (Argo 2023).

The Vital Signs web site (NASA 2023b) offers databases on ocean temperature and heat energy stored in the ocean in zettajoules (ZJ) or 10^{21} J. Measurements of December 2021 indicated 337 zJ added since 1955 integrated over the top 2km of the ocean. Stored heat causes ocean water to expand, which is responsible for about 30%–50% of global sea level rise. Most of the added energy is stored at the surface, at a depth of zero to 700 m.

NOAA provides a daily Optimum Interpolation SST, an analysis constructed by combining observations from different platforms (satellites, ships, buoys, and Argo floats) on a regular global grid, and interpolating to fill in gaps (NOAA 2023a). Satellite data are from the Advanced Very-High-Resolution Radiometer, which provide high temporal-spatial coverage since 1981.

GROUNDWATER MONITORING

As with surface water, groundwater monitoring may relate to quantity, represented by water levels and flows, and quality, related to constituents in the water. The major means of observing water levels and quality is by a network of wells located in the aquifers to be monitored. Most aquifers will show variation in level due to pumping and recharge, while others may show a declining trend with time when pumping is much more significant than recharge.

Groundwater quality is related to a combination of natural factors, mostly the geologic formation of the aquifer, and anthropogenic activities contributing to pollution of the aquifer. Natural factors are related to the minerals dissolved from the rocks in the basin draining to the aquifer; for example, an aquifer receiving water from a watershed with an abundance of limestone, dolomite, and gypsum will have a high content of calcium Ca, magnesium Mg, and sulfate SO_4. Anthropogenic activity may contribute to sodium Na, potassium K, and chloride Cl. A method used to differentiate the various sources of constituents in groundwater is based on the chemical ratios of constituents in the water. One example is SAR, which we discussed earlier in this chapter. Other ratios include Na/Cl, Ca/Cl, Mg/Cl, SO_4/Cl, (Na+Cl)/TDS, and (Ca+Mg)/(Na+K) (Alderman 2001).

In the United States, the USGS keeps groundwater monitoring data available online (USGS 2023d) with current conditions at selected sites based on automated recording equipment at a fixed interval, transmitted to the USGS every hour, and reported daily. For the sake of an example, we now look at site USGS 315712106361803 MBOWN-238 near Santa Teresa, New Mexico, and El Paso, Texas, which has water level data since June 10, 1985, and water quality data, including inorganics, nutrients, organics, pesticides, and radio isotopes, at selected years. One can download the water level time series for analysis. Figure 12.19 is a monthly average of the average water level showing a decreasing trend with a period of higher recharge in the early 2000s.

Also provided by the USGS, the Hydrologic Analysis Package (HASP) is an R package that can retrieve groundwater level and groundwater quality data from USGS, aggregate these data, plot them, and generate basic statistics (USGS 2023b). HASP also allows analysis of groundwater level trends in major aquifers. We will practice this package in the lab session companion of this chapter.

Another comprehensive source of groundwater monitoring data for the USA is the National Groundwater Monitoring Network (USGS 2023a) that encompasses groundwater monitoring wells from Federal, State, and local groundwater monitoring networks across the nation. Its data portal contains current and historical data, including water levels, water quality, lithology, and well construction. Water level data are useful to calibrate and evaluate groundwater hydrologic models that predict head and flow. There is an increased recognition that monitoring and modeling the entire watershed by including surface hydrology in conjunction with groundwater leads to a better understanding of the system. The Next-Generation Water Observing System (USGS 2023c) will provide real-time data on water quantity and quality in a more comprehensive manner in more locations.

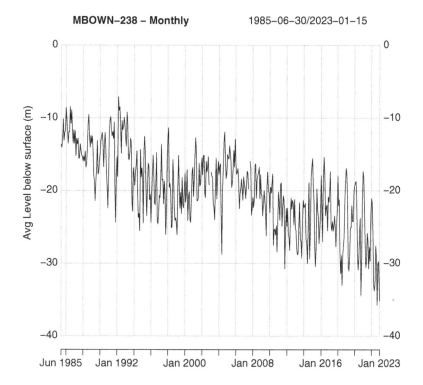

FIGURE 12.19   Depth to water level for well USGS 315712106361803 MBOWN-238.

AUTOREGRESSIVE ANALYSIS OF TIME SERIES

We have emphasized throughout the book the importance of time series analysis for environmental modeling. In this chapter, we present prediction models based on time series and regression. An *Autoregressive* (AR) process of order p, denoted AR(p), is such that

$$x(t) = a_1 x(t-1) + a_2 x(t-2) + \cdots + a_p x(t-p) + e(t) \qquad (12.13)$$

that is at time t, $x(t)$ is a linear combination of terms x lagged up to p plus some noise (or residual variability) $e(t)$. This noise is Gaussian white noise $N(0, \sigma)$ and $x(t)$ is a zero-mean process.

For example, an AR(1) process is simply

$$x(t) = a_1 x(t-1) + e(t) \qquad (12.14)$$

Note that of course for $a_1 = 0$, we have a white noise process, and scatter plots of lagged values would not indicate a potential relationship between lagged values. As an example, consider $a_1 = -0.5$, we would notice a slight negative relationship for values lagged by one Δt in Figure 12.20, while no relationship is suggested for other lags.

The *Yule-Walker* (YW) equations are relationships in terms of covariance or of correlation. Assume $\rho(h)$ is an autocorrelation of an AR(p) process $x(t)$, at lag h; then the a_i coefficients of the AR(p) satisfy the YW equations

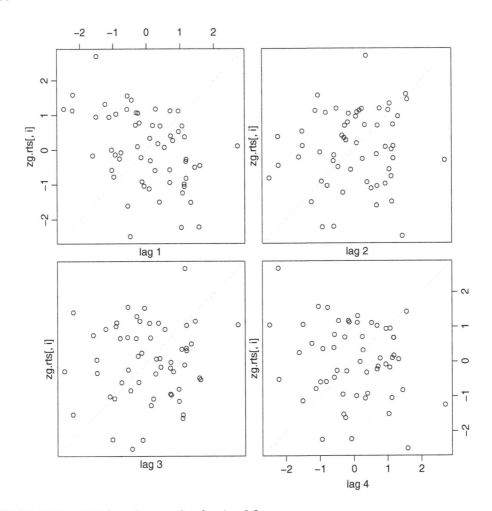

FIGURE 12.20 AR(1) lagged scatter plots for $a1 = -0.5$.

$$\rho(h) = a_1\rho(1-h) + a_2\rho(2-h) + \cdots + a_p\rho(p-h) \quad \text{where } h = 1, 2, \cdots, p \tag{12.15}$$

this is a system of p linear equations (recall that autocorrelation is even, $\rho(k) = \rho(-k)$)

$$\rho(1) = a_1\rho(0) + a_2\rho(1) + \cdots + a_p\rho(p-1)$$
$$\rho(2) = a_1\rho(1) + a_2\rho(0) + \cdots + a_p\rho(p-2)$$
$$\cdots \tag{12.16}$$
$$\rho(p) = a_1\rho(p-1) + a_2\rho(p-2) + \cdots + a_p\rho(0)$$

Which we write in matrix form (recall that $\rho(0) = 1$)

$$\begin{bmatrix} \rho(1) \\ \rho(2) \\ \cdots \\ \rho(p) \end{bmatrix} = \begin{bmatrix} 1 & \rho(1) & \cdots & \rho(p-1) \\ \rho(1) & 1 & \cdots & \rho(p-2) \\ \cdots & \cdots & \cdots & \cdots \\ \rho(p-1) & \rho(p-2) & \cdots & 1 \end{bmatrix} \begin{bmatrix} a_1 \\ a_2 \\ \cdots \\ a_p \end{bmatrix} \qquad (12.17)$$

Or for short using vector and matrix notation

$$\rho = \Phi a \qquad (12.18)$$

Matrix Φ is symmetric and has an inverse. Therefore, we can solve for coefficients $a = \Phi^{-1}\rho$.
For illustration, when AR(1) there is only one equation with an obvious solution

$$\rho(1) = a_1 \qquad (12.19)$$

For example, consider the series in Figure 12.21, the autocorrelation at lag 1 is $\rho(1) = -0.48$ and relatively smaller for higher lags. Thus, we could model the series as AR(1), then solving Equation (12.19), we have $a_1 = \rho(1) = -0.48$.

Increasing p from 1 to 2 we now have an AR(2)

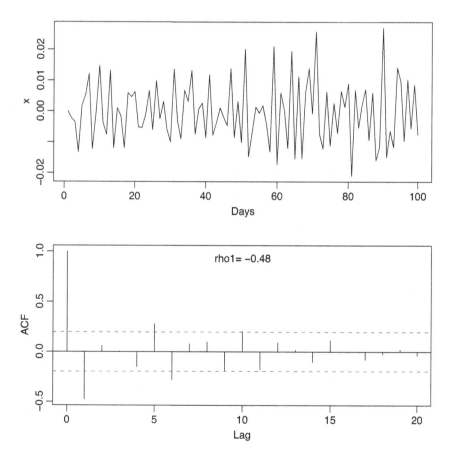

FIGURE 12.21 AR(1) Estimating coefficients from autocorrelation.

$$x(t) = a_1 x(t-1) + a_2 x(t-2) + e(t) \tag{12.20}$$

And the YW equations would be

$$\begin{bmatrix} \rho(1) \\ \rho(2) \end{bmatrix} = \begin{bmatrix} 1 & \rho(1) \\ \rho(1) & 1 \end{bmatrix} \begin{bmatrix} a_1 \\ a_2 \end{bmatrix} \tag{12.21}$$

Consider Figure 12.22, assume AR(2) and use the correlation values 0.3923 and −0.2735 to substitute in Equation (12.21)

$$\begin{bmatrix} 0.3923 \\ -0.2735 \end{bmatrix} = \begin{bmatrix} 1 & 0.3923 \\ 0.3923 & 1 \end{bmatrix} \begin{bmatrix} a_1 \\ a_2 \end{bmatrix} \tag{12.22}$$

Solving this equation yields $a_1 = 0.59$ and $a_2 = -0.51$. Note that we decided to ignore the correlation at lag=3. This series could be modeled as AR(3) by solving one more equation. An important question is then how to identify the order p of the AR(p).

The *partial autocorrelation function* (PACF) is the ACF truncated after lag p (Box and Jenkins 1976). Partial autocorrelation is obtained by recursion: fit AR(p) models successively from $p=1$ to the maximum lag solving the YW equations. The structure of the equations makes them solvable

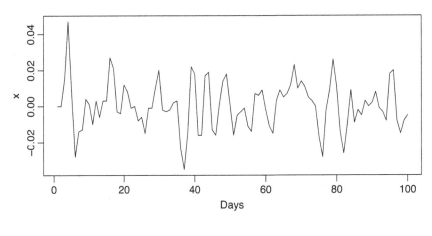

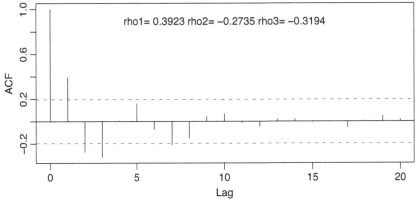

FIGURE 12.22   AR(2) Estimating coefficients from autocorrelation.

by a recursive method. For each step in the recursion, keep the pth coefficient. We try several values of p until the partial autocorrelation value for the next p is very low or within a confidence interval. For example, the AR(2) process estimated above yields PACF 0.39 and -0.51 which correspond to the ACF at lag 1 and the a_2 estimated by solving YW shown in equation (12.22).

For evaluation, consider that the residual time series should behave like Gaussian white noise; red flags to look for are outliers, trends, and drift. A handy tool to check that the residuals behave like white noise is to do an autocorrelation of the residuals. All spikes except lag=0 should be within the confidence interval. To identify the order p, we can use the Akaike Information Criterion (AIC) already studied in previous chapters. Its objective is to balance the reduction of estimated error variance with the number of estimated parameters; this is accomplished by minimizing

$$AIC(p) \sim \log(\sigma_X^2) + 2p \tag{12.23}$$

Figure 12.23 illustrates the results for the simple example just discussed. We can see the plot of PACF with only two prominent spikes (top left panel), the AIC plot dropping to zero at $p=3$ (top right panel), and the Gaussian noise behavior of the residuals (bottom panels). In the lab session, we will learn how to employ these techniques for a general value of p.

We use this model to forecast 10 days ahead from day 50 of the data series and calculate upper and lower limits based on the double of the standard error. Figure 12.24 shows the results where we

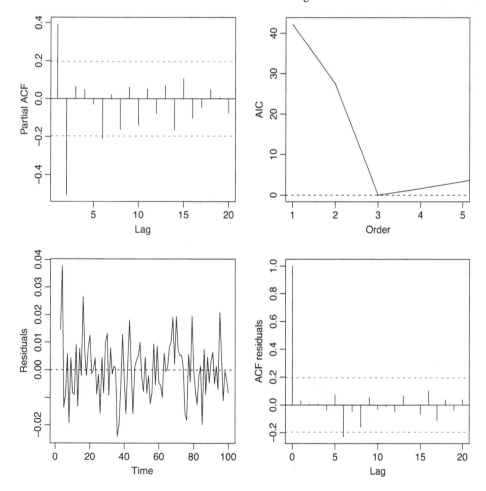

FIGURE 12.23 Results of AR(2): PACF, AIC, residual, and its ACF.

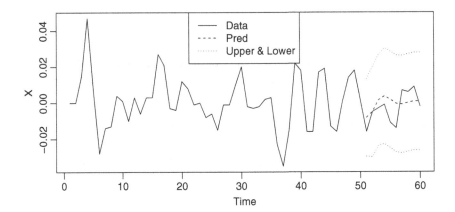

FIGURE 12.24   Ten days ahead forecast using AR(2) for the first 50 days of data and comparing to observed data.

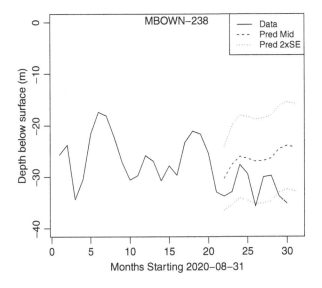

FIGURE 12.25   Groundwater monthly level predicted from an AR(11) model. Data from station USGS 315712106361803 MBOWN-238.

plot the forecast together with the existing values in the interval 50–60 to evaluate the prediction. You can see that the mean of the prediction follows the general pattern of the data.

As an application example, we take the time series of groundwater levels discussed in the previous section, use the monthly average, calibrate an AR model that yields 11 coefficients, select a segment for prediction and one for testing, and compare the predictions to the existing test data. The result is shown in Figure 12.25 that can be interpreted in a similar manner to the previous figure.

EXERCISES

Exercise 12.1

Assume we have an acoustic water level sensor looking at the surface of a stream from a height of 6 m above the streambed. Assume that the speed of sound in dry air is $v_s = 331.4 + a \times T$ m/s where T is the air temperature in °C, and $a = 0.6 \frac{m/s}{°C}$. At time 16:00:00, the sound travel time of the sound wave is 20 ms, the air temperature is 5°C, and RH is almost 0. Calculate water level or stream height at 16:00:00.

Exercise 12.2

Suppose we measure EC 800 µS/cm at 30°C. Assume that the temperature coefficient is 0.020 1/°C. What would be the reference conductivity at 25°C?

Exercise 12.3

Suppose we measure EC 800 µS/cm at 25°C. What would be the corresponding TDS on a 500 scale.

Exercise 12.4

Suppose water analysis of well water yields sodium 359 mg/L, calcium 301 mg/L, and magnesium 187 mg/L. What is the SAR?

Exercise 12.5

Visit the Vital Signs web site and explore the data offered for the increase of heat energy in ZJ as a function of time. Calculate an approximate rate of increase for three periods in the curve.

Exercise 12.6

Consider the graph of Figure 12.19 and calculate an approximate overall rate at which level is decreasing since the 2000s.

Exercise 12.7

Write the YW equation for an AR(3) process.

REFERENCES

Acevedo, M.F. 2024. *Real-Time Environmental Monitoring: Sensors and Systems, Second Edition – Lab Manual*. Boca Raton, FL: CRC Press, Taylor & Francis Group. 463 pp.

Acevedo, M. F., T. W. Waller, D. Smith, D. Poage, and P. McIntyre. 1995. Cladoceran population responses to stress with particular reference to sexual reproduction. *Non Linear World* 2:97–129.

Alderman, J. H. 2001. *Development of a Procedure to Evaluate Groundwater Quality and Potential Sources of Contamination in the East Texas Basin*. PhD Dissertation (Environmental Science), Denton, TX: University of North Texas.

Allen, J., W. Franz, D. Macke, and S. Panguluri. 2014. *Source Water Monitoring and Biomonitoring Systems*. EPA, accessed 2014. http://www.horsleywitten.com/sourcewater/pdf/Allen.pdf.

Allen, J. H., W. T. Waller, M. F. Acevedo, E. L. Morgan, K. L. Dickson, and J. H. Kennedy. 1996. A minimally-invasive technique to monitor valve movement behavior in bivalves. *Environmental Technology* 17:501–507.

Argo. 2023. *How do floats work*. accessed January 2023. https://argo.ucsd.edu/how-do-floats-work/.

Artiola, J. F. 2004. "Monitoring Surface Waters." In *Environmental Monitoring and Characterization*, edited by J. F. Artiola, I. L. Pepper and M. L. Brusseau, 141–161. Burlington, MA: Elsevier Academic Press.

Box, G. E. P., and G. M. Jenkins. 1976. *Time Series Analysis: Forecast and Control*. San Francisco, CA: Holden-Day.

Decagon Devices. 2014. *ES-2 Electrical Conductivity and Temperature Sensor*. Pullman, WA: Decagon Devices Inc.

Fondriest Environmental. 2014. *Fluorometers*. accessed January 2015. http://www.fondriest.com/environmental/water-quality/fluorometers.htm.

Fricker, P. 2014. *Analyzing and Visualizing Flows in Rivers and Lakes with MATLAB*. MathWorks, accessed April 2020. https://www.mathworks.com/company/newsletters/articles/analyzing-and-visualizing-flows-in-rivers-and-lakes-with-matlab.html.

García-Iturbe, S. 2005. *Simulation of Physical and Chemical Processes in Reservoirs: Two Case Studies*. PhD Dissertation (Environmental Science), Denton, TX: University of North Texas.

Gertz, E., and P. Di Justo. 2012. *Environmental Monitoring with Arduino: Building Simple Devices to Collect Data about the World Around Us*. Sebastopol, CA: O'Reilly Media Inc.

Glover, C. R. 1996. *Irrigation Water Classification Systems*. Las Cruces, NM: New Mexico State University.

Gruber, D., C. H. Frago, and W. J. Rasnake. 1994. Automated biomonitors - first line of defense. *Journal of Aquatic Ecosystem Health* 3:87–92.

Iridium. 2023. *Iridium*. accessed January 2023. https://www.iridium.com/.

Korver, R. M., and J. B Sprague. 1988. "A Real-Time Computerized Video Tracking System to Monitor Locomotor Behavior." In *Automated Biomonitoring: Living Sensors as Environmental Monitors*, edited by D. S. Gruber and M. Diamond, 157–171. Chichester: Ellis Horwood Ltd.

NASA. 2023a. *Missions: Argo*. accessed January 2023. https://sealevel.nasa.gov/missions/argo.

NASA. 2023b. *Vital Signs Ocean Warming*. accessed January 2023. https://climate.nasa.gov/vital-signs/ocean-warming/.

NOAA. 2017. *Harmonic Constituents - Station Selection*. NOAA, accessed October 2017. https://tidesandcurrents.noaa.gov/stations.html?type=Harmonic+Constituents.

NOAA. 2023a. *Ocean Temperatures*. accessed November 2022. https://www.nesdis.noaa.gov/ocean-temperatures.

NOAA. 2023b. *Tides and Currents*. accessed January 2023. https://tidesandcurrents.noaa.gov/.

Ong, K. G., J. Wang, R. S. Singh, L. G. Bachas, and C. A. Grimes. 2001. Monitoring of bacteria growth using a wireless, remote query resonant-circuit sensor: application to environmental sensing. *Biosensors and Bioelectronics* 16 (4–5):305–312.

Pallás-Areny, R., and J. G. Webster. 2001. *Sensors and Signal Conditioning*. New York: Wiley.

Radiometer Analytical. 2004. *Conductivity Theory and Practice*. Villeurbanne Cedex, France: Radiometer Analytical SAS.

Redfearn, H. 2005. *Rainfall-Runoff Changes due to Urbanization: A Comparison of Different Spatial Resolutions For Lumped Surface Water Hydrology Models Using HEC-HMS*. MS Thesis, Environmental Science, Denton, TX: University of North Texas.

Sadar, M. 2014. *Turbidity Measurement: A Simple, Effective Indicator of Water Quality Change* Hach Hydromet, accessed 2014. http://www.ott.com/en-us/products/download/turbidity-white-paper/.

Sanders, P. 2014. *An Introduction to Algae Measurements Using In Vivo Fluorescence*. accessed 2014. http://www.ott.com/download/fluorescence-white-paper/.

Turner Designs. 2014. *C3 Submersible Fluorometer. User's Manual*. Turner Designs, accessed 2015. http://www.turnerdesigns.com/t2/doc/manuals/998-2300.pdf.

U.S. EPA. 2014a. *NPDES Home*. US EPA, accessed 2014. http://water.epa.gov/polwaste/npdes/.

U.S. EPA. 2014b. *STORET/WQX*. accessed 2014. http://www.epa.gov/storet/.

U.S. EPA. 2014c. *Summary of the Clean Water Act*. US EPA, accessed 2014. http://www2.epa.gov/laws-regulations/summary-clean-water-act.

U.S. EPA. 2014d. *Water Monitoring and Assessment. 5.5 Turbidity*. U.S. EPA, accessed 2014. http://water.epa.gov/type/rsl/monitoring/vms55.cfm.

U.S. EPA. 2014e. *Water Monitoring and Assessment. 5.9 Conductivity*. US EPA, accessed January 2014. http://water.epa.gov/type/rsl/monitoring/vms59.cfm.

USGS. 2023a. *Groundwater Monitoring*. accessed January 2023. https://www.usgs.gov/programs/groundwater-and-streamflow-information-program/groundwater-monitoring.

USGS. 2023b. *HASP*. accessed January 2023. https://rconnect.usgs.gov/HASP_docs/.

USGS. 2023c. *Next Generation Water Observing System (NGWOS)* accessed January 2023. https://www.usgs.gov/mission-areas/water-resources/science/next-generation-water-observing-system-ngwos.

USGS. 2023d. *USGS Groundwater Data for the Nation*. accessed January 2023. https://waterdata.usgs.gov/nwis/gw.

USGS. 2023e. *WaterWatch*. accessed January 2023. https://waterwatch.usgs.gov/index.php.

Waller, W. T., M. F. Acevedo, E. L. Morgan, K. L. Dickson, J. H. Kennedy, L. P. Ammann, H. L. Allen, and P. R. Keating. 1994. "Biological and Chemical Testing in Storm Water." In *Stormwater NPDES Related Monitoring Needs*, edited by H.C. Torno. Colorado: Mount Crested Butte.

Yang, H., J. Kong, H. Hu, Y. Du, M. Gao, and F. Chen. 2022. A review of remote sensing for water quality retrieval: Progress and challenges. *Remote Sensing* 14:1770.

13 Terrestrial Ecosystems Monitoring

INTRODUCTION

This chapter is devoted to monitoring a variety of characteristics and processes in terrestrial ecosystems, including the *vadose* zone (i.e., the terrestrial subsurface from the surface to the groundwater table, including the soil) and vegetation. We already covered surface water and groundwater in Chapter 12, and we postpone discussion of terrestrial animals to Chapter 14. After discussing soil monitoring, we will focus on those variables related to plant productivity together with water use, gas fluxes, and relationships with environmental factors, including evapotranspiration models. Plant productivity and leaf area are described at several scales from plant to canopy. We also discuss ground penetrating radar, lysimeters, and remote sensing. This chapter ends with an overview of remote sensing applications to terrestrial systems, emphasizing vegetation and soil moisture. Although the sensor material is presented thinking of natural environments, some of the techniques contained here are applicable as well to greenhouses, nurseries, or plantations. These concepts are further elaborated by computer exercises in Lab 13 of the companion Lab Manual (Acevedo 2024).

SOIL MOISTURE

Measuring water content in the soil by electronic devices is based on soil electrical properties, such as dielectric constant (i.e., relative permittivity) which has been shown to be an effective estimator of soil moisture, and several techniques are used to relate the response of an electric circuit to relative permittivity (Robinson et al. 1998, 1999). Among these techniques, time domain reflectometry (TDR) (Kelleners et al. 2005; Kallioras et al. 2016; Robinson et al. 1999) and capacitance probes (Robinson et al. 1998, 1999; Fares et al. 2009; Kargas and Soulis 2012) are prevalent soil relative permittivity sensors.

TDR

We mentioned in Chapter 4 that the dielectric properties of soil vary with soil water content, this principle is used in TDR to measure soil moisture (Nielsen et al. 1995; Jones et al. 2002) by relating waveform properties of an electromagnetic (EM) pulse in the microwave range (2–3 GHz) injected to a waveguide (steel rods) inserted into the soil (Figure 13.1). The waveguide is an extension of a coaxial cable; one rod connects to the wire inside the coaxial cable and the other to the cable shield (Figure 13.2).

Pulse wave velocity depends on the soil dielectric properties and therefore, the travel time of the reflected signal from the open end of the waveguide depends on soil water content. Increased moisture increases permittivity, and this translates into an increase in travel time. A microcontroller infers travel time from the waveform making it amenable for datalogging and real-time continuous monitoring. The travel time T_p is related to the pulse velocity V_p and the travel path length given by twice the length L_p of the waveguide

$$V_p = \frac{2L_p}{T_p} \tag{13.1}$$

DOI: 10.1201/9781003425496-13

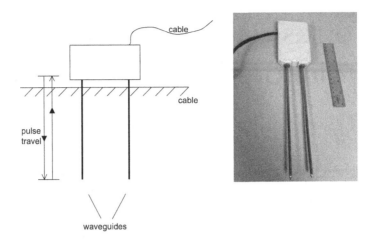

FIGURE 13.1   TDR probe to measure soil moisture.

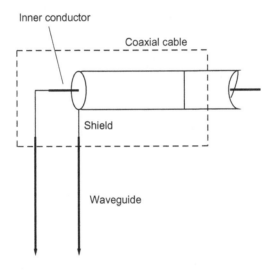

FIGURE 13.2   TDR probe: from coaxial cable to waveguide.

The velocity of the received reflected pulse V_p is related to permittivity by

$$V_p = \frac{c}{\sqrt{\varepsilon_r \mu_r}} \tag{13.2}$$

where c is the speed of light, ε_r and μ_r are the relative electrical permittivity and magnetic permeability of the soil, respectively. Relative magnetic permeability μ_r of soil is equal to 1 since typically there is no presence of magnetic material. Therefore, from equations (13.2) and (13.1), once we measure T_p, we get permittivity from

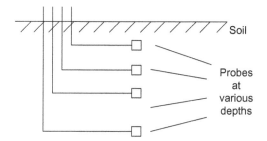

FIGURE 13.3   Obtaining a profile of soil moisture using sondes at various depths.

$$\varepsilon_r = \left(\frac{c}{V_p} \right)^2 = \left(\frac{cT_p}{2L_p} \right)^2 \tag{13.3}$$

Permittivity is a complex number, i.e., it has real and imaginary parts, and the above value is just the real part. The imaginary part is proportional to the soil conductivity and inversely proportional to frequency. Thus, it is only relevant for high values of soil conductivity; furthermore, the probe minimizes its effect by using high pulse frequency (greater than 3 GHz).

Once we estimate permittivity by Equation (13.3), we relate water content to the permittivity by using one of several equations proposed in the literature. Although manufacturers provide a generic calibration curve relating dielectric constant to moisture, calibration may be required for specific soil types and conditions. Likewise, deployment may be more difficult for some soil conditions, since this probe requires direct contact between the waveguide rods and the soil; for example, we need to make sure that there is no air in between the rods and the soil. The TDR probes provide a local measurement of soil moisture because their zone of influence is small (in order of inches). Therefore, we must use several probes to obtain a soil moisture profile (Greco and Guida 2008), or a more global view of soil moisture conditions (Figure 13.3).

CAPACITANCE PROBES

Capacitance probes are less expensive than TDR probes; therefore, they have become popular for developing low-cost soil moisture sensor systems (Fares et al. 2011; Kargas and Soulis 2012; Visconti et al. 2014). Typically, the output of an off-the-shelf capacitance probe requires only a low DC excitation voltage (e.g., 2–5 V), producing a signal in the order of mV, which can be converted to a digital value and calibrated to soil moisture, as volumetric water content (VWC) for various soil types. Out of several off-the-shelf capacitance probes, the Decagon's electrical conductivity (EC)-5 probes are widely used (METER 2020), while Campbell Scientific TDR devices are commonly installed in weather and soil monitoring stations (Campbell Scientific Inc. 2020a) and also integrated as portable handheld units such as the HS-2 Hydrosense II (Campbell Scientific Inc. 2020b). Calibration equations are developed by experiments collecting voltage output concurrently with soil moisture obtained from a controlled experiment measuring water content by weight.

SOIL TENSION: TENSIOMETER

A tensiometer transducer is based on a bridge circuit with strain gages or piezoresistive sensors. Examples are UMS (Cobos 2007) and TEROS 31 Tensiometers (Meter 2022). The UMS Tensiometer

FIGURE 13.4 UMS Tensiometer.

(Figure 13.4) uses a thin wafer piezoresistive pressure sensor to measure soil water tension; the sensor resistance changes when deformed by the pressure difference of water tension and atmosphere. The signal is obtained placing the sensor in a Wheatstone full bridge (Cobos 2007).

INFILTROMETERS

An important process determining soil moisture dynamics is the hydraulic properties of soil controlling how fast water goes into the soil or *infiltration rate* and therefore becomes an important factor in terrestrial ecosystems. Infiltration rate changes as a function of soil water content. An important condition to establish infiltration rate is the *hydraulic conductivity* when the soil is saturated, denoted by K_s or K_{sat}.

An infiltrometer is an apparatus to perform this measurement by using the rate of change of water level in a reservoir that provides water to the soil. Simultaneously, it monitors the VWC of the soil to perform a balance. The simplest design is a single-ring infiltrometer, which consists of partially inserting a ring in the soil (Reynolds and Elrick 1990; Prieksat et al. 1992). The ring will hold ponded water acting as an open reservoir (Figure 13.5). We can operate it a constant head, i.e., constant water level, or at falling head, i.e., decreasing water level. In the former, infiltration rate is the rate at which we must supply water to keep the level constant, which is a process that can be automated using the water level sensor to control a valve to supply water as the head tries to decrease. When operating at falling head, infiltration rate is the rate of water level decrease.

A double-ring infiltrometer (Touma and Albergel 1992) adds a second ring, larger and concentric to the measuring ring, to reduce the effect of lateral movement of water (Figure 13.6). The water flow from the inner ring to the soil would then be mostly vertical. This infiltrometer can also be used at constant head or falling head.

A more sophisticated design is a disk infiltrometer (Angulo-Jaramillo et al. 2000), which has a Mariotte column to serve as a reservoir and a disk to make good contact between the water in the column and the surface of the soil (Figure 13.7). We can use various types of sensors to monitor water level in the column and soil water content. For example, we can use TDR to measure soil water content and a resistance-based level sensor to measure water in the reservoir, or TDR for both, water level and soil content (Moret et al. 2004).

Once we collect data, it can be processed by a variety of software tools and inform models of soil water dynamics (Ankeny et al. 1993; Lepore et al. 2009; Chen 2008). Flow sensors, peristaltic pumps, capacitance probes, and air pressure control can be used as the basis of building a continuous automated infiltrometer capable of producing variable hydraulic head conditions without

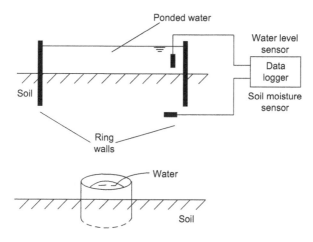

FIGURE 13.5 Single-ring infiltrometer with sensors to monitor water level change and soil moisture.

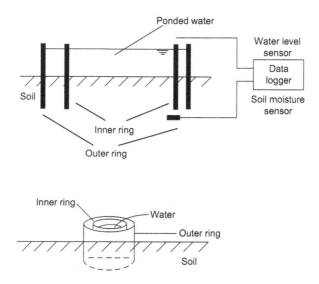

FIGURE 13.6 Double-ring infiltrometer with sensors to monitor water level change and soil moisture.

requiring user intervention (Wacker et al. 2013). Such a device enables to run multiple infiltrometers getting a better quantification of spatial variability.

SOIL EC

In addition to soil moisture, it is of great interest to measure soil EC because it provides an estimate of soil salinity as discussed in Chapter 12, and of course EC must be accompanied by sensing temperature so that we can properly correct EC for temperature. Examples are the TEROS 12 that is a transducer equipped with an SDI-12 interface and the JXBS-3001-EC which interfaces with other

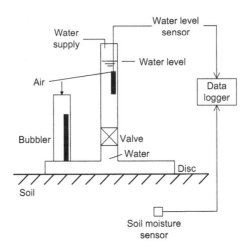

FIGURE 13.7   Disk infiltrometer using a Mariotte column as a reservoir and sensors to monitor water level in reservoir and soil water content.

sensors and dataloggers using RS-485. We described these sensors in Chapter 5 in the context of interfaces and presented an example of using the JXBS-3001-EC in a WSN in Chapter 6.

To further illustrate the use of TEROS 12, consider Figure 13.8 that shows VWC, temperature, and EC, measured by a TEROS 12 with address "A" buried at 6 inches below the surface in an agronomic experimental plot irrigated with brackish groundwater of ~1600 µS/cm. The spikes in VWC and EC correspond to irrigation events, the oscillation of temperature corresponds to the diurnal cycle of insolation. This sensor "A" is part of a network of 16 TEROS 12 sensors, installed one in each of 16 plots, treated with either well water or desalinated water, and amended or not by compost inoculation. This 16-sensor network is divided in 8 clusters of 2 sensors each to facilitate troubleshooting in the field. A Python program running on a Raspberry Pi collects data from the sensor network and creates a datalog file that is processed by an R script to produce graphs like Figure 13.8. This network is deployed at the Brackish Groundwater National Desalination Research Facility (Alamogordo, NM) and is part of a project aiming to improve crop yield and soil salinity funded by the INFEWS (Innovations at the Nexus of Food, Energy and Water Systems) program of the National Science Foundation and US Department of Agriculture.

EVAPOTRANSPIRATION

Soil water balance for a terrestrial ecosystem is determined by input from precipitation and demand from evapotranspiration (ET). This last term is a combination of evaporation and transpiration by plants. Evaporation occurs from surface water, soil surface, or water intercepted from rain by the canopy. Transpiration by plants draws water from the soil and thus ET represents a demand for soil water, which becomes part of water balance for a terrestrial ecosystem. The *actual* ET is a fraction of the *potential* ET (PET) demand estimated from weather conditions, because the soil may not have enough moisture to supply the PET.

A reference ET, denoted by ET_0, is calculated by combining several variables: air temperature, solar radiation, relative humidity (RH), and wind speed. These are variables typically measured by a weather station (Chapter 11). There are several models to calculate ET_0. For example, the Penman method requires solar radiation, air temperature, wind speed, and RH as inputs. It assumes that the total evaporation is due to two terms: an energy term (driven by net solar radiation) and an aerodynamic term (driven by wind speed and RH). Air temperature affects both terms.

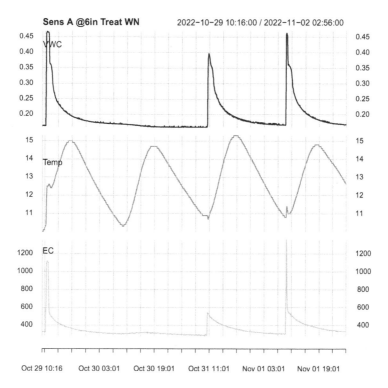

FIGURE 13.8   Example of VWC, temperature (in °C), and EC (in μS/cm) measurements using a TEROS 12 node part of a network.

We will study the Penman-Monteith equation following calculations based on the Food and Agriculture Organization (FAO) method, as described for example in Chapter 4 of Allen et al. (1998). Applying the Penman-Monteith method requires solar radiation, air temperature, wind speed, and RH as inputs. It assumes that the total evaporation is due to two terms: an energy term (driven by net solar radiation) and an aerodynamic term (driven by wind speed and RH). Air temperature affects both terms. The equation can be applied in several time scales; in the following, we calculate hourly values in mm/h.

The forcing weather variables are hourly values of T average air temperature in °C, Q total solar energy radiation in MJm^{-2}, average RH (%), average P barometric pressure in kPa, and average u wind speed at a height of 2 m in m/s. It requires values for α=surface albedo, which is about 0.23 for vegetation. Recall that Pa and mbar are units of pressure and are related by 1000 mbar= 1bar= 100kPa.

Water vapor pressure at saturation in kPa as a function of T in °C

$$e_s(T) = 0.6108 \times \exp\left[\frac{17.27 \times T}{T + 237.3} \right] \tag{13.4}$$

Vapor pressure is a fraction RH/100 of the vapor pressure at saturation,

$$e(T,P,RH) = \frac{e_s(T,P)RH}{100} \tag{13.5}$$

Using this estimate and air temperature allows to calculate the slope of saturation vapor pressure curve (kPa/°C)

$$\Delta(T) = 4096 \left[\frac{e_s}{(T + 237.3)^2} \right] \tag{13.6}$$

The γ=psychrometric constant (kPa/°C) is a function of P=barometric pressure in kPa, c_p= specific heat capacity of the air at constant pressure, latent heat of vaporization, and molecular ratio of wet and dry air. For simplicity, we assume that the psychrometric constant be approximated by

$$\gamma \simeq 0.7 \times 10^{-3} P \tag{13.7}$$

Net radiation (MJm^{-2}) is incoming radiation (discounting reflected radiation given by albedo).

$$R_n = Q(1 - \alpha) \tag{13.8}$$

Aerodynamic resistance (s/m) is a function of wind speed at a height of $z = 2\,\text{m}$, but here, we will use a factor, where 37 would be replaced by 900 if calculating daily

$$f(u) = \frac{37}{T + 273} u \tag{13.9}$$

Latent heat of vaporization in MJ/kg will be simplified to

$$L(T) = 2.5 - 0.0022 \times T \tag{13.10}$$

Now we combine radiation and aerodynamic terms using weighting factors for each,

$$W_R(T) = \frac{\Delta(T)}{\Delta(T) + \gamma \times (1 - 0.34u)} \tag{13.11}$$

$$W_A(T) = \frac{\gamma}{\Delta(T) + \gamma \times (1 - 0.34u)} \tag{13.12}$$

Using G=soil heat-flux the contribution from radiation is

$$E_r = \frac{R_n - G}{L(T)} W_R(T) \tag{13.13}$$

and from aerodynamics is

$$E_a = f(u)(es - e)W_A(T) \tag{13.14}$$

after adding up, we get ET$_0$

$$ET_0 = E_r + E_a \tag{13.15}$$

The ET_0 reference values are adjusted by a crop coefficient in order to estimate ET specific to the crop cover. Given the importance of ET for agriculture, many meteorological networks report estimates of ET_0 along with weather variables and crop coefficients. For example visit websites managed by Colorado State University (2023).

SAP FLOW

Sap flow sensors measure water consumption by plants. These are energy balance sensors that measure the heat carried by the sap, which is converted into sap flow in grams per hour based on heat convection principles. The sensors are non-intrusive if the plants heat up only by a small amount (~1°C). This sensing technology allows to measure water use by plants of agricultural, economic, and ecological importance. To accommodate a variety of applications, sap flow sensors come in several sizes, adapted to stems and trunks (Figure 13.9). Examples are sap flow sensors by Dynamax, which has a full range of sensors from 2 up to 125 mm (DYNAMAX 2014).

Electrical power input P_{in} (in W) balances with tissue heat flow Q_v in the vertical direction, heat flow Q_r in the radial direction, and by heat convection Q_f by the sap flow

$$Q_f = P_{in} - Q_v - Q_r \tag{13.16}$$

if we can determine Q_f from measurements and physical principles, then sap flow rate q_s (in g/s) would be calculated as

$$q_s = \frac{Q_f}{C_p \times dT} \tag{13.17}$$

where C_p is the specific heat of water and dT is measured by thermocouples. In order to determine Q_f, we would need to determine Q_v and Q_r from geometry, thermal gradients, and tissue thermal conductivity. This is difficult for large diameter and an open stem environment, but more feasible for thin stems and by isolating the stem into a closed compartment (Herzog et al. 1997).

Other methods are based on estimating sap-flux density ($cm^3 cm^{-2} h^{-1}$), either by continuous heating (such as the thermal dissipation and heat field deformation methods) or by heat pulses (Vandegehuchte and Steppe 2013). Water content and tissue thermal properties are assumed constants, however these may vary and therefore generate difficulty in the estimation of sap-flow density (Lopez-Bernal et al. 2014; Vergeynst et al. 2014).

LYSIMETERS

In its basic form, a lysimeter is an instrument designed to measure water movement in the vadose zone and has been expanded to include monitoring of other variables, including ET, nutrients and leachates, and contaminants. Therefore, there are several types of lysimeters, such as weighing, drainage, percolation, sampling, and pan lysimeters. We will discuss just a couple of these. Long-term data from lysimeters contribute to understand soil water and temperature dynamics of great value for agriculture (Seyfried et al. 2001).

A *weighing* lysimeter is a contained parcel of soil sitting on a weigh scale that measures the total water balance (input minus output) by the change in weight over time. The amount of soil on top of the scale and the desired accuracy of weight determine the complexity of the lysimeter. Some lysimeters contain a relatively small amount of soil (e.g., 0.25 m^2 of area and 0.35 m of depth) and can be installed temporarily at different field sites, e.g., Freebairn et al. (1986), whereas others have a very

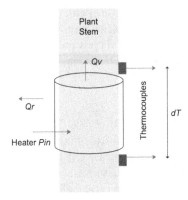

FIGURE 13.9   Sap-flow sensor.

large amount of soil, are permanently installed at a site, and have underground access for operation and maintenance, e.g., Andales et al. (2018).

Drainage lysimeters measure the water exiting the soil profile and allow to measure nutrients, salts, or contaminants, leached with that water. They work by collecting the water that infiltrates the soil by a divergence control tube and a wick and storing it into a reservoir that stores the water until its volume is measured and the water is analyzed for chemical constituents, thus water flow and chemical flux are determined.

PRODUCTIVITY

In this section we describe methods to estimate primary productivity based on gas flux, which can be employed at the leaf level and the canopy level.

GAS EXCHANGE

Measuring gas exchange allows to measure photosynthesis of individual leaves, whole plants, or even plant canopies. Gas exchange is a direct measure of net rate of photosynthetic carbon assimilation and is instantaneous and non-destructive. At the leaf level, CO_2 exchange systems can be closed or open. In closed systems, we place the leaf entirely in a closed transparent chamber, whereas in an open system, we partially cover the leaf by the chamber. The latter is the more common method now employed. The rate of CO_2 fixed by the leaf is given by the change in the CO_2 concentration of the chamber's airflow. Due to the low values of ambient atmospheric CO_2 concentration (~400 ppm), its measurement requires sensitive sensors.

We already explained in Chapter 12 some of the methods available to measure the concentration of gases in atmospheric air. Recall that CO_2 absorbs in the infrared (IR) wavelength range and we can use absorption spectroscopy to measure CO_2 concentration. In photosynthetic applications, we need to correct for water vapor presence (which also absorbs IR) by drying the air or by measuring vapor separately for correction.

CHLOROPHYLL FLUORESCENCE COMBINED WITH GAS EXCHANGE

Fluorescence techniques allow the measurement of how absorbed light is actually used within the leaf. At this time, we need to review that photosynthesis entails two major processes: firstly, light energy is converted to chemical energy by electron transport; secondly, CO_2 is taken from the air and assimilated. The first process encompasses the light reactions, and the second one consists of the Calvin reactions. Fluorescence, in combination with gas exchange applied to the same portion of a leaf, allows to determine quantum or light efficiency ψ of each one of these processes, i.e., electron transport ψ_{PSII} and quantum efficiency of carbon assimilation ψCO_2.

When both efficiencies match, we conclude that most of the chemical energy produced by the light reactions goes to CO_2 assimilation; however, a larger difference indicates that some of the energy produced goes into other processes different from assimilating CO_2.

CANOPY GAS EXCHANGE

Once we measure leaf gas exchange, we can extrapolate to the canopy level using models. However, it is also possible to measure gas exchange of the entire canopy to estimate the instantaneous productivity of a plant community. Flux of CO_2 is a measure of ecosystem metabolism and a key to understand the relationship between these ecosystems and climate. A chamber is placed over the canopy to monitor CO_2 flux. However, the chamber itself affects canopy conditions, since it alters air temperature, wind speed conditions, and radiation balance.

MICROMETEOROLOGICAL FLUX MEASUREMENTS

We can avoid the potentially problematic effects of closing the canopy by estimating fluxes in the open canopy based on micrometeorological measurements. An often used method is eddy covariance (Baldocchi 2014), which allows to estimate net ecosystem exchange by continuous high-frequency measurements of fluxes of CO_2, water, and energy (Xiao et al. 2014).

COVARIANCE: A REVIEW OF BASIC CONCEPTS

In its simplest case, we can calculate the relationship between two variables. i.e., *bivariate*. Suppose we have two random variables X and Y. Each one of these variables has first and second moments (mean and variance). Covariance is the *joint variation* or the expected value of the product of the two variables, where each one is centered at the mean. This is to say

$$\text{cov}(X,Y) = E[(X - \mu_X)(Y - \mu_Y)] \tag{13.18}$$

As given here, this is a population concept since the expectation operator implies using the distribution of the product (Acevedo 2013). Therefore, we require the joint probability density function of X and Y. When applying the same idea to a sample x_i, y_i of size n leads to sample covariance

$$\text{cov}(X,Y) = \frac{1}{n-1} \sum_{i=1}^{n} \left(x_i - \bar{X}\right)\left(y_i - \bar{Y}\right) \tag{13.19}$$

When we have two time series $X(t)$ and $Y(t)$, we extend this concept to

$$\langle X(t)Y(t) \rangle = \frac{1}{T} \int_{0}^{T} \left[X(t) - \langle X \rangle\right]\left[Y(t) - \langle Y \rangle\right] dt \tag{13.20}$$

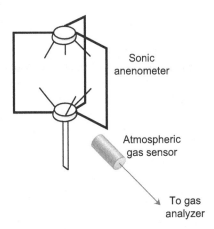

FIGURE 13.10   Open canopy gas exchange: eddy covariance.

where the triangular braces $\langle X \rangle$ indicates time average or the integral of the series over a given period T divided by the integration period.

EDDY COVARIANCE

Eddy covariance allows to measure vertical turbulent fluxes within atmospheric boundary layers by combining high-frequency measurement of the wind field and atmospheric gas (Figure 13.10). A sonic anemometer measures the wind field (see Chapter 12), while simultaneously a gas analyzer measures the gas concentration (see Chapter 12). These two instruments are mounted together in proximity on a tower. The method of eddy covariance is general and thus can be applied to monitor fluxes of water vapor, momentum, heat, and methane. A similar technique is used underwater to measure flux from the seafloor in benthic systems. The flux footprint is the area around the tower that generates the upwind field that applies to the measured flux. The fetch is the distance from the tower that determines this area.

You can think of turbulent airflow as a combination of multiple rotating eddies of various sizes and having horizontal and vertical components. The eddy covariance method attempts to measure net vertical movement from the tower (Figure 13.11). At time t, an eddy moves a parcel of air down at the speed $W(t)$. Then, at time $t + dt$, an eddy moves a parcel up at the speed $W(t + dt)$. Assume that each parcel of air has a gas concentration, pressure, temperature, and humidity. If we know these variables, along with the wind speed, then we can determine the flux (Burba 2013). For example, vertical flux of water vapor would be determined by vapor concentration X in the eddies; using how much goes down $X(t)$ at t and how much goes up $X(t + dt)$ at $t + dt$. So, vertical flux is estimated by the covariance cov(W, X) of the vertical wind velocity $W(t)$ and the concentration $X(t)$ of the entity of interest. The instruments used should detect very small changes at high frequency (in a meteorological scale), ranging from minimum of 5 to 40 Hz.

Using the mixing ratio of density of substance to density of air $X(t) = (\rho_x / \rho_a)$ the flux F is the time average of $\rho_a(t)W(t)X(t) = \rho_x(t)W(t)$, in other words

$$F = \left\langle \rho_a(t)W(t)X(t) \right\rangle \tag{13.21}$$

Now, using decomposition of $W(t)$ and $X(t)$ in terms of means $\langle W \rangle$, $\langle Y \rangle$ and deviations with respect to the mean $w(t) = W(t) - \langle W \rangle$ and $x(t) = X(t) - \langle X \rangle$ a series of assumptions discussed below, this equation is simplified to

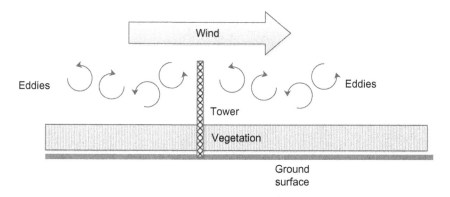

FIGURE 13.11 Schematic representation of airflow with eddy components.

$$F \approx \langle \rho_a(t) \rangle \langle w(t)x(t) \rangle \qquad (13.22)$$

which means that flux is approximately proportional to the covariance of the deviations $w(t)$ and $y(t)$.

To arrive at this expression, the eddy covariance method uses several assumptions, including: the measurements represent an upwind area inside the boundary layer of interest, flux footprint is adequate, flux is turbulent with zero average fluctuations, density fluctuations are negligible, convergence or divergence are negligible, the terrain is horizontal, and its cover is uniform (Burba 2013).

Implementing this method requires synchronization of the gas analyzer and anemometer output, air density corrections, and software to compute the covariance. Open source options are available, such as EddyPro (LI-COR 2015).

Open-path methane gas analyzers are used in eddy covariance systems to measure methane emissions from ecosystems (Detto et al. 2011). The open-path system appears more versatile for unattended stations. Synchronization between wind speed and methane data and air density corrections have impacts on the computed covariance. These systems require intensive data processing.

Flux measurement is an area where networks are making considerable effort to standardize and unify technical approaches. For example, FLUXNET (ORNL DAAC 2015), AmeriFlux (US DOE 2015), ICOS (ICOS 2015), iLEAPS (iLEAPS 2015), and NEON (NEON 2015).

TREE GROWTH, DENDROMETERS

A dendrometer is a device that measures the growth and size of plant stems. This instrument allows monitoring changes of diameter over time. In its simplest form, we attach it to the stem and record changes over time by reading the scale. For automated electronic recording, we can install a full-bridge strain gage attached to a flexible arm. The output signal from the bridge would be low (millivolts) and requires conditioning. Once conditioned, we can record in real time by a datalogger. Applications of dendrometers include monitoring tree stands and testing plants under a variety of conditions, such as water stress, elevated ozone, and other atmospheric pollutants. The device requires temperature compensation, calibration parameters, and zero offset. These parameters and computations are loaded to the datalogger. For examples, see the DEX series of Dynamax (DYNAMAX 2014).

LEAF AREA

In this section, we study leaf area, its role to determine light capture by canopies, and its relationships with gas exchange, energy, and photosynthesis in terrestrial ecosystems. Instead of absolute leaf area, it is common to work with the Leaf Area Index (LAI) in m²/m² representing how much leaf area in m² corresponds to 1 m² of area on the ground.

Our starting point to understand light interception by canopies is the Beer-Lambert law, which predicts sunlight transmission through a canopy. Assume that layer of leaves absorbs solar radiation I uniformly, and that a layer at height h in the canopy has LAI $l(h)$. The rate of light extinction is

$$\frac{dI(h)}{dh} = k \times l(h) \times I(h) \tag{13.23}$$

where $I(h)$ = solar radiation at height h (W/m²), h = height with respect to ground level (m), and k is a constant to adjust units; because LAI is a dimensional, k should be 1/m. Besides determining light interception by canopies, LAI participates in regulating carbon dioxide, water, and energy exchanges between plants and atmosphere (Ryu et al. 2012).

Moving $I(h)$ to the denominator of the left-hand side and dh to the numerator of the right-hand side $\frac{dI(h)}{I(h)} = kl(h)dh$. Integrating from the top of the canopy at height h_t to height h

$$\ln(I(h) - I(h_t)) = \ln\left(\frac{I(h)}{I(h_t)}\right) = k \int_{ht}^{h} l(s)\,ds \tag{13.24}$$

where $s > h$ is used for all heights above h. The integral of LAI is denoted by l_h.

Many canopy models calculate light extinction adding leaf area for all trees higher than h, thereby approximating the integral. A hypothetical special case is when leaf area is constant for all h and equal to l_a, i.e., a vertically homogeneous canopy above h,

$$\ln\left(\frac{I(h)}{I(h_t)}\right) = kl_a(h - h_t). \tag{13.25}$$

LEAF LEVEL

We can measure leaf size of individual leaves using a portable scanner and measuring length, width, perimeter, and geometrical features. Specific hardware and software add capabilities to a basic scanner to make it a leaf area meter. An example is the CI-203 Handheld Laser Leaf Area Meter (CID-Bioscience 2015) (Figure 13.12). These measurements are time stamped by an RTC and geo-referenced by GPS. For example, the data below are a segment of a data file obtained by a CI-203.

FIGURE 13.12   Leaf size measurements by portable scanners.

```
       Date     Time      Area   Length    Width   Perim   Factor
 Ratio Void          Lat            Lon           Alt

 04/24/2013 22:36:03     6.70,    4.66,    2.29,   12.47,   2.04,
 0.54,   1,   3322.7243N, 09701.0783W, 194.0M
 04/24/2013 22:41:55     8.72,    4.40,    2.90,   13.42,   1.52,
 0.61,   3,   3322.7276N, 09701.0790W, 206.8M
 04/24/2013 22:43:05     7.42,    5.79,    2.33,   13.74,   2.48,
 0.49,   0,   3322.7283N, 09701.0768W, 204.7M
 04/24/2013 22:46:29     9.02,    5.62,    2.53,   13.17,   2.22,
 0.65,   0,   3322.7286N, 09701.0755W, 198.1M
 04/24/2013 22:49:38    28.76,    7.99,    6.16,   28.86,   1.30,
 0.43,   4,   3322.7284N, 09701.0771W,199.8M
```

Measurements of leaf size could accompany the measurements of leaf weight and of photosynthesis at leaf levels using gas exchange. To scale up from leaf to canopy, the challenge is then to integrate these detailed measures of individual leaf size to layers in the canopy. For this purpose, we sample the layers to estimate the number of leaves of the various species considered.

CANOPY ANALYZER

A common method to estimate LAI is by optical means, such as upward-pointing digital hemispheric and wide angle photographs as illustrated in Figure 13.13 (Acevedo et al. 2003). Image analysis is embedded in Plant Canopy Analyzers, e.g., LAI-2200 by LI-COR, or CI-110 by CID (CID-Bioscience 2022a). The latter combines hemispherical canopy photography and image analysis with light measurement to calculate LAI and other canopy parameters (Figure 13.14). For monitoring purposes, the methods we have mentioned so far require regular field visits to measure LAI. This approach is of course labor-intensive and has coarse temporal frequency, and we may miss phenological and disturbance events.

An alternative is to combine upward-pointing digital cameras with electronic processing to monitor changes in LAI continuously and obtain seasonal dynamics (Ryu et al. 2012). A camera would point toward the zenith and a program would identify pixels as gap (sky) or cover (vegetation). With these data, we can calculate gap fraction (Welles and Cohen 1996), LAI, and clumping index. The

FIGURE 13.13 Example of upward-pointing photographs in a tropical cloud forest (Acevedo et al. 2003).

FIGURE 13.14 Left: example of image captured by canopy analyzer in an open canopy oak-elm forest. Right: example of canopy analyzer, CI-110 by CID Biosciences.

clumping index indicates departures from random distribution of foliage areas and can be used to correct the estimation of LAI. All cover pixels in the image give plant area index (PAI) instead of LAI because it includes woody contribution of branches and trunks. Thus, there is a need to convert PAI to LAI.

SOLAR RADIATION AND SPECTRAL MEASUREMENTS

We introduced pyranometers and net radiometers in Chapter 11, and quantum sensors in Chapter 12. These sensors have application to monitor solar radiation in terrestrial environments to relate its measurements to ET processes, and to photosynthesis.

For example, differences in canopy and understory light conditions are fundamental factors in forest dynamics. Differences in solar radiation from the ground to the top of the canopy are determined in part by the optical properties of the leaves, that is, absorption, reflection, and transmission of light. As described in Acevedo et al. (2001), at a point in the understory, light depends on a combination of (1) light coming directly through canopy gaps, (2) light reflected by leaves, tree branches and trunks, and epiphytes, and (3) light transmitted through the leaves (Figure 13.15). These may change with layers in the canopy.

A commonly used measure of the radiation field is the proportion of area occupied by canopy gaps and its geometry. This measure has importance in determining light environment, predicting light available for photosynthesis, and estimations of LAI.

Light spectral characteristics constitute a determinant factor in many biological processes. A portable fiber-optic-based spectrometer can be used in a variety of ways to monitor light quality in a forest environment. For example, take a spectrometer as described in Chapter 11, designed to measure the spectrum between the ultraviolet and the near IR (NIR), in a range of 200–850 nm, using an array of 2048 diodes with an aperture slit of 100 μm. The dispersion is $(850–200)/2048 = 0.32$ nm/diode, with a resolution of $12 \times 0.32 = 3.8$ nm (FWHM).

For canopy measurements (Acevedo et al. 2003), one fiber collects light incident to the canopy by placing it in the open (reference spectrum) and the other collects light in the understory environment (sample spectrum). For leaf measurements (Acevedo and Ataroff 2012), two optical fibers conduct reflected and transmitted light from the leaf to a spectrometer. The fiber end to measure reflectance looks at the leaf upper surface and the one for transmittance is placed near the lower surface of the leaf. The other end of each fiber connects to an optical switch that allows blocking light to obtain dark response, which is subtracted from the signals to correct for electronic noise of the instrument.

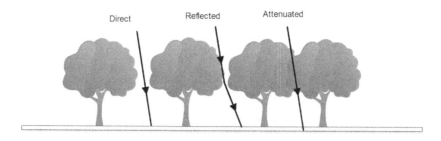

FIGURE 13.15 Schematic representation of the light field in a forest canopy (Acevedo et al. 2001).

Signals are analyzed in a range (400–750 nm) of interest in photosynthesis and plant responses to far-red light.

Spectral signals, reference (lamp) as well as reflected and transmitted, are smoothed and divided into the reference signal to obtain ratios of reflectance $R(\lambda)$ and transmittance $T(\lambda)$ as functions of wavelength λ (Acevedo and Ataroff 2012). Absorbance $A(\lambda)$ for each leaf is calculated as $A(\lambda) = 1 - R(\lambda) - T(\lambda)$. As an example, Figure 13.16 shows these signals for leaves of tree species in the canopy, sub-canopy, and understory.

Variables can be extracted from the spectrum, some of them correspond to averages over wide wavelength bands and other are averages over a narrow band. For example, the contribution of the spectrum in the far-red (705–750 nm) relative to the one in the full photosynthetic range (400–705 nm) is calculated as the ratio of the integral in the first range over the integral in the second range. As an example of a narrower band (618–622 nm), consider absorbance A_{620} at 620 nm, which is the wavelength at which maximum relative quantum efficiency occurs.

Leaf spectra measurements can be integrated in a single device to measure leaf spectra directly in the field. For example, the CI-710 Miniature Leaf Spectrometer (CID-Bioscience 2022b) is an integrated leaf probe, with dual light source (Halogen/LED) and bifurcated optical fiber (Figure 13.17). The CI-710 Miniature Leaf Spectrometer measures the transmission, absorption, and reflection of light by biological substances within a wide range of wavelengths that cover visible and NIR light. Figure 13.18 shows examples of spectra collected using leaves of two species in a forest plot in the Ray Roberts Lake State Park, near Denton, Texas. This device would still require an operator to perform periodic measurements. An automated version that could measure the leaf spectra by non-contact would allow for continuous real-time monitoring.

IR THERMOMETER

An IR thermometer is a radiometer that monitors surface temperature by measuring the thermal energy radiated from any surface within its field of view (Apogee Instruments 2014). The thermometer provides non-contact measurements (Figure 13.19) according to distance to target and its diameter; it is necessary to measure the temperature of the sensor in order to correct for its temperature sensitivity. Applications include monitoring water use and employing it to estimate leaf stomatal conductance and in turn estimate ET, and use canopy temperature to estimate vapor pressure deficit. These measurements can be used as an indication of plant water stress because under high atmospheric demand for water when the soil is dry, plants regulate transpiration by modifying stomatal aperture. Transpiration thus controls leaf temperature.

The Apogee IR thermometer has a built-in microcontroller and is an example of sensor producing output data in SDI-12 format (Chapter 5). By virtue of using SDI-12, the thermometer has three wires, one for power, one for ground, and one for serial data.

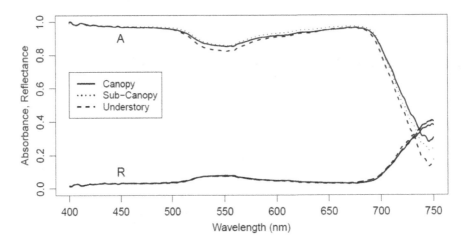

FIGURE 13.16 Example of leaf absorbance (A) and reflectance (R) spectrum for species in canopy, sub-canopy, and understory (Acevedo and Ataroff 2012).

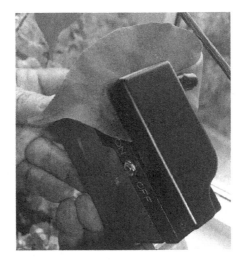

FIGURE 13.17 Portable leaf spectrometer.

The radiometer operates based on the Stefan-Boltzmann law (Figure 13.19) which states that the total power emitted per unit area of a blackbody is proportional to the fourth power of temperature $E = \sigma T^4$ where E is the power per unit area, σ is the Stefan-Boltzmann constant 5.6704×10^{-8} W m^{-2} K^{-4}, and T is the temperature of the blackbody (in K). This law results from integrating Planck's distribution over solid angle and frequency.

The IR radiometer generates a voltage V (in mV range) proportional to the balance between target (σT_T^4) and detector-radiated power (σT_D^4). This balance is $T_T^4 - T_D^4 = aV + b$ where a is the slope and b is the intercept of a regression line that adjust sensor output V to the measurand or target temperature T_T.

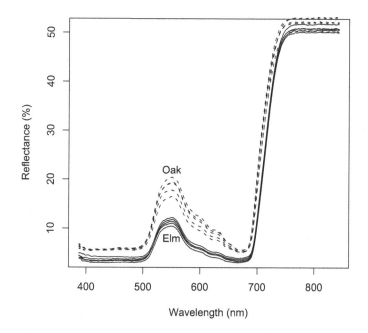

FIGURE 13.18 Examples of leaf spectra acquired by portable leaf spectrometer.

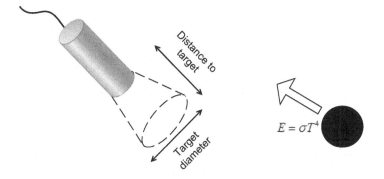

$$E = \sigma T^4$$

FIGURE 13.19 IR radiometer used as IR thermometer to infer surface temperature.

GROUND PENETRATING RADAR

A *Ground Penetrating Radar* (GPR) is an EM device used to collect underground data to about 10 m in depth based on changes in dielectric permittivity, EC, and magnetic permeability of the subsurface. A source (*Tx*) transmits an RF pulse (10 MHz to 2.6 GHz) into the ground, the RF wave is distorted because of the soil and subsurface EM properties. Where the subsurface EM properties change abruptly, RF signals can reflect or refract. A receiver (*Rx*) measures the amplitude and travel time of distorted RF signals, and these data are then used to image discrete targets and physical boundaries (EM GeoSci 2022).

GPR surveys can provide useful information for environmental monitoring of the vadose or unsaturated zone. For instance, using GPR one can map a contaminant plume, determine the direction of contaminant movement, and map hydraulic conductivity. These results then can provide insight into the deployment of real-time monitors, sensor networks, and WSN. Another good example is mapping soil moisture that has direct applications to monitoring. Permittivity is estimated by an equation similar to Equation (13.2) when magnetic permeability is 1 $\varepsilon_r = \left(\dfrac{c}{v}\right)^2$ and v is the propagation velocity of the ground. Once permittivity is mapped, it can be converted to soil moisture by the same principles that are applied to soil moisture sensors (Clement and Ward 2003).

REMOTE SENSING OF TERRESTRIAL ECOSYSTEMS

Many aspects of terrestrial ecosystem monitoring benefit from remote sensing in multiple technologies and platforms, including passive and active sensors. We already discussed in detail the analysis and classification of optical images for vegetation and land cover (Chapters 8 and 9) to help the reader understand the principles of remote sensing. Many other characteristics of vegetation are investigated by remote sensing, to name a few: foliage cover, tree density, tree height, vegetation health, LAI, and aboveground biomass; regarding soil, remote sensing is applied to soil type, soil properties, and soil moisture (Lechner et al. 2020). We will discuss only a couple of topics in this vast field.

Synthetic aperture radar (SAR) systems are active sensors, i.e., emit a microwave pulse and measure the backscatter reflecting to the sensor, and can differentiate land features by surface roughness and water content. From our discussion on TDR and GPR, you would suspect that spaceborne microwave sensors would prove useful for remote sensing of soil moisture. In fact, microwave remote sensing has been successful for estimating soil dielectric properties and thereby allowing soil moisture estimation (Mohanty et al. 2017). Nevertheless, optical and thermal remote sensing has also proven effective to estimate soil moisture when supported by ground validation data, and new algorithms to derive soil moisture from geostationary weather satellites may prove useful (Wang et al. 2018).

When using microwave, it is important to distinguish between passive sensors, such as radiometers, and active sensors, such as radar. Soil Moisture Active Passive (SMAP) (NASA 2023) is designed to carry two instruments to map soil moisture, the radiometer allows spatial resolution of 36 km every 2–3 days, whereas a combination of radar and radiometer measurements would have mapped soil moisture at a spatial resolution of 9 km. Although the SMAP radar failed in 2015, the radiometer has been providing global monitoring of near-surface (0–5 cm) soil moisture. Higher spatial resolution (~1 km) is provided by the European Space Agency (ESA) Sentinel-1 (S-1) platform (ESA 2023a). The C-band S-1 has two satellites (S-1A & B), is based on SAR, and has frequent revisit and large geographical coverage, allowing for monitoring of near-surface soil moisture. Validation studies have demonstrated the usefulness of S-1 data at 1-km resolution (Balenzano et al. 2021).

Light detecting and ranging (LiDAR) systems are also active sensors, emitting a laser pulse and measuring distance to a target and the reflected light is used to make digital 3D representations of the target in the form of *point clouds*. In forests, tree height and stand structure are needed for many purposes, such as forest inventories and carbon storage assessments, and wildlife habitat characteristics. To quantify vertical structure of forest stands, LiDAR point cloud density is related to tree height, the first object hit by the laser is the canopy, whereas the last object hit by the laser is the ground surface. A point cloud can be processed to create a high-resolution raster, which after subtracting values from a DEM raster, provides a tree height raster.

Hyperspectral remote sensing provides many more bands than Landsat, for example Hyperion covered the 0.4–2.5 μm spectrum in more than 200-nm resolution bands, and in raster format with 30 m × 30 m spatial resolution. Historical images are still available via the USGS Earth Explorer (USGS 2022). Unsupervised classification of hyperspectral images can provide information on vegetation type in broad categories such as grass/herbaceous, bottomland hardwood forest, and oak/elm forests. Moreover, when combined with LiDAR-derived tree height, more refined class can be derived such as distinguishing between secondary and mature bottomland hardwood forests (Matsubayashi 2013).

The ESA Sentinel-2 platform consists of two polar-orbiting satellites placed in the same sun-synchronous orbit, but with a phase difference of 180°, which shortens the revisit time to half when using 2 satellites, i.e., 5 days at low latitude and 2–3 days at mid-latitudes (ESA 2023b). A multispectral scanner instrument is on board and capable of scanning a 290-km swath width and detecting 13 spectral bands from 442 nm to 2202 nm, many of these having 15–30-nm bandwidth.

EXERCISES

Exercise 13.1

When measuring VWC (θ) of soil by TDR, we determine that the pulses return to the head of the probe in 4 ns. The probe rods are 12-cm long. For this soil, we have a calibration of $\theta = 0.115 \times \sqrt{\varepsilon_r} - 0.176$. Calculate the VWC.

Exercise 13.2

A sap-flow monitor measures 26.5°C and 25°C at the output and input, respectively. The heater produces 20 W, the radial heat loss is 1 W, and the heat flow at the output is 17 W. Specific heat capacity of water is 4.187 J/gK. Calculate the sap-flow.

Exercise 13.3

Downwelling photosynthetic active radiation (PAR) at the forest floor is 20% of the corresponding one above the canopy. The canopy is 12-m tall, and the PAR attenuation coefficient is 0.15 m^{-1}. Calculate the LAI at the ground assuming it is constant from the ground to the top of the canopy. Hint: use equation 13.25 and assume $h=0$ at ground level.

Exercise 13.4

Assume we maintain constant head at an infiltrometer of cross-section 400 cm^2 by adding 180 ml of water per minute while the soil is saturated. Calculate the hydraulic conductivity of the soil at saturation.

Exercise 13.5

An IR thermometer for monitoring leaf temperature measures 400 W m^{-2}. What is the estimated leaf temperature?

Exercise 13.6

At a flux tower, located at sea level, measurements of CO_2 by a gas analyzer and wind speed by a sonic anemometer yield a covariance of 10.0 (m × μmol)/(s × kg). Assume dry air at sea level is 1.225 kg/m^3. At a higher location, where air is 1.097 kg/m^3, another tower yields a covariance of 15.0 (m × μmol)/(s × kg). Calculate the difference in CO_2 flux estimate.

REFERENCES

Acevedo, M. F. 2013. *Data Analysis and Statistics for Geography, Environmental Science & Engineering. Applications to Sustainability*. Boca Raton, FL: CRC Press, Taylor & Francis Group.

Acevedo, M.F. 2023. *Real-Time Environmental Monitoring: Sensors and Systems, Second Edition – Lab Manual*. Boca Raton, FL: CRC Press, Taylor & Francis Group. 463 pp.

Acevedo, M. F., and M. Ataroff. 2012. Leaf spectra and weight of species in canopy, sub-canopy, and under-story layers in a Venezuelan Andean cloud forest. *Scientifica* 2013:14.

Acevedo, M. F., M. Ataroff, S. Monteleone, and C. Estrada. 2003. Heterogeneidad estructural y lumínica del sotobosque de una selva nublada andina de Venezuela. *Interciencia* 28 (7):394–403.

Acevedo, M. F., S. Monteleone, M. Ataroff, and C. A. Estrada. 2001. Aberturas del dosel y espectro de la luz en el sotobosque de una selva nublada andina de Venezuela. *Ciencia* 9 (2):165–183.

Allen, R. G., L. S. Pereira, D. Raes, and M. Smith. 1998. *Crop Evapotranspiration - Guidelines for Computing Crop Water Requirements*. Rome: FAO.

Andales, Allan A., Dale Straw, Thomas H. Marek, Lane H. Simmons, Michael E. Bartolo, and Thomas W. Ley. 2018. Design and Construction of a Precision Weighing Lysimeter in Southeast Colorado. *Transactions of the ASABE* 61 (2): 509–521. doi: https://doi.org/10.13031/trans.12282.

Angulo-Jaramillo, R., J.-P. Vandervaere, S. Roulier, J.-L. Thony, J.-P. Gaudet, and M. Vauclin. 2000. Field measurement of soil surface hydraulic properties by disc and ring infiltrometers: A review and recent developments. *Soil and Tillage Research* 55 (1–2):1–29.

Ankeny, M. D., K. M. Noh, T. C. Kaspar, and M. A. Prieksat. 1993. Flowdata: Software for analysis of infiltration data from automated infiltrometers. *Agronomy Journal* 85 (4):955–959.

Apogee Instruments. 2014. *Owner's Manual Infrared Radiometer*. Logan Utah: Apogee Instruments.

Baldocchi, D. 2014. Measuring fluxes of trace gases and energy between ecosystems and the atmosphere - the state and future of the eddy covariance method. *Global Change Biology* 20 (12):3600–3609.

Balenzano, A., F. Mattia, G. Satalino, F. P. Lovergine, D. Palmisano, J. Peng, P. Marzahn, U. Wegmüller, O. Cartus, K. Dąbrowska-Zielińska, J. P. Musial, M. W. J. Davidson, V. R. N. Pauwels, M. H. Cosh, H. McNairn, J. T. Johnson, J. P. Walker, S. H. Yueh, D. Entekhabi, Y. H. Kerr, and T. J. Jackson. 2021. Sentinel-1 soil moisture at 1 km resolution: A validation study. *Remote Sensing of Environment* 263:112554.

Burba, G. 2013. *Eddy Covariance Method for Scientific, Industrial, Agricultural and Regulatory Applications*. Lincoln, Nebraska: LI-COR Biosciences.

Campbell Scientific Inc. 2020a. *CS655, 12 cm Soil Moisture and Temperature Sensor*. accessed Apriol 2020. https://www.campbellsci.com/cs655.

Campbell Scientific Inc. 2020b. *HS2- HydroSense II Handheld Soil Moisture Sensor*. accessed April 2020. https://www.campbellsci.com/hs2.

Chen, L. 2008. *Soil Characteristics Estimation and Its Application in Water Balance Dynamics*. Denton, TX: University of North Texas.

CID-Bioscience. 2015. *CI-203 Handheld Laser Leaf Area Meter*. accessed March 2015. http://www.cid-inc.com/products/leaf-area-lai/handheld-laser-leaf-area-meter.

CID-Bioscience. 2022a. *CI-110 Plant Canopy Imager*. accessed April 2022. https://cid-inc.com/blog/product/ci-110-plant-canopy-imager/.

CID-Bioscience. 2022b. *CI-710s SpectraVue Leaf Spectrometer*. accessed Aprl 2022. https://cid-inc.com/plant-science-tools/leaf-spectroscopy/ci-710-miniature-leaf-spectrometer/.

Clement, W., and A. Ward. 2003. *Using Ground Penetrating Radar to Measure Soil Moisture Content*. ASAE Annual Meeting Las Vegas, NV. American Society of Agricultural and Biological Engineers

Cobos, D. R. 2007. *Measuring UMS Tensiometers with Non-UMS Control and Data Acquisition Systems*. Pullman, WA: Decagon Devices.

Colorado State University, CoAgMET Colorado's Mesonet. 2023. *CoAgMET Regional ETR Summaries*. accessed May 2023. https://coagmet.colostate.edu/etr_form.php.

Detto, M., J. Verfaillie, F. Anderson, L. K. Xu, and D. Baldocchi. 2011. Comparing laser-based open- and closed-path gas analyzers to measure methane fluxes using the eddy covariance method. *Agricultural and Forest Meteorology* 151 (10):1312–1324.

DYNAMAX. 2014. *Products*. DYNAMAX, accessed February 2015. http://www.dynamax.com/Products.htm#6.

EM GeoSci. 2022. *Ground Penetrating Radar*. accessed April 2022. https://em.geosci.xyz/content/geophysical_surveys/gpr/index.html.

ESA. 2023a. *Sentinel-1 for Surface Soil Moisture*. accessed January 2023. https://eo4society.esa.int/projects/s1-for-surface-soil-moisture/.

ESA. 2023b. *Sentinel-2.* accessed January 2023. https://sentinel.esa.int/web/sentinel/missions/sentinel-2.

Fares, A., M. Safeeq, and D. M. Jenkins. 2009. Adjusting temperature and salinity effects on single capacitance sensors. *Pedosphere* 19 (5):588–596.

Fares, A., F. Abbas, D. Maria, and A. Mair. 2011. Improved calibration functions of three capacitance probes for the measurement of soil moisture in tropical soils. *Sensors* 11 (5):4858–4874.

Freebairn, D. M., N. H. Hancock, and S. C. Lott. 1986. Soil Evaporation Studies Using Shallow Weighing Lysimeters: Techniques and Preliminary Results. *Conference on Agricultural Engineering 1986: Preprints of Papers*, Barton, ACT.

Greco, R., and A. Guida. 2008. Field measurements of topsoil moisture profiles by vertical TDR probes. *Journal of hydrology* 348 (3–4):442–451.

Herzog, K. M., R. Thum, R. Zweifel, and R. Häsler. 1997. Heat balance measurements - to quantify sap flow in thin stems only?. *Agricultural and Forest Meteorology* 83:75–94

ICOS. 2015. *Integrated Carbon Observation System.* accessed 2015. http://www.icos-infrastructure.eu/.

iLEAPS. 2015. *Integrated Land Ecosystem - Atmosphere Processes Study.* accessed 2015. http://www.ileaps.org/.

Jones, S. B., J. M Wraith, and D. Or. 2002. Time domain reflectometry measurement principles and applications. *Hydrological Processes* 16:141–153.

Kallioras, A., A. Khan, M. Piepenbrink, H. Pfletschinger, F. Koniger, P. Dietrich, and C. Schuth. 2016. Time-domain reflectometry probing systems for the monitoring of hydrological processes in the unsaturated zone. *Hydrogeology Journal* 24 (5):1297–1309.

Kargas, G., and K. X. Soulis. 2012. Performance analysis and calibration of a new low-cost capacitance soil moisture sensor. *Journal of Irrigation & Drainage Engineering* 138 (7):632–641.

Kelleners, T. J., M. S. Seyfried, J. M. Blonquist, Jr., J. Bilskie, and D. G. Chandler. 2005. Improved interpretation of water content reflectometer measurements in soils. *Soil Science Society of America journal* 69 (6):1684–1690.

Lechner, A. M., G. M. Foody, and D. S. Boyd. 2020. Applications in remote sensing to forest ecology and management. *One Earth* 2 (5):405–412.

Lepore, B. J., C. L. S. Morgan, J. M. Norman, and C. C. Molling. 2009. A mesopore and matrix infiltration model based on soil structure. *Geoderma* 152 (3–4):301–313.

LI-COR. 2015. *EddyPro.* accessed 2015. http://www.licor.com/env/products/eddy_covariance/software.html.

Lopez-Bernal, A., E. Alcantara, and F. J. Villalobos. 2014. Thermal properties of sapwood of fruit trees as affected by anatomy and water potential: errors in sap flux density measurements based on heat pulse methods. *Trees-Structure and Function* 28 (6):1623–1634.

Matsubayashi, S. 2013. *Quantifying Forest Vertical Structure to Determine Bird Habitat Quality in the Greenbelt Corridor.* PhD Dissertation, Environmental Science, Denton, Texas: University of North Texas.

METER. 2020. *EC-5 | Soil Moisture Sensor | METER Environment.* accessed April 2020. https://www.metergroup.com/environment/products/ec-5-soil-moisture-sensor/.

Meter. 2022. *TEROS 31 Lab Tensiometer.* accessed April 2022. https://www.metergroup.com/en/meter-environment/products/teros-31-lab-tensiometer.

Mohanty, B. P., M. H. Cosh, V. Lakshmi, and C. Montzka. 2017. Soil moisture remote sensing: state-of-the-science. *Vadose Zone Journal* 16 (1):201.

Moret, D., J. L. Arrue, and M. V. Lopez. 2004. TDR application for automated water level measurement from Mariotte reservoirs in tension disc infiltrometers. *Journal of hydrology* 297 (1–4):229–235.

NASA. 2023. *SMAP Soil Moisture Active Passive.* accessed January 2023. https://smap.jpl.nasa.gov/.

NEON. 2015. *National Ecological Observatory Network.* NEON, accessed 2015. http://www.neoninc.org/.

Nielsen, D. C., H. J. Lagae, and R.L. Anderson. 1995. Time-domain reflectometry measurements of surface soil water content. *Soil Science Society of America Journal* 59 (1):103–15.

ORNL DAAC. 2015. *The FLUXNET project.* ORNL, accessed 2015. http://daac.ornl.gov/FLUXNET/fluxnet.shtml.

Prieksat, M. A., T. C. Kaspar, and M. D. Ankeny. 1992. Design for an automated, self-regulating, single-ring infiltrometer. *Soil Science Society of America journal* 56 (5):1409–1411.

Reynolds, W. D., and D. E. Elrick. 1990. Ponded infiltration from a single ring: Analysis of steady flow. *Soil Science Society of America journal* 54:1233–1241.

Robinson, D. A., C. M. K. Gardner, and J. D. Cooper. 1999. Measurement of relative permittivity in sandy soils using TDR, capacitance and theta probes: Comparison, including the effects of bulk soil electrical conductivity. *Journal of hydrology* 223 (3):198–211.

Robinson, D. C., G. J. Evans, J. Cooper, M. Hodnett, and J. Bell. 1998. The dielectric calibration of capacitance probes for soil hydrology using an oscillation frequency response mode. *Hydrology and Earth Systems Science* 21:111–198.

Ryu, Y., J. Verfaillie, C. Macfarlane, H. Kobayashi, O. Sonnentag, R. Vargas, S. Ma, and D. D. Baldocchi. 2012. Continuous observation of tree leaf area index at ecosystem scale using upward-pointing digital cameras. *Remote Sensing of Environment* 126:116–125.

Seyfried, M. C. H. M. Murdock, and S. Vactor. 2001. Long-term lysimeter database, reynolds creek experimental watershed, Idaho, United States. *Water Resources Research* 37:2853–2856.

Touma, J., and J. Albergel. 1992. Determining soil hydrologic properties from rain simulator or double ring infiltrometer experiments: A comparison. *Journal of hydrology* 135 (1–4):73–86.

US DOE. 2015. *Ameriflux*. US Department of Energy, accessed 2015. http://ameriflux.lbl.gov/.

USGS. 2022. *EarthExplorer*. accessed December 2022. https://earthexplorer.usgs.gov/.

Vandegehuchte, M. W., and K. Steppe. 2013. Sap-flux density measurement methods: Working principles and applicability. *Functional Plant Biology* 40 (3):213–223.

Vergeynst, L. L., M. W. Vandegehuchte, M. A. McGuire, R. O. Teskey, and K. Steppe. 2014. Changes in stem water content influence sap flux density measurements with thermal dissipation probes. *Trees-Structure and Function* 28 (3):949–955.

Visconti, F., J. Miguel de Paz, D. Martínez, and M. José Molina. 2014. Laboratory and field assessment of the capacitance sensors Decagon 10HS and 5TE for estimating the water content of irrigated soils. *Agricultural Water Management* 132:111–119.

Wacker, K., G. Campbell, and L. Rivera. 2013. *An Automated Dual-Head Infiltrometer for Measuring Field Saturated Hydraulic Conductivity*. Tampa, FL: Food, Energy and Innovation for a Sustainable World.

Wang, Y., J. Peng, X. Song, P. Leng, R. Ludwig, and A. Loew. 2018. Surface soil moisture retrieval using optical/thermal infrared remote sensing data. *IEEE Transactions on Geoscience and Remote Sensing* 56 (9):5433–5442.

Welles, J. M., and S. Cohen. 1996. Canopy structure measurement by gap fraction analysis using commercial instrumentation. *Journal of Experimental Botany* 47 (302):1335–1342.

Xiao, J. F., S. V. Ollinger, S. Frolking, G. C. Hurtt, D. Y. Hollinger, K. J. Davis, Y. D. Pan, X. Y. Zhang, F. Deng, J. Q. Chen, D. D. Baldocchi, B. E. Law, M. A. Arain, A. R. Desai, A. D. Richardson, G. Sun, B. Amiro, H. Margolis, L. H. Gu, R. L. Scott, P. D. Blanken, and A. E. Suyker. 2014. Data-driven diagnostics of terrestrial carbon dynamics over North America. *Agricultural and Forest Meteorology* 197:142–157.

14 Wildlife Monitoring

INTRODUCTION

Wildlife monitoring entails animal tracking, by radio and acoustic signals that provide not only location of the animals but also information such as environmental and physiological conditions of the animal. Satellite technology, prominently GPS and data satellites, provides extensive and long-term monitoring in remote areas and particularly in the ocean environment. Together with animal tracking, remote sensing helps to monitor terrestrial wildlife habitat, by classifying images in terms of factors that make land suitable as habitat for a particular species. Analyzing the changes on habitat, for example by time series of fragmentation metrics, allows to monitor changes in habitat. This chapter provides a brief introduction to spatial analysis using point patterns and kriging interpolation, since spatial distributions are of importance in many types of wildlife population distribution as well as habitat structure. These concepts are elaborated further by computer exercises in Lab 14 of the companion Lab Manual (Acevedo 2024).

RADIO MONITORING

An important area of wildlife monitoring is tracking individual animals by using EM in radio frequency (RF) or acoustic signals. In this section, we focus on RF and provide examples of tracking in the terrestrial and aquatic environment.

TERRESTRIAL

In its simplest form, to track an animal with RF, you attach a transmitter and antenna to the animal, and as it moves, its attached transmitter's signal is received and processed at a base station. At a minimum, the receiving system can indicate the presence of the signal, however more information could be carried by the transmission as we learned in Chapter 6. To track more than one individual, we need to give each one a unique signal that you can identify. Uniqueness of radio signals is accomplished by varying the carrier frequency, the modulated signal characteristic (say beeping rates), or much better by a digital code (Lotek 2023c).

Another important aspect of monitoring wildlife with telemetry is to be able to tell not just the presence but also the position. This requires an array of receivers with multiple antennae; a system of towers with receivers that can geo-locate the tags by the characteristics of the reception. For example, one could deploy three towers with receivers at the site and four antennae at each tower (Taylor et al. 2011). Every few seconds the towers receive signals to determine presence, location, and activity levels of tagged individuals within a given range (say 10 km). Receivers cycle through antennae at regular intervals. When the antennae are directional, each pointed in a cardinal direction, we can use signal strength received at each antenna to determine position (Figure 14.1). A datalogger or computer processes the signals so that the digital identification (ID) signal encoded on each tag can be determined along with signal strength in each antenna. Such a system allows for monitoring of survival, activity patterns, spatial patterns, and migratory departure from a site. Small individuals, such as birds, can be monitored using very small tags; for example, "nanotags" (Lotek 2023a) that are very small and unobtrusive.

Depending on the type of movement (e.g., flight height) of monitored individuals and habitat characteristics (e.g., foliage and canopy height), antennae should be located at a minimum height above the ground (say 10–20 m). Messages contain tag number, date, location, antenna number,

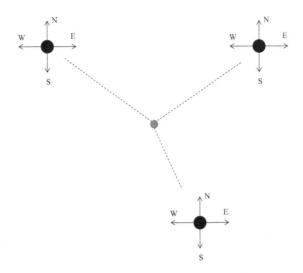

FIGURE 14.1 Array of three towers with four antennae each to monitor radio tags. A set of receiver calculates position based on each antenna's signal strength.

signal strength (related to the log of the received signal in dB), and time. Real time requires coordination between receivers, and this may entail using an RTC at each receiver. An important consideration is battery life for the transmitters, which is maximized by limiting transmission power, either by reducing duty cycle of the pulses or transmitting only in bursts. In these cases, the burst rate determines the time that the receiver allocates to each antenna.

Suppose we know coordinates (x_i, y_i) of three towers, and distance d_i from an animal to each tower is measured by signal strength. What are the coordinates (x, y) of the animal? Start relating coordinates to distance

$$(x - x_1)^2 + (y - y_1)^2 = d_1^2$$
$$(x - x_2)^2 + (y - y_2)^2 = d_2^2 \quad (14.1)$$
$$(x - x_3)^2 + (y - y_3)^2 = d_3^2$$

Expand each equation

$$x^2 - 2xx_1 + x_1^2 + y^2 - 2yy_1 + y_1^2 = d_1^2$$
$$x^2 - 2xx_2 + x_2^2 + y^2 - 2yy_2 + y_2^2 = d_2^2 \quad (14.2)$$
$$x^2 - 2xx_3 + x_3^2 + y^2 - 2yy_3 + y_3^2 = d_3^2$$

Subtract the second from the first equation, and the third from the second equation

$$x^2 - 2xx_2 + x_2^2 + y^2 - 2yy_2 + y_2^2 - x^2 + 2xx_1 - x_1^2 - y^2 + 2yy_1 - y_1^2 = d_2^2 - d_1^2$$
$$x^2 - 2xx_3 + x_3^2 + y^2 - 2yy_3 + y_3^2 - x^2 + 2xx_2 - x_2^2 - y^2 + 2yy_2 - y_2^2 = d_3^2 - d_2^2 \quad (14.3)$$

Simplify

$$2x(x_1 - x_2) + 2y(y_1 - y_2) = (d_2{}^2 - d_1{}^2) + (x_1{}^2 + y_1{}^2) - (x_2{}^2 + y_2{}^2)$$

$$2x(x_2 - x_3) + 2y(y_2 - y_3) = (d_3{}^2 - d_2{}^2) + (x_2{}^2 + y_2{}^2) - (x_3{}^2 + y_3{}^2)$$

(14.4)

and relabel terms to obtain the simple system of equations

$$a_1 x + b_1 y = c_1$$

$$a_2 x + b_2 y = c_2$$

(14.5)

Passive integrated transponder tags (PIT-tags) are small (~10 mm range) and lightweight (fraction of a gram) electronic tags based on RF identification (RFID); i.e., a microchip or transponder provides identification (ID) to an RF scanner or receiver. These tags are called passive because they do not have a battery, whereas the other tags described in this chapter are called active tags, because they have their own battery. The PIT-tag is inactive until a scanner powers the tag circuitry by RF induction and reads the code. Note that this convention is opposite from the sensor terminology of Chapter 2 (and that we have used so far in the book); recall that we defined an active sensor as one requiring external energy to generate the response signal, and a passive sensor one that generates its own signal. In addition to small size, PIT-tags have a long life (possibly decades); however, they must be near the reader (fraction of a meter) to become active and therefore require the installation of antenna at a constriction area that forces the animals to go through and approach the reader. There can be a variety of materials in between the tag and the receiver, but ferrous metals are to be avoided. Previously, a frequency of 400 kHz was used, which has a limited read range, so now the 125- and 134.2-kHz tags are more common and can read from longer distances.

AQUATIC

Radio signals would also propagate in fresh water near the surface, but not through sea and brackish water because the dissolved salts attenuate the radio waves. In this case, we can use acoustic telemetry for wildlife monitoring in marine habitats, such as salmon, trout, cod, crabs, and sea turtles (Bloor et al. 2013).

Wave propagation in water is different from the one in air because of high permittivity ($\varepsilon_r \sim 80$) and conductivity (value depends on dissolved salts) of water. Attenuation in water is higher than the one in air and increases with frequency (Jiang and Georgakopoulos 2011). Therefore, refraction at the water/air interface is high and given by the refractive index $n = \sqrt{\varepsilon_r \mu_r}$; however, permeability of water is $\mu_r = 1$ and thus the refractive index is purely a function of relative permittivity ε_r.

Snell's law of refraction $n_1 \sin\theta_1 = n_2 \sin\theta_2$ allows us to predict that the transmission angle at the interface from water to air is given by $\sin\theta_t = \sqrt{\varepsilon_r} \sin\theta_i$ or that the transmission angle from air to water is given by $\sin\theta_t = \dfrac{\sin\theta_i}{\sqrt{\varepsilon_r}}$ (Figure 14.2). In the companion lab guide, we write a script to calculate transmission angle θ_t for a variety of incidence angle θ_i between 0° and 90° for both scenarios producing Figure 14.3, which shows that due to water's high relative permittivity, the transmission angle from air to water is low and below 6.41° meaning that waves penetrate almost normal to the surface. Likewise, the transmission from water to air is such that waves with incidence angles higher than 6.41° emerge almost parallel to the water surface.

Consequently, in freshwater, the radio waves from a transmitter tag propagate in all directions (omnidirectional propagation) but only waves in a narrow range (~6.4°) of directions emerge close to the normal (Figure 14.4). Therefore, even though we can receive the waves from many directions,

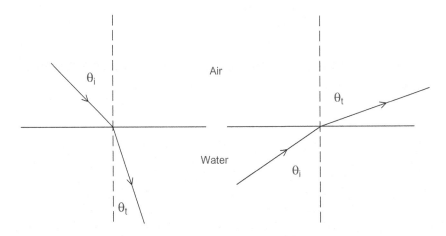

FIGURE 14.2 Radio wave refraction at the water surface or water/air interface.

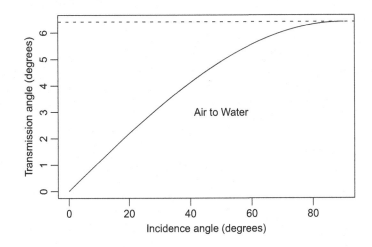

FIGURE 14.3 Transmission angle as a function of incidence angle for both scenarios.

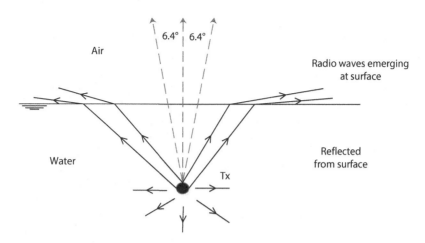

FIGURE 14.4 Radio wave propagation in water and emerging at the surface. Only the directions closest to normal to the surface emerge close to this direction.

we can only tell tag location by signal strength when positioning the antenna near the tag (Thorstad et al. 2013).

Radio telemetry has limitations in terms of depth range. At greater depths, we must resort to acoustic telemetry. It is also possible to submerse an antenna to detect the radio waves directly under water and connect this antenna to a receiver above the surface. A simple arrangement is a coaxial cable connected at the receiver and stripped at the submersed end (Thorstad et al. 2013).

Radio transmitters for aquatic animals use 30–225 MHz; at the lower end we have lower attenuation, making it better for greater depths and higher water conductivity. However, these lower frequencies have longer wavelengths and require larger antennae making it less practical for tag transmitters. Thus, higher frequencies in the order of 130–170 MHz are more practical; particularly for lower water conductivity and shallow water. Since attenuation increases at these higher frequencies, their use is less practical for brackish/saline water or deeper water. In these circumstances, it is better to resort to acoustic tagging.

ACOUSTIC MONITORING

A sound wave consists of oscillating high- and low-pressure regions across a medium and corresponding to regions of compression and rarefaction of the medium. Sound propagation can be modeled by a sine wave, with maxima for compression regions, minima for rarefaction, and zero for neither. Therefore, an acoustic sensor is in essence a pressure sensor, and the most common example is a microphone. As with RF, we can relate the speed of sound v to frequency f and wavelength λ using $v = f\lambda$; however, the speed of sound varies with the medium with large differences between air and water.

TERRESTRIAL

Acoustic monitoring for terrestrial animals is convenient when the animals emit sound and ultrasound as part of their behavior, such that it occurs for birds, frogs, and some insects. In this section, we provide an example for birds.

Bird populations and communities have been monitored for a long while by human observation or by recording using microphones. However, acoustic interactions among songbirds can be complex;

for example, birds may vocalize at random or divide a *soundscape* so that they avoid overlapping their songs with those of other birds. Therefore, monitoring involves collecting spatiotemporal data in which multiple individuals and species are singing sometimes simultaneously. Male birds may produce long songs to advertise their territory or attract females in the breeding season, or shorter calls may occur in other circumstance such as flight, threat, and alarm (Susuki et al. 2017).

Advances in acoustic engineering and microphone arrays make possible to develop new tools to monitor species of interest for a long period of time without human intervention. Compared to single microphones, microphone arrays (System in Frontier 2023) can detect the direction of arrival (DOA) of the sound event. A microphone array is a set of microphones arranged in a pattern to detect signals in several directions; for example, six microphones arranged in 60° separation. Using DOA of sound events acquired from multiple microphone arrays, a system could determine the position of the sound source using the same localization methods described in Figure 14.1 and Equations (14.1–14.5) at the beginning of this chapter.

Research to develop systems that employ this approach involves, for example *robot audition* and *microphone arrays* (Susuki et al. 2017); the HARKBird system consists of a portable system for robot audition, HARK (HRI 2023), together with a low-cost and commercially available microphone array. Experiments have been conducted in two different types of forests in the USA and Japan and have determined that this system can automatically estimate the DOA of the songs of multiple birds and separate them as different signals in the recordings as well as provide insight about asymmetries among species in their tendency to partition temporal resources.

Difficulty in implementing this type of monitoring depends on the pitch and intensity of bird species songs, how often they sing, and complexity of the habitat. Loud bird songs in the 2–5 kHz range and in low vegetation or forest gaps are easy to monitor, but infrequent songs in the 0.5–1 kHz range in dense vegetation are more difficult to monitor.

Aquatic

Sound speed in water varies with temperature and salinity, for instance at 20°C in freshwater, the speed of sound is ~1500 m/s and in seawater is ~1520 m/s. These values are about 4.4 times the speed of sound in air, 343 m/s, and therefore wavelength for the same frequency is very different in water compared to air; for example, at 1 kHz in freshwater, the wavelength would be $\lambda = \dfrac{1500\,\text{m s}^{-1}}{1000\,\text{s}^{-1}} = 1.5\text{m}$, whereas in air $\lambda = \dfrac{343\,\text{m s}^{-1}}{1000\,\text{s}^{-1}} = 0.34\text{m}$. In similar manner to radio waves, acoustic waves propagate in all directions under water; but contrastingly to radio waves, acoustic waves do not emerge at the surface, because the air interface behaves as an almost perfect reflector of sound for frequencies above ~1 kHz. Thus, for acoustic monitoring, we need to submerse an acoustic sensor, referred to as a *hydrophone*.

A hydrophone is based on a piezoelectric sensor that responds to changes in pressure relating these to sound waves and able to be submersed to withstand expected higher pressures underwater. A hydrophone array, like a microphone array, is a series of hydrophones arranged to detect acoustic signals in several directions. An acoustic monitoring system may be connected by wire to a receiver station above the surface equipped with the datalogger (Figure 14.5). This station may be at the shore, floating on a buoy, or on a boat. Acoustic signals have a lower propagation speed than radio waves and this can generate undesirable latency or delay in the telemetry system.

Acoustic transmitters for aquatic animals usually operate in the frequency range of 30–400 kHz. The size of the tag is a function of the frequency; the higher the frequency, the smaller the resonant elements are. However, higher frequencies have a shorter range of distance. Therefore, small transmitters (for small fish) have a shorter maximum range of distance than larger transmitters (Thorstad et al. 2013).

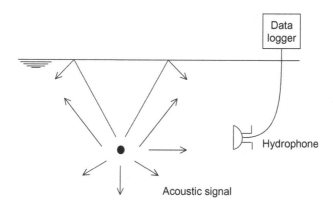

FIGURE 14.5 Acoustic waves propagate in the water and do not emerge at the surface therefore must be received by a microphone under water (hydrophone) connected by wire to a datalogger.

In addition to presence or position, acoustic transmitters can integrate with sensors of environmental and physiological parameters. These can include hydrostatic pressure (to provide water depth), temperature, salinity, DO (for water quality conditions), accelerometers (for animal activity and body position), electromyogram (for muscle activity), heartbeat rate, and differential pressure (for coughing, breathing, and feeding activity) (Thorstad et al. 2013).

SATELLITE

Radio and acoustic systems typically communicate over relatively short distances but tracking capabilities can be extended by using satellites such as GPS, Iridium, Globalstar, or Argos.

GLOBAL POSITIONING SYSTEM

Global positioning system (GPS) allows determining geographical location (latitude, longitude, and elevation) using a network of satellites that orbit the Earth. A GPS receiver must have a clear reception of at least three satellites to determine position. To do this, the receiver determines distance to each satellite from the time taken for the signal from that satellite to reach the receiver. Then it calculates the position by triangulation using these distances. To determine signal travel time, the receiver must synchronize its clock with the satellite constellation clocks. The receiver may need a fourth satellite to synchronize the clock, particularly when restarting the process or changing elevation.

Each satellite transmits a stream of data segments lasting 30 seconds that contain status of the transmitting satellite, clock time, location (ephemeris), and orbits (almanac). When the receiver starts, it must "acquire" the satellites, i.e., find each satellite's signal and read the clock and ephemeris data. This process may take some time (a minute or two) but once the satellites are acquired, the receiver can update location several times in a second, even though the transmit data rate is low, 50 bits per second. Almanac data allow calculating a satellite approximate location in orbit. A receiver uses almanac data when finding satellites; it searches for the satellites that are nearby and skips those that are below the horizon.

GPS devices are essential for mapping and geospatial analysis such as geographical information systems (GIS) and remote sensing. GPS receivers have become very popular in everyday life because now it is affordable and applied to vehicle navigation, outdoor sports (such as hiking, hunting, and fishing), geocaching, and many other activities that involve location.

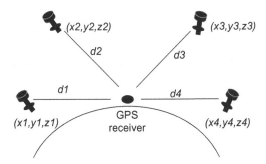

FIGURE 14.6 GPS coordinates.

The method to calculate coordinates involves a system of equations on unknown coordinates x, y, z of the receiver and known x_i, y_i, z_i coordinates of the satellites, and distances d_i to each satellite $i = 1, 2, 3, 4$

$$\sqrt{(x - x_i)^2 + (y - y_i)^2 + (z - z_i)^2} + ct_c = d_i \tag{14.6}$$

c is the speed of light and t_c is a clock time offset between the satellite and the receiver clocks (Figure 14.6). This offset is the same for all equations if the clocks of the satellites are synchronized, and it is zero if the receiver clock is synchronized to the satellite clocks. We can linearize this system by squaring both sides of each equation and pairwise subtraction. We can see that the GPS receiver is not just a simple receiver but must be able to perform calculations.

Naturally, we can use GPS to track an animal location by attaching a collar with receiver on large animals or in a small backpack for small animals. GPS collars are equipped with a battery pack and a GPS receiver that can be placed on relatively large animals to track their location (Figure 14.7). Collars also have a VHS RF beacon in the 140–170 MHz range to locate the animal and retrieve the collar when needed. In addition, the beacon can transmit battery status and alert of whether the animal has died. Some collars have a temperature sensor for cross reference and study animal behavior. Practical issues are to make the system waterproof and impact resistant.

Pop-Up Satellite Archival Tags

Pop-up satellite archival tags (PSAT) are used to monitor fish, data are stored on the tag and both the pop-up position along with the stored data are transferred to satellites (see Chapter 12) when the

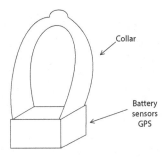

FIGURE 14.7 Example of GPS collar.

tag loosens from the fish and pops up to the water surface (Thorstad et al. 2013; Wildlife Computers 2015). From these tags, we can generate diving behavior and obtain time series of depth and time-at-depth charts.

PROXIMITY SENSORS

Of great interest now in wildlife biology is to gather data on animal interaction, including the duration and frequency of encounters, i.e., how often and for how long are animals in proximity to each other (Lotek 2023b). A two-collar proximity sensor system has a *Master* collar, which senses the presence or absence of the *SUB* ("Slave UHF Beacon") collar.

DATA STORAGE TAGS

Data storage tags (DSTs) or *archival tags* record the information on movement and other environmental, e.g., temperature, and physiological, e.g., heart rate, information on the tag, but the device does not transmit the data real time; thus, one must recapture the animal to retrieve the data. In essence, it is a datalogger attached to the animal to collect data as it moves. For marine life, DST may datalog temperature, salinity, depth, light, pitch, and roll. They are designed to be small size, e.g., 2–4 cm, store a lot of data, e.g., 500,000 points with time stamp, and remain operational for a very long time, e.g., nine years by using a battery and low power consumption. When used for marine life, DSTs may have a depth switch to turn off data collection for preset thresholds (Star ODDI 2023).

For recapture purposes, DST can have text on the housing with return information when the tags are used on commercially fished species, so that fishermen can return a tag to the researcher. Other options are for the DST to float or come to shore upon death of the animal allowing then to retrieve data. For terrestrial animals, options are to provide a transmitter that would allow retrieving data periodically but infrequently as the animal passes a station.

Using small tags, researchers can study fish movement allowing to understand seasonal patterns, for example movement between spawning and feeding areas employing tidal streams, or to calibrate models giving a probability distribution of the position of the fish, which can be used to predict path of the most probable movement (Metcalfe and Arnold 1997; Pedersen et al. 2008).

Methods to attach tags to fish are surgical implantation, external attachment, and ingestion. Because the tag is a foreign object in the animal body, there are potential negative handling effects such as inflammations and infections, altered behavior and swimming, reduced feeding, and consequently reduced growth and increased mortality. Therefore, the catch, handling, and tagging procedures should have minimal effects on the fish to monitor unaltered behavior and not the one potentially induced by the tags themselves. We must consider ethical standards for the use of experimental animals and to ensure fish survival and welfare (Thorstad et al. 2013).

CAMERA AND VIDEO

Advances in camera technology and devices to record video have made possible to monitor wildlife in their habitats and nesting sites generating a wealth of information on behavior (Cox et al. 2012; Brown and Gehrt 2009). Camera technology is used for wildlife monitoring in a variety of habitats, such as tree top bird nesting sites and polar bears in the tundra. Nest cameras have been used to observe nest predation studies, feeding ecology, and adult behavior (Cox et al. 2012).

In the simplest form, we can deploy still cameras operated by batteries that will trigger upon movement and capture images of the animal. Such cameras are widely used by hunters and wildlife lovers and offer the possibility of collecting visual data that can be analyzed to infer movement and behavior. A more demanding application is to use a video camera that would continually monitor animals. Two approaches of adding video recording to monitor wildlife are common: vendor-built

FIGURE 14.8 Example of a burrowing owl video image captured from monitoring cameras. Screenshot captured from a YouTube video (TPWD 2015).

systems made specifically for wildlife monitoring, and user-built systems with outdoor cameras (Cox et al. 2012).

There are commercially available, self-contained digital video recording systems used for wildlife monitoring (SeeMore Wildlife Systems 2015). Other alternatives are to build your own system using standard off-the-shelf components, such as video surveillance equipment and outdoor cameras, and presenting an opportunity to improve design and implementation (Golightly and Schneider 2011).

An example of user-built system is a system to monitor burrowing owls in their burrows using standard and off-the shelf Digital Video Recorders (DVR) and security video cameras (Williams 2014) (Figure 14.8). Such a system operates off- grid using solar panels (see Chapter 7) and is adapted successfully to harsh desert environment. The DVR system uses the H.264 coding standard to compress the video to save it to the internal hard disk drive. This makes the system more efficient when storing the video files and yielded less data volume to transfer per hourly video segment. This system used high-quality outdoor infrared (IR) cameras required a minimum illumination of zero lux, which was necessary for the den camera. In order to save disk space and power, the system records only when interesting events are occurring; for this purpose, a detector was implemented to activate the DVR. Movement (or change between frames in the video) was one of the most obvious events that triggered a recording response. Adjusting the sensitivity of motion detection is important to record only important events, such as the bird entering or exiting the burrow or feeding the offspring. The motion detection for this surveillance system was done by the DVR unit using digital processing, instead of a PIR sensor at the camera like many game-camera designs (Brown and Gehrt 2009). Unfortunately, that meant the cameras ran continuously, thus being less efficient in power utilization (Williams 2014).

An interesting user-built approach would be the development of customized low power processing boards for video data acquisition and storage (see Chapters 2–6 for information on microcontroller, SBC, and WSN). As an example, a USB camera and an SBC have been used to capture images at a specified frequency (Kandula 2011). In addition, this SBC setup allows interacting with environmental sensors on a monitoring station. The development includes two modes of image acquisition, including a basic activity recognition algorithm (Kandula 2011). Capturing audio information together with video would allow for multimedia environmental monitoring.

AUTONOMOUS VEHICLES

Autonomous vehicles contribute to monitor wildlife in both the aquatic and terrestrial environments. Besides contributing to monitor water environmental conditions, *autonomous underwater vehicles* equipped with camera for video monitoring as well as stations to receive signals from tags attached to animals can provide a wealth of data on animal behavior and physiological conditions (Hawkes et al. 2020). Unmanned aerial vehicles (UAV) or drones provide data on terrestrial animals facilitating counts in remote locations and for species that are difficult or sensitive to survey from the ground. Onboard thermal IR cameras, sensitive to heat emitted by the animals, are of great value to capture data on species difficult to detect by optical sensors based on light in the visible range (Witczuk et al. 2018). There are multiple concerns regarding the use of UAV to monitor wildlife, including affecting animal behavior and impacting their habitat.

HABITAT MONITORING

A valuable support to monitoring wildlife is to monitor changes in the habitat and its suitability. For animals of restricted movement and thus habitats occupying small area, it is possible to define variables that define the habitat and deploy sensor networks and other tools as video tracking to monitor the various variables that determine the habitat. Take an example from aquatic organism in a stream reach and habitat that is characterized by depth, water velocity, temperature, DO, sedimentation, and turbidity. These variables are all amenable to track using sensors, and long-term time series of these variables can help us predict habitat changes.

For large animals with habitat occupying large areas at landscape and regional scales, remote sensing and GIS are of great value to monitor and map habitat changes (Oeser et al. 2021). In these cases, of particular importance is building time series of a *habitat suitability index* (HSI) by images based on remote sensing combined with GIS analysis. HSI models have been used effectively as a snapshot in time using Landsat images and GIS; for example, to evaluate the habitat of One-horned Rhinoceros in Nepal (Thapa et al. 2014), and the Red-cockaded Woodpecker near the Big Thicket National Preserve (Thapa and Acevedo 2016). For monitoring, a time series of HSI would be needed to assess changes in suitability due, for example, to urbanization or natural disasters.

Habitat Suitability Index

Dettki et al. (2003) discuss two approaches to HSI modeling, process-oriented models are based on functional processes underlying habitat use thus providing a conceptual framework, while empirical models analyze data on habitat use and habitat characteristics collected at specific sites. Process-oriented HSI models use habitat requisites or parameters such as food, shelter or cover, and water as input variables to a function providing a dimensionless 0.0–1.0 index, where 0 and 1 indicate unsuitable and optimum habitat conditions, respectively.

To evaluate the habitat of One-horned Rhinoceros in Nepal (Thapa et al. 2014), the HSI was derived from the literature and field observations. Landsat images were classified to identify grass, agriculture, sal forest, and mixed forest, which were used to calculate the HSI for a target pixel based on a neighborhood of 70×70 pixels around of a target pixel as input variables to calculate the HSI value for that pixel. Recall neighborhood and zonal calculations from Chapter 10. The neighborhood size was determined according to the animal's average mean annual home range of $\sim 4\,km^2$. Food (grass and agriculture) and cover (sal forest and mixed forest) were evaluated as proportions in the neighborhood; suitability for those factors was defined as a function with increasing values for the factors until reached a saturation point where the value is 1. These saturation points are calculated from food consumption rates and cover needs. To combine those factors, seasonality was accounted for by calculating the weighted geometric mean when both factors are important or the maximum when one factor is dominant. To account for uncertainty in these estimations, the results are evaluated by sensitivity analysis. Once a raster with HSI values is derived, it is subject to pattern analysis or calculation of fragmentation metrics to determine the overall quality of the habitat.

HABITAT FRAGMENTATION ANALYSIS

In addition to a habitat suitability of a cell or pixel, one must consider the spatial configuration of those cells across the habitat. For example, it is not the same to have a large area of contiguous high HSI cells that serve as large habitat patch than a collection of isolated small patches of high HSI separated by large expanse of low HSI cells. These spatial features are quantifiable by *fragmentation indices* or metrics that have evolved since their inception in the field of landscape ecology in the late 1980s (O'Neill et al. 1988).

There is an abundance of fragmentation metrics and are classified whether they apply at the patch level, the class level, or the landscape level. A class is understood as the collection of a single type of patch (e.g., forest) on the landscape; this is to say, the cells identified as that type of patch on a raster layer. To explain this better, consider the simple raster with 30×30 cells shown in Figure 14.9 which is the sample dataset in R package `landscapemetrics` (Hesselbarth et al. 2019). This "toy" landscape has three classes labeled 1–3. Observe how you can follow all contiguous cells of one class and separate them on the landscape into patches; your observations should match the three maps of Figure 14.10, one for each class showing all the patches you can trace for each. Think of each class as a patch type, then each one of the three maps of Figure 14.10 is a *class* meaning is the layer representation for that patch type, and it shows the patches in that class. You can count the patches for each class and think of the distribution of patch size, its edges, shape, and a few more things about these patches; that is what landscape metrics are. For example, for class 1, there are only two large core area patches. Relating this to animal habitat, you can see how this species would be restricted to only two major areas. Compare this to class 3 which has fewer but larger patches and would be better habitat if this were the class of importance for the species.

Now there are some important technical details to understand. You would notice that there is a *boundary* around each patch, and cells inside the boundary, which are defined as *core area*; this has to do with the rule that assigns a cell to a patch, which entails several decisions. First, a cell may look in all eight directions from it to its neighbors, like a queen moves in chess, or only in four directions from it, as in the rook move in chess. A decision is made as to how many sides of a cell

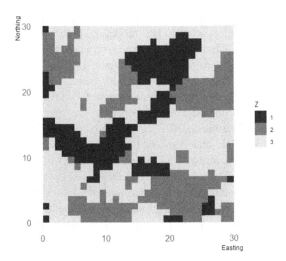

FIGURE 14.9 Sample example from package landscapemetrics.

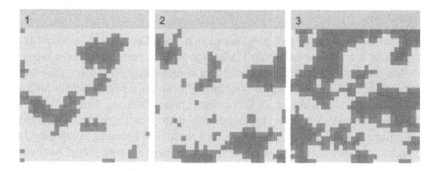

FIGURE 14.10 Classes by patch type.

are boundary and how far away from the patch would a cell be considered part of the patch. A cell is considered part of the core area if all the neighbors are of the same class. Another technical detail is that calculations are done in meters with area in hectares (ha). Recall that 1 ha = 10,000 m². This is convenient when the image is in UTM coordinates. A 30×30 pixel is $\dfrac{30 \times 30 \text{ m}^2}{10,000 \text{ m}^2/\text{ha}} = 0.09$ ha or nearly a tenth of a hectare.

Per this explanation, one would understand that class-level metrics are of interest and typically identified with a habitat. It is important to use class metrics that relate to pattern and are independent of total abundance on the landscape (Wang et al. 2014), because after all having a lot of total area is not desirable if it is fragmented. However, class-level metrics depend on metrics defined at the patch level.

To better understand these concepts, consider the calculation of a metric based on area. For a patch, one can define a metric called *core area index* (CAI) that is defined as the proportion of core area of the patch in relation to total patch area expressed in percent. For example, if the total area of one of the large patches of class 1 is 0.0148 ha and its core area is 0.0071, then $CAI = \dfrac{0.0071}{0.0148} \times 100 = 48\%$.

Now at the class level, we can use the metric Mean CAI as the average of the *CAI* of all patches in that class. In this example, the mean *CAI* for class 1 is 6.92%, which tells us that on the average these patches have little core area compared to the patch size. The same calculation yields 14.8% for class 3, confirming our intuition that this class would be a better habitat. Looking beyond the average, one can calculate the coefficient of variation of the patches *CAI* for the class, which is the variability scaled by the mean, allowing a better comparison. In this case, we get 233 for class 1 vs. 158 for class 3.

One *shape* metric of the patches is their *fractal dimension*, which expresses a ratio of perimeter to area, thus measuring the patch complexity. Fractal dimension formulas involve a function of the natural log of perimeter and area. At the class level, metrics may consist of the mean and coefficient of variation of the fractal dimension of all patches in the class. Computing the mean fractal dimension of the example, we get 1.15, 1.23, and 1.30 for classes 1, 2, and 3, indicating an increase of complexity of the patches from class 1–3. In terms of *edge* of patches, one metric is *edge density* calculated by the sum of all edges of a class divided by the landscape area and given in m/ha or km/ha. In the example, class 1 has edge density of 2 km/ha, whereas class 3 has 3.6 km/ha. Finally, we mention an *aggregation* metric, defined by the proportional deviation of the proportion of like adjacencies (of the class) from that expected under a spatially random distribution. In the example, this metric for class 1 would have a value of 0.732, while for class 3, the value is lower 0.649, indicating more aggregation in class 1.

Vertical Structure

As you likely noticed reading the two previous sections, once we classify a cell of land cover as a particular class of habitat or a particular HSI value, we can perform multiple raster operations to determine habitat quality for the study area under consideration. It is also common to find that habitat suitability depends on the vertical structure of cover at a pixel, such as bird habitat as a function of canopy layers in a forest. In these cases, remote sensing analysis must account for a way of inferring vertical structure. This is possible with high spatial resolution radar images, hyperspectral imagery, LiDAR, or a combination of these. For example, combining LiDAR data with hyperspectral imagery allowed to infer vertical structure in a bottomland hardwood forest as a metric of habitat quality for birds (Matsubayashi 2013). This study demonstrated that 3D-habitat descriptions show how the distribution of bird species relates to forest composition and structure at various scales. Accuracy assessments showed that integrated LiDAR-hyperspectral increased the overall classification accuracy.

SPATIAL ANALYSIS

Point pattern analysis is used in many monitoring applications such as spatial distribution of plants and trees on the landscape, sampling sites in a waste field, sampling sites in a lake, locations of intense events such as quakes or tornadoes. We encountered point patterns in Chapter 10 when we studied GIS vector layers. In this chapter, we study the basics of point pattern analysis as it can also be used to study animal species distribution; in particular, for territorial species such as songbirds, and species that flock or congregate in resource patches (Matsubayashi 2013). It is also useful to interpolate from a spatial point pattern obtained by sampling population density to infer values at unsampled points; one effective technique is *kriging* that we will study in this section. In addition, when adding other environmental variables, e.g., obtained by remote sensing, to the kriging estimate using *cokriging*, it is possible to make predictions of a species spatial distributions. This approach has been used for fish populations, e.g., Amiri et al. (2017).

Point patterns are typically a collection of points placed over a spatial 2D domain with coordinates *x*, *y* as illustrated in Figure 14.11, where at the top points are uniformly distributed and at the bottom the pattern is non-uniform, either following a gradient or by aggregation. The key is the variation of the *density of points*, i.e., the spatial variability of the number of points per unit area. Note that a regular pattern is uniform, but a uniform pattern is not necessarily regular.

Testing Spatial Patterns: Cell Count Methods

One way to determine the spatial pattern is to divide the domain in a grid of T cells or tiles of equal size, count the total number of points, m, and the number of points o_i in each tile. If the distribution is uniform, then the expected number e_i of points per cell is $E = m/T$, thus we can use a chi-square test with $df = T-2$. When there are large departures from the expected, the chi-square value is sufficiently large, and we can reject the null and conclude that the pattern is non-uniform. For example, Figure 14.12 shows 90 points distributed over 9 tiles, we would expect $e_i = 90/9 = 10$ points per cell for all cells i. We obtain a chi-square of 5.8, which for $df = 9-2 = 7$ has a p-value of 0.56. Therefore, we should not reject H0 and we conclude that the pattern may be uniform.

Spatially random patterns follow a Poisson distributed random variable (RV), which is a binomial with very small probability of an event of interest, i.e., a good model for rare events. Define λ density (number of points per unit area) by $\lambda = \dfrac{m}{A}$ where A is the total area of the domain, and divide the total area in T cells or tiles, then the mean number of points per cell is m/T and should be equal to the density multiplied by the area a of a cell, therefore $a\lambda = \dfrac{A}{T}\dfrac{m}{A} = \dfrac{m}{T}$. A Poisson RV with rate $a\lambda = mT$ can be used to calculate the probability of having $r = 0$, 1, 2, ... points in

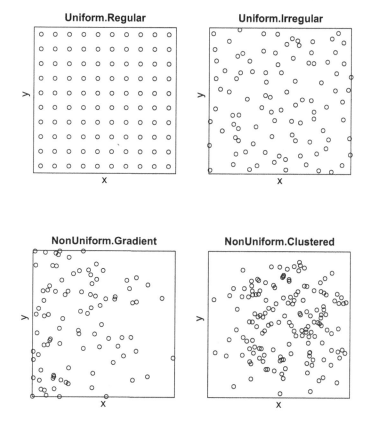

FIGURE 14.11 Examples of two-dimensional spatial point patterns.

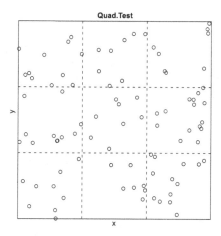

FIGURE 14.12 Spatial pattern of 90 points within a 3×3 grid or 9 cells.

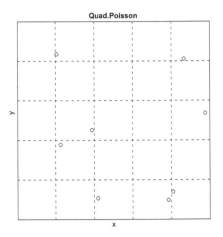

FIGURE 14.13   A potentially random point pattern.

a cell and apply χ^2 to compare to observed values. For example, take the pattern in Figure 14.13, where we have $m = 8$ points in a grid of $5 \times 5 = 25$ cells over a domain of unit area $A = 1$. The rate is $a\lambda = \dfrac{m}{T} = \dfrac{8}{25} = 0.32$. The probabilities of a Poisson with this rate are 0.73, 0.23, and 0.04 for $r = 0$, 1, 2. The expected values of number of cells that have r points are 18, 6, and 1. The observed are 17 cells with 0 points, 8 cells with 1 point, and 0 cells with 2 points. The chi-square values are 0.055, 0.666, and 1.00. These add up to 1.72 and we have $df = 3 - 2 = 1$ because we used three categories or bins. The p-value is 0.19 and cannot reject, therefore, the pattern could be random.

A clustered pattern is a set of points closer together than expected by chance alone. One way of testing for cluster patterns using cell counts is with the negative binomial $P(r) = \begin{pmatrix} k+r-1 \\ r \end{pmatrix} \left(\dfrac{p}{1+p} \right)^r \left(\dfrac{1}{1+p} \right)^k$ where r is the number of points in a cell, k is a clustering or shape parameter $k = \dfrac{(m/T)^2}{s^2 - (m/T)}$, and p is the probability that a cell contains a point $p = \dfrac{\lambda}{k} = \dfrac{\frac{m}{T}}{k}$. The mean is the rate m/T. We then use chi-square to test if pattern was distributed according to a negative binomial.

NEAREST NEIGHBOR ANALYSIS

In this section, we cover methods based on nearest neighbor analysis, based on Euclidian distance, which we discussed in Chapter 10 as a technique for GIS vector layer analysis. Consider the nearest neighbor distance d_i for each point i. If the points follow a Poisson distribution, with density λ, then the PDF of distances is

$$p(d) = 2\lambda \pi d \exp(-\lambda \pi d^2) \qquad (14.7)$$

which has mean $\mu_d = \dfrac{1/2}{\sqrt{\lambda}}$ and variance $\sigma^2_d = \dfrac{4-\pi}{4\pi\lambda}$. We perform the analysis by comparing an empirical CDF (ECDF) of d with the theoretical CDF, which is

$$G(d) = 1 - \exp(-\lambda \pi d^2) \tag{14.8}$$

To calculate the ECDF, for each d, we count of all distances to nearest neighbors less or equal than the value d

$$\widehat{G}(d) = \frac{1}{m} \sum_1^n \delta_i(d_i, d) \tag{14.9}$$

Here we use an indicator function

$$\delta_i(d_i, d) = \begin{cases} 1 & \text{when } d_i \leq d \\ 0 & \text{otherwise} \end{cases} \tag{14.10}$$

Now we can compare the ECDF $\widehat{G}(d)$ with the theoretical CDF $G(d)$.

For example, consider the point pattern in Figure 14.14 which looks clustered. The process above would yield a plot of empirical and theoretical CDFs for visual comparison as in Figure 14.15. The circles correspond to the ECDF raw or uncorrected for edge effects. There are several schemes to correct for edge effect, e.g., the Kaplan-Meier estimates. When we include this correction for this example (Figure 14.15), we visually appreciate that the corrected ECDF departs only slightly from the ECDF and mostly for larger values of d. In conclusion, this visual exploration tells us that the ECDF departs substantially from the theoretical CDF, reinforcing the intuition of clustered pattern of the points.

Another more refined approach is to look at the distance to the second closest neighbor, the third neighbor, the kth neighbor. A possible approach is the *Ripley's K* function, an estimator of the second-order properties. The K function is the cumulative distribution of points within a distance interval. The theoretical $K(d)$ is

$$K(d) = \frac{N(d)}{\lambda} \tag{14.11}$$

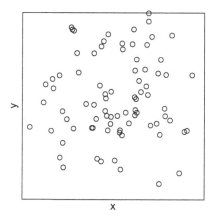

FIGURE 14.14 A spatial point pattern for distance analysis.

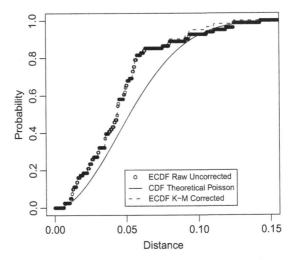

FIGURE 14.15 Empirical (raw and corrected) and theoretical CDF of distance.

where $N(d)$ is the expected number of points within a distance d. The ECDF is

$$\widehat{K}(d) = \frac{1}{\lambda} \sum_i^m \sum_j^m \delta_{ij}(d_{ij}, d)/m \tag{14.12}$$

for $i \neq j$ and where d_{ij} is distance between points i and j. Here we use an indicator function

$$\delta_{ij}(d_{ij}, d) = \begin{cases} 1 & \text{when } d_{ij} \leq d \\ 0 & \text{otherwise} \end{cases} \tag{14.13}$$

Recall that $\lambda = m/A$ and substitute in Equation (14.12) to get

$$\widehat{K}(d) = \frac{A}{m^2} \sum_i^m \sum_j^m \delta_{ij}(d_{ij}, d) \tag{14.14}$$

Again, we can compare theoretical (Poisson) with the empirical. There are several schemes to correct for edge effect such as the border and isotropic estimates. When we include isotropic correction for this example (Figure 14.16), we visually appreciate that the corrected departs only slightly from the raw and mostly for larger values of d. In conclusion, this visual exploration tells us that the empirical K departs substantially from the theoretical K, reinforcing the suspicious clustered pattern of the points.

As an aid to visualization, we can calculate

$$\hat{L}(d) = \sqrt{\frac{K(d)}{\pi}} \tag{14.15}$$

and plot the $L(d)$ function vs. d and check whether it follows a straight line. The reason is that a completely random pattern will have

$$K(d) = \pi d^2 \tag{14.16}$$

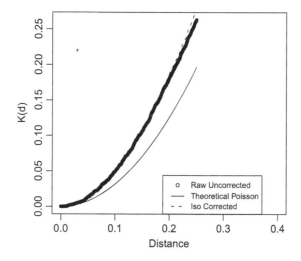

FIGURE 14.16   Empirical K raw and corrected vs. theoretical comparison.

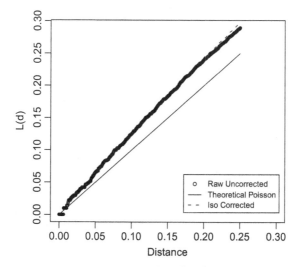

FIGURE 14.17   Empirical L raw and corrected vs. theoretical comparison.

and therefore if the pattern is close to random

$$\hat{L}(d) = \sqrt{\frac{K(d)}{\pi}} = \sqrt{\frac{\pi d^2}{\pi}} = d \tag{14.17}$$

As illustrated in Figure 14.17, the empirical $L(d)$ departs substantially from the 1:1 line given by the theoretical Poisson (1:1 line). We compare these also with Monte Carlo simulation as we did for $G(d)$. See Figure 14.18.

GEOSTATISTICS

In addition to its location, each point may have a mark or associated value, yielding a *marked point pattern*. The mark may be a value from a categorical variable, e.g., species, or a continuous variable,

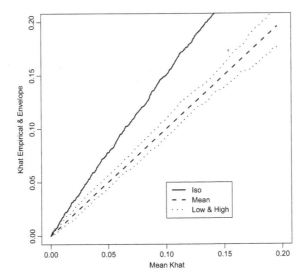

FIGURE 14.18 Empirical K departing from a band obtained by simulation.

e.g., size. *Geostatistics* consists of predicting values of variables in non-sampled points of a spatial domain using a collection of sampled points, i.e., a marked point pattern. We calculate a measure of similarity using the covariance between points separated at given distances then interpolate values at non-sampled points using the covariance structure to predict the values of the variables at the non-sampled points, this is called *kriging*, as we mentioned at the beginning of this section. Geostatistics assumes that a variable is continuous over the domain and modeled as an RV $Z(x, y)$ where location x, y are coordinates of a point. The technique is also applicable to 3D and in this case, Z would have values at three coordinate axes. In this book, we limit the presentation to the spatial case of an stationary variable, i.e., the distribution of variable $Z(x, y)$ does not depend on location x, y, and then the expected value of $Z(x, y)$ at all points is a constant μ_Z, $E[Z(x,y)] = \mu_Z$ and the variance at all points is also constant σ_Z^2 $E[(Z(x, y) - \mu_Z)^2] = \sigma_Z^2$.

The covariance of Z between points (x_1, y_1) and (x_2, y_2) separated by distance h, or *lag*, is the same for any two points separated by this distance h

$$\text{cov}(h) = E[Z(x_1, y_1)Z(x_2, y_2)] - \mu_Z^2 \tag{14.18}$$

This expression would remind you of the autocorrelation of an RV $X(t)$, where the lag is time. However, here the points occur in 2D, and the lag h is the Euclidian distance between points. Note that $\text{cov}(h)$ is maximum at zero lag ($h = 0$) and equal to the variance σ_Z^2 because at $h = 0$ it is the same sample point $Z(x_1, y_1) = Z(x_2, y_2)$. Covariance decreases as h increases because the similarity of values will tend to decrease as points are increasingly apart. The *empirical covariance* is the average of products of values z at all point pairs (x_i, y_i) and (x_j, y_j) separated by distance in the interval $h \leq d_{ij} < h + \Delta h$ minus the square of the sample mean

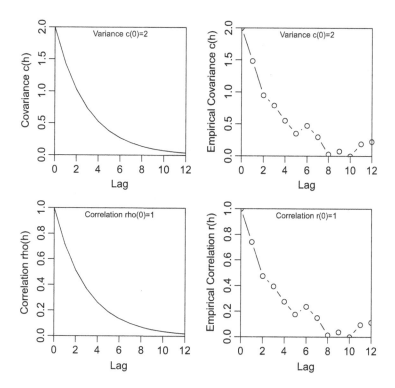

FIGURE 14.19   Variograms: covariance and correlation coefficient.

$$c(h) = \frac{\displaystyle\sum_{h \le d_{ij} < h + \Delta h} z(x_i, y_i) z(x_j, y_j)}{n_h} - \overline{Z}^2 \qquad (14.19)$$

Here n_h is the number of point pairs at distance h.

The top panels of Figure 14.19 show examples of a theoretical covariance (left) and an empirical covariance (right). A plot of covariance as a function of h is called a *covariogram*. Dividing $cov(h)$ by $c(0)$ or σ_Z^2, we obtain the correlation coefficient $\rho(h) = \dfrac{cov(h)}{cov(0)} = \dfrac{cov(h)}{\sigma_Z^2}$ which has a value of 1 at $h = 0$ and decays to 0 with increasing h. The empirical correlation coefficient $r(h)$ is the ratio of empirical covariance over variance. The bottom panels of Figure 14.19 show correlation coefficients corresponding to the examples in the top panels. On the left, theoretical $\rho(h)$ and on the right the empirical $r(h)$ coefficient. The plot of correlation coefficient vs. lag is called the *correlogram*.

The semivariance is a measure of dissimilarity denoted with the Greek letter gamma γ. It is defined as half of the average of the squares of the differences between points (x_i, y_i) and (x_j, y_j) separated by distance in the interval $h \le d_{ij} < h + \Delta h$

$$\gamma(h) = \frac{\displaystyle\sum_{h \le d_{ij} < h + \Delta h} \left(z(x_i, y_i) - z(x_j, y_j) \right)^2}{2 n_h} \qquad (14.20)$$

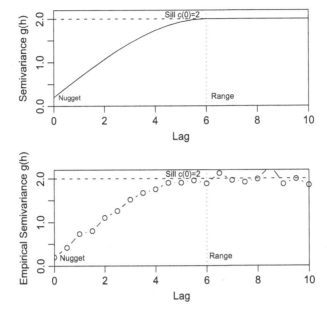

FIGURE 14.20 Semivariance: nugget, sill, and range.

The prefix "semi" reminds us that this is half of the total squared difference at a given h. A plot of semivariance vs. lag is a *semivariogram*. Note that if $h = 0$, the difference should be zero and therefore we would expect that $\gamma(0) = 0$. However, the empirical semivariance is not calculated at $h = 0$, but at a small distance representing a minimum lag in the dataset. Then $\gamma(h)$ is extrapolated from the calculated values back to the $h = 0$ axis. The non-zero intercept found by this extrapolation is called the *nugget effect* and represents the fine-scale or micro-scale variation (Figure 14.20). Many times, as points become more and more distant, the dissimilarity reaches a plateau called the *sill* at a distance called the *range* (Figure 14.20).

The semivariance is equal to the sample variance s_z^2 minus the covariance $c(h)$, and the variance is equal to the covariance at zero lag $c(0)$, therefore

$$\gamma(h) = s_Z^2 - c(h) = c(0) - c(h) \tag{14.21}$$

because $c(h)$ goes to zero for large h, the asymptotic value of the semivariance or sill is equal to the variance $sill = \gamma(\infty) = c(0) = s_Z^2$. Figure 14.21 shows another example of semivariogram and illustrates the relationship with the covariogram. We have ignored the nugget for simplicity. In practice, the term variogram is used to refer generically to the covariogram, correlogram, and semivariogram. When searching for points separated at given distances h, we can move along lines with given directions, for example north to south or east to west. When we search all directions, then we have an *omnidirectional* variogram.

A variogram model is an equation defining the semivariance or the covariance and several mathematical functions are available and employed to model variograms. The exponential

$$c(h) = c(0)\exp\left(\frac{h}{a}\right) \tag{14.22}$$

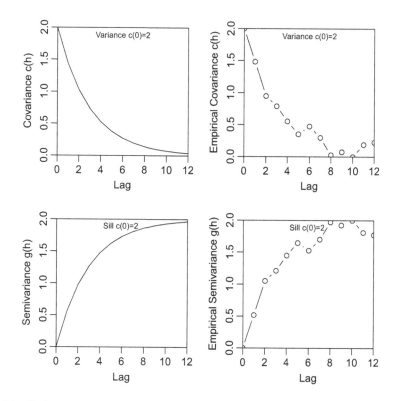

FIGURE 14.21   Variograms: covariance and semivariance.

is the simplest model prescribing a decay of covariance from the maximum $c(0) = \sigma_Z^2$ to zero at a rate $1/a$. An example is shown in Figure 14.22. Then the exponential semivariance is

$$\gamma(h) = c(0) - c(h) = c(0) - c(0)\exp\left(\frac{-h}{a}\right) = c(0)\left(1 - \exp\left(\frac{-h}{a}\right)\right) \qquad (14.23)$$

It asymptotically reaches a maximum value (sill) of $c(0)$ when h/a is large. Recall that $c(0) = \sigma_Z^2$, and that the sill is the variance of Z. From the last equation, also note that $\gamma(0) = 0$. Thus, in this model, the nugget is zero. To account for a non-zero nugget effect, we can use a discontinuity $\gamma(0^+) = nugget$ where 0^+ is a value of lag slightly larger than 0.

We can fit a model variogram to an empirical variogram by finding the values of the model parameters, e.g., the values of $c(0)$ and a in Equation (14.22) that produce the least difference between the model and the empirical variogram. For this purpose, we can employ a numerical fitting routine; it requires an initial guess for values of model parameters, and other conditions such as the number of iterations and tolerance for convergence. A convenient procedure is non-linear least squares regression.

Kriging consists of using the covariance model to predict the values of the variables at the non-sampled points, using a regular grid for the target points for prediction. This way, we have values distributed in the entire domain; some of these are measured values, while others are kriging estimates. The result can be visualized as a grid or raster image with overlaid contour lines (Figure 14.23). By decreasing step size between the prediction grid points, we can produce a higher resolution image (Figure 14.24).

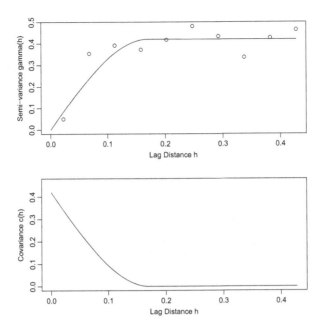

FIGURE 14.22 Exponential variogram model with nugget $= 0.2$, $a = 3$, $c(0) = 1$.

There is a variety of kriging procedures, in this book, we will mention only *ordinary* kriging which assumes that there is not a trend, i.e., Z is stationary. Using a compact vector notation for x, y coordinates, denote x_0 as a point x where we want to estimate the value of Z. The kriging estimate $\widehat{Z}(x_0)$ is obtained by the linear combination of n known values $Z(x_i)$ around the target location x_0 using a set of weights or coefficients λ_i to form the equation

$$\widehat{Z}(x_0) = \sum_{i=1}^{k} \lambda_i Z(x_i) \tag{14.24}$$

the kriging error is the difference between the theoretical and the estimated values

$$e = Z(x_0) - \widehat{Z}(x_0)$$

Thus, the expected value of the estimate should be equal to the mean or expected value of Z at any point and therefore the expected value of the kriging error should be zero. This is to say

$$E[e] = E\left[Z(x_0) - \widehat{Z}(x_0) \right] = \mu_Z - E[\widehat{Z}(x_0)] = 0 \tag{14.25}$$

We must find the weights λ_i by minimizing the expected value of the square of the error

$$E[e^2] = E\left[\left(Z(x_0) - \widehat{Z}(x_0) \right)^2 \right] \tag{14.26}$$

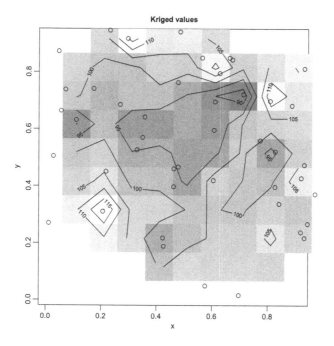

FIGURE 14.23 Kriged regionalized variable using 10×10 grid.

which happens to be equal to the variance of the kriging error because $E[e]=0$.

$$\sigma_e^2 = E[e^2] - E[e]^2 = E[e^2] - 0 = E[e^2]$$

It is important to generate an image of this variance to see how the error varies with location. Figure 14.25 illustrates this variance for the higher resolution grid.

In other words, the kriging procedure consists of finding the λ_i such that

$$\min_{\lambda_i} \sigma_e^2 = \min_{\lambda_i} E\left[\left(Z(x_0) - \widehat{Z}(x_0) \right)^2 \right] \tag{14.27}$$

Equation (14.25) means that we want an unbiased estimator.

$$E[\widehat{Z}(x_0)] = \mu_Z \tag{14.28}$$

Note that

$$E[\widehat{Z}(x_0)] = E\left[\sum_{i=1}^{k} \lambda_i Z(x_i) \right] = \sum_{i=1}^{k} \lambda_i E[Z(x_i)] = \mu_Z \sum_{i=1}^{k} \lambda_i \tag{14.29}$$

Because the expected value at all points x_i is equal to the constant mean. Therefore, to obtain Equation (14.28), we need to require that the weights sum up to 1

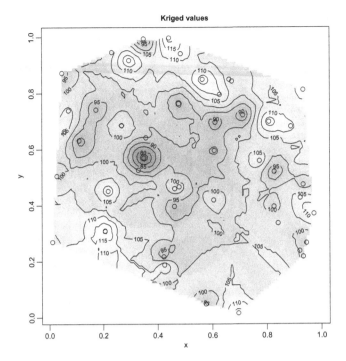

FIGURE 14.24 Kriged regionalized variable using 100×100 grid.

$$\sum_{i=1}^{k} \lambda_i = 1 \qquad (14.30)$$

to obtain the unbiased estimates. The estimates at the measured points are the same as the observed values. This means that kriging is an exact interpolator.

In essence, this implies solving the optimization problem Equation (14.27) subject to a constraint Equation (14.30). This type of problem is solved by taking partial derivatives of the objective function (variance of the kriging error in this case) with respect to the coefficients λ_i and using an extra parameter called a Lagrange multiplier μ. Using this method, we form the matrix equation

$$
\begin{bmatrix}
c(h_{11}) & c(h_{12}) & \dots & c(h_{1k}) & 1 \\
c(h_{21}) & c(h_{22}) & \dots & c(h_{2k}) & 1 \\
\dots & \dots & \dots & \dots & \dots \\
c(h_{k1}) & c(h_{k2}) & \dots & c(h_{kk}) & 1 \\
1 & 1 & \dots & 1 & 0
\end{bmatrix}
\begin{bmatrix}
\lambda_1 \\
\lambda_2 \\
\dots \\
\lambda_k \\
-\mu
\end{bmatrix}
=
\begin{bmatrix}
c(h_{01}) \\
c(h_{02}) \\
\dots \\
c(h_{0k}) \\
1
\end{bmatrix}
\qquad (14.31)
$$

where $c(h_{ij})$ denotes the covariance model evaluated at lag h_{ij}. This is to say, the covariance of $Z(x_i)$ and $Z(x_j)$ for points x_i and x_j separated by lag h_{ij}. As you can see, all the known quantities of this

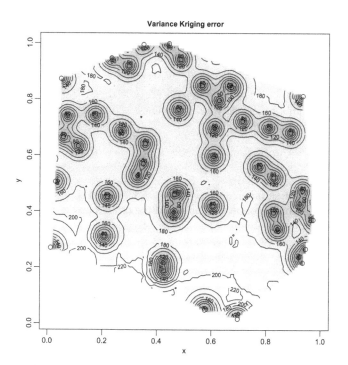

FIGURE 14.25   Variance of kriging error using 100 × 100 grid.

equation are obtained by evaluating the covariance model, or equivalently the semivariance model since $c(h) = \sigma_Z^2 - \gamma(h)$.

POPULATION AND COMMUNITY MODELING

Animal tracking, habitat monitoring, and spatial distributions provide useful information to develop models of wildlife at the ecological *population*, *metapopulation*, and *community* levels. Mechanistic or process-based models emphasize dynamics of population density (individuals per unit area) using difference and differential equations of various degrees of complexity based on reproduction, recruitment, and mortality rates (Acevedo 2012). Empirical models emphasize data-driven prediction of population numbers or density and include time series, regression models, and mixed effects models (Zuur et al. 2009).

Population models consider one species at a locale or a contiguous area or habitat, whereas metapopulation models consider one species at a set of locations or patches of the landscape. In this latter case, spatial distribution of the population and habitat spatial patterns are very important. Community models consider multiple interacting species at a locale or contiguous area, as well as across patches of the landscape. While the concepts of habitat fragmentation presented in this chapter are intuitive for terrestrial and coastal environments, it is harder to conceptualize metrics for aquatic organisms in pelagic environments. In this case, population and community dynamics modeling benefits from data on animal movement, environmental conditions, and spatial distributions.

EXERCISES

Exercise 14.1

Suppose the coordinates (x_i, y_i) of three towers are $(x_1, y_1) = (0,0); (x_2, y_2) = (1,0); (x_3, y_3) = (0,1)$ and the distance d_i from a beacon to each tower are $d_1 = 0.361; d_2 = 0.854; d_3 = 0.728$. What are coordinates (x, y) of the beacon?

Exercise 14.2

A drone flies at $y = 6\,m$ of elevation above a lake searching for a submersed radio emitter. How close in horizontal distance x to the emitter would the drone need to be for receiving a strong radio signal from the emitter?

Exercise 14.3

Compare acoustic signal wavelength in freshwater and the air for 10 and 40 kHz.

Exercise 14.4

CAI metrics, for example in Figure 14.10, are 6.92, 11.2, and 14.8 for the mean, 233, 160, and 158 for the CV, and 16.1, 17.9, and 23.4 for the standard deviation. Explain these values in terms of the patterns you observe in Figure 14.10.

Exercise 14.5

Use the spatial point pattern of Figure 14.12 and the chi-square test to determine that the pattern is uniform (homogeneous). What is the number of cells, T, the total number of points, m, and the expected number of points for each cell if the distribution were uniform? What is the null hypothesis for a chi-square test? How many degrees of freedom df? Calculate the chi-square value. Use the 1-pchisq(x, df) expression in R to calculate the p-value. Can you reject the null hypothesis? What is your conclusion?

Exercise 14.6

Uniform point patterns can be either regular or random, but the chi-square test does not specify which. Does the point pattern of the previous exercise appear more regular or random? What distribution and test could you use to determine the nature of a uniform pattern?

Exercise 14.7

Examine the spatial point pattern of Figure 14.13 for randomness. What is the total area, A, the total number of points, m, the density, λ, the total number of cells, T, and the rate, $\lambda a = m/T$? Use the probabilities that a cell will have r number of points. Calculate the expected number of cells that will have 0, 1, 2, or 3 points. How do these values compare to the point pattern? What is the chi-square value? Could the pattern be random?

Exercise 14.8

Consider four points on the *x-y* plane, with following coordinates A = (0,1), B = (3,3), C = (2,0), and D = (2,1). For each point, identify its nearest neighbor and calculate that distance NND.

Exercise 14.9

Suppose a semivariogram has a range of 4, sill of 20, and a nugget of 2. Write an expression for the variogram using the exponential model as a function of lag *h*. Sketch a graph of the semivariogram. Write an expression for the covariance as a function of *h*. Draw a graph of the covariogram.

REFERENCES

Acevedo, M. F. 2012. *Simulation of Ecological and Environmental Models*. Boca Raton, FL: CRC Press, Taylor & Francis Group. 464 pp.

Acevedo, M.F. 2024. *Real-Time Environmental Monitoring: Sensors and Systems, Second Edition – Lab Manual*. Boca Raton, FL: CRC Press, Taylor & Francis Group. 463 pp.

Amiri, K., N. Shabanipour, and S. Eagderi. 2017. Using kriging and co-kriging to predict distributional areas of Kilka species (Clupeonella spp.) in the southern Caspian Sea. *International Journal of Aquatic Biology* 5:108–113.

Bloor, I. S. M., V, J. Wearmouth, S. P. Cotterell, M. J. McHugh, N. E. Humphries, E. L. Jackson, M. J. Attrill, and D. W. Sims. 2013. Movements and behaviour of European common cuttlefish *Sepia officinalis* in English Channel inshore waters: First results from acoustic telemetry. *Journal of Experimental Marine Biology and Ecology* 448:19–27.

Brown, J., and S. D. Gehrt. 2009. *The Basics of Using Remote Cameras to Monitor Wildlife, Fact Sheet: Agriculture and Natural Resources*. Columbus, OH: The Ohio State University.

Cox, A. W., M. S. Pruett, T. J. Benson, S. J. Chiavacci, and F. R. Thompson. 2012. "Development of Camera Technology for Monitoring Nests." In *Video Surveillance of Nesting Birds*, edited by C. A. Ribic, F. R. Thompson and P. J. Pietz, 185–209. Berkeley, CA: University of California Press.

Dettki, H., R. Lofstrand, and L. Edenius. 2003. Modeling habitat suitability for moose in coastal northern Sweden: Empirical vs process-oriented approaches. *Ambio* 32 (8):549–556.

Golightly, R. T., and S. R. Schneider. 2011. *Years 9 and 10 of a Long-Term Monitoring Effort at a Marbled Murrelet Nest in Northern California*. Arcata, CA: National Park Service.

Hawkes, L. A., O. Exeter, S. M. Henderson, C. Kerry, A. Kukulya, J. Rudd, S. Whelan, N. Yoder, and M. J. Witt. 2020. Autonomous underwater videography and tracking of basking sharks. *Animal Biotelemetry* 8 (1):29.

Hesselbarth, M. H. K., M. Sciaini, K. With, K.A. Wiegand, and J. Nowosad. 2019. Landscapemetrics: An open-source R tool to calculate landscape metrics (ver.0). *Ecography* 42:1648–1657.

HRI. 2023. *Open-Source Robot Audition Software—HARK*. accessed January 2023. https://www.jp.honda-ri.com/en/activity/hark/.

Jiang, S., and S. Georgakopoulos. 2011. Electromagnetic wave propagation into fresh water. *Journal of Electromagnetic Analysis and Applications* 3:261–266.

Kandula, H. 2011. *Development of a Testbed for Multimedia Environmental Monitoring*. MS Thesis, Electrical Engineering. Denton, TX: University of North Texas.

Lotek. 2023a. *Nanotags (Coded VHF) for Birds and Bats*. accessed January 2023. https://www.lotek.com/products/nanotags/.

Lotek. 2023b. *Counter Proximity*. accessed January 2023. https://www.lotek.com/products/ncounter-proximity/.

Lotek. 2023c. *Radio*. Lotek, accessed January 2023. https://www.lotek.com/technology/radio/.

Matsubayashi, S. 2013. *Quantifying Forest Vertical Structure to Determine Bird Habitat Quality in the Greenbelt Corridor, Denton, Texas*. PhD Dissertation, Denton, TX: University of North Texas.

Metcalfe, J. D., and G. P. Arnold. 1997. Tracking fish with electronic tags. *Nature* 387 (6634):665–666.

O'Neill, R. V., J. R. Krummel, R. H. Gardner, G. Sugihara, B. Jackson, D. L. DeAngelis, B. T. Milne, M. G. Turner, B. Zygmunt, S. W. Christensen, V. H. Dale, and R. L. Graham. 1988. Indices of landscape pattern. *Landscape Ecology* 1 (3):153–162.

Oeser, J., M. Heurich, C. Senf, D. Pflugmacher, and T. Kuemmerle. 2021. Satellite-based habitat monitoring reveals long-term dynamics of deer habitat in response to forest disturbances. *Ecological Applications* 31 (3):e2269.

Pedersen, M. W., D. Righton, U. H. Thygesen, K. H. Andersen, and H. Madsen. 2008. Geolocation of North Sea cod (Gadus morhua) using hidden Markov models and behavioural switching. *Canadian Journal of Fisheries and Aquatic Sciences* 65(11):2367–2377.

SeeMore Wildlife Systems. 2015. *Remote Wildlife Viewing Systems.* accessed March 2015. http://www.seemorewildlife.com/.

Star ODDI. 2023. *Aquatic Animals.* accessed January 2023. https://www.star-oddi.com/products/archival-tags.

Susuki, R., S. Matsubayashi, R. W. Hedley, K. Nakadai, and H. G. Okuno. 2017. HARKBird: Exploring acoustic interactions in bird communities using a microphone array. *Journal of Robotics and Mechatronics* 29(1):213–223.

System in Frontier. 2023. *Advanced Sound Technology Products: Microphone Array.* accessed January 2023. https://sifi.co.jp/en/product/microphone-array/.

Taylor, P. D., S. A. Mackenzie, B. G. Thurber, A. M. Calvert, A. M. Mills, L. P. McGuire, and C. G. Guglielmo. 2011. Landscape movements of migratory birds and bats reveal an expanded scale of stopover. Invitation for re-review. *PLOS One* 6(11): e27054.

Thapa, V., and M. F. Acevedo. 2016. Habitat quantity of red-cockaded woodpecker picoides borealis (Aves: Piciformes: Picidae) in its former historic landscape near the Big Thicket National Preserve, Texas, USA. *Journal of Threatened Taxa* 8 (1):8309–8322.

Thapa, V., M. F. Acevedo, and K. P. Limbu. 2014. An analysis of the habitat of the Greater One-horned Rhinoceros Rhinoceros unicornis (Mammalia: Perissodactyla: Rhinocerotidae) at the Chitwan National Park, Nepal. *Journal of Threatened Taxa* 6 (10):6313–6325.

Thorstad, E. B., A. H. Rikardsen, A. Alp, and F. Okland. 2013. The use of electronic tags in fish research - An overview of fish telemetry methods. *Turkish Journal of Fisheries and Aquatic Sciences* 13:881–896.

TPWD. 2015. *Checking Out the New Home.* accessed April 2015. https://www.youtube.com/watch?v=Asqr1 P0cHcM.

Wang, X. F., G. Blanchet, and N. Koper. 2014. Measuring habitat fragmentation: An evaluation of landscape pattern metrics. *Methods in Ecology and Evolution* 5 (7):634–646.

Wildlife Computers. 2015. *Pop-Up Tags.* accessed March 2015. http://www.wildlifecomputers.com/.

Williams, J. 2014. *An Application of Digital Video Recording and Off-Grid Technology to Burrowing Owl Conservation Research.* MS Thesis, Electrical Engineering. Denton, TX: University of North Texas.

Witczuk, J., S. Pagacz, A. Zmarz, and M. Cypel. 2018. Exploring the feasibility of unmanned aerial vehicles and thermal imaging for ungulate surveys in forests - preliminary results. *International Journal of Remote Sensing* 39:5503–5520.

Zuur, A. F., E. N. Ieno, N. J. Walker, A. A. Saveliev, and G. M Smith. 2009. *Mixed Effects Models and Extensions in Ecology with R.* New York: Springer.

Index

1-sun 169, 173–177, 184; *see also* incoming solar radiation; solar; solar radiation
2.4 GHz 38–39, 139, 144, 148–149, 158; *see also* Wi-Fi; wireless; wireless fidelity; wireless sensor network
4-20 mA 105, 108; *see also* signal conditioning
5 GHz 38, 139–140, 154; *see also* Wi-Fi; wireless; wireless fidelity; wireless sensor network

abiotic 1–2; *see also* ecosystem
absorption 23, 70, 74, 132, 185–186, 284, 291, 293–296, 299, 307–308, 319, 322, 327, 348, 354–355
accuracy 1, 21–22, 25, 60–61, 63, 123, 220, 233–236, 246–247, 315, 327, 347, 376; *see also* precision
acoustic 312, 335, 363, 365, 367–369, 390–392
action 2, 23, 41, 224
active 53, 62–63, 115, 120, 147, 150, 185, 323, 358–359, 361, 365
   active sensors 53, 358 (*see also* remote sensing)
actual 16, 68–73, 80–81, 83, 85, 90, 99, 105, 148, 344
aerosols 279, 301, 306
aggregation 375–376
air mass 159, 170, 184, 281, 298–300, 304, 307
   AM1.5 170, 184 (*see also* PV cell)
air monitoring stations 289
albedo 186, 282, 302, 306, 345–346
alternating current 55, 101, 316
AC 42, 55, 57, 60–62, 101, 140, 218–221, 248, 316
ALU 29, 41
ammeter 60, 62; *see also* current
amplifier 59, 78–79, 95–97, 100–101, 106, 108, 294–297, 308
   amplification 95
   differential amplifier 95–97, 100
   gain 4, 73, 93, 96–97, 100–101, 107–108, 132–134, 137, 140, 148, 153, 174, 193, 224, 238, 264, 291, 309, 318
   in-amp 100–101, 104, 107–108
   instrumentation amplifier 100, 108
   inverting amplifier 96
   non-inverting amplifier 96
   operational amplifier 79, 95
amplitude 89, 91, 101–102, 105, 129–130, 138, 357
analog-to-digital conversion 60
   A/D converter 30, 41, 104, 109
   ADC 30, 39–40, 42–43, 49, 51, 60, 68–70, 77, 104–105, 108–110, 118, 138, 143–144, 187
   analog channel 110–111 (*see also* datalogger)
anemometers 178, 302; *see also* sonic anemometer; ultrasonic sensors
animal tracking 363, 389; *see also* archival tags; data storage tags; passive integrated transponder tags; pop-up satellite archival tags
ANOVA 240
antennas 129–130, 136–140, 148, 150, 153
aquifers 328; *see also* groundwater
archival tags 370–371; *see also* animal tracking

Arduino 29, 33, 42–45, 50–51, 115, 144, 147, 315, 336
ARM 39, 86–89, 96–97, 100, 351
aspect 1, 3, 25, 182, 241, 267–270, 275–276, 297, 322, 358, 363; *see also* slope
asynchronous 32, 109, 117
atmosphere 2, 5, 24, 27, 141, 169–170, 178, 186, 279–281, 283–284, 288, 297–299, 301, 304, 306, 309, 327, 342, 352, 360–361
   atmospheric carbon 4–5, 26, 279, 308
   atmospheric effects 23, 169, 185
   atmospheric processes 159, 170, 279, 288
attenuated 193, 293, 319–320, 322
auto-covariance 11
   autocorrelation 11–12, 329–333, 382 (*see also* autoregressive)
autonomous underwater vehicles 327, 373
autoregressive 309, 329

backscatter 23, 185, 319, 358
backup 111, 174, 274, 277
balanced source 79, 81–82, 85–86, 100, 107
bands 174, 185–189, 192–196, 200, 207, 236, 247, 267, 306–307, 355, 358–359
bandwidth 13, 38, 105, 137–139, 147, 186, 359
barometric 120–121, 125, 152, 309, 345–346
base rate 219, 222
base reflectivity 305
base station node 141, 143, 146
base velocity 305
Bayes' 215, 220–223, 236, 248, 324
BCD 111–113, 127
beam splitter 291–292, 295
bias 21, 277
binary PSK 138
binary-coded decimal 111
binomial 228, 230, 376, 378
biomonitor 324–325, 336
biosensing 223
biosphere 2, 277
biotic 1–2
biplot 203–204, 206–208, 214
bipolar 104, 113
bipolar junction transistor 113
birds 363, 367–368, 376, 391–392
bit rate 32
bivariate 10, 17, 228, 232, 349
BJT 113
Bluetooth Low Energy 39, 141
Boolean 264, 266–267
boundary 178, 189–190, 270, 274, 350–351, 357, 374–375
boxplot 12–14
   box and whiskers 12
   hinges 12
broker 147–148, 152–153, 252–253
buck-boost converter 166–167
bulk 35, 260, 361
bus 29, 31–32, 39, 115, 117–118, 152

canopy 3, 149, 284, 339, 344, 348–356, 358–360,
362–363, 376
capacitance probes 339, 341–342, 361–362
capacity 1, 41, 138, 173–176, 184, 302, 327, 346, 359
capacity factor 175
carbon dioxide 2, 4–5, 26, 279, 281, 308, 323, 352
carrier 34, 138, 140, 148, 363
cascading style sheets 43
categorical data 232
categories 231–232, 359, 378
CCD 289–290
cell constant 316
central processing unit 29
chemiluminescent analyzer 294
chi-square 230–231, 233–234, 376, 378, 390
chi-square test 231, 376, 390
cladding 292
class c 37, 246
classical 12, 15–16, 297, 300
classification 23, 185, 198, 207, 209, 212, 214–215, 219–224,
230, 233–235, 242–249, 336, 358–359, 376
CART 215, 242–247
classification and regression trees 215, 243, 249
classification tree 230, 242–244
regression tree 215, 242–245, 249
client 38, 115, 146–148, 153, 251–253, 274
climate 1–2, 4, 24, 26–27, 184, 279, 284, 286, 289, 308,
327, 336, 349
cloud 30, 186–187, 279, 282, 301, 306, 353, 358, 360
cluster 141, 198, 207, 209–214, 344, 378
cluster analysis 198, 207, 209–210, 213–214
agglomerative 198, 209, 212
cluster heads 141
clustered pattern 378–380
CMF 226–227, 248
CMR 100–101
coaxial cable 103–104, 137, 339–340, 367
coefficient of variation 8, 375
Cohen's kappa 235
cold-junction compensation 76, 78, 104; *see also*
thermocouple
collinearity 215, 239, 241
column metric preserving 204
combinations 216–217
comma separated values 46
common-mode rejection 79, 100
community 43, 230, 263, 300, 349, 367, 389, 392
complement 42, 104, 215–216, 223, 248
complexity parameter 246
composite reflectivity 305
computer organization 29, 51
conditional probability 218–222, 229
confidence interval 14–15, 20, 333
confusion matrix 215, 220, 222, 233–234, 246–247, 249
error of commission 235–236
error of omission 236
contingency table 215, 232
continuous 1–3, 5–6, 12, 23, 27, 30, 39, 60, 101, 115, 174,
188, 242–244, 289, 318, 320, 327, 339, 342, 347,
349, 355, 362, 381–382
control 2, 20, 29–31, 35–36, 39, 41, 43–46, 49, 102,
108–110, 113–115, 118, 127–128, 141, 143,
146–147, 149, 152, 277, 288, 328, 336, 342, 348,
355, 360

control structures 43
convolution 267–268
coordinate reference system 264
core area 374–375
correlation coefficient 10–11, 16–17, 19, 239–240, 383
correlation test 16
correlogram 383–384
cosine correction 323
cosine response 302
counters 111
covariance 10–11, 18, 200–205, 210, 213, 239, 329,
349–351, 359–360, 382–386, 388–389, 391
covariogram 383–384, 391
CPU 29–30, 39–40, 126, 144
CR1000 120–121, 124, 126–128
critical region 14–15
cross-sectional 20, 56, 150, 178, 311, 313
cross-tabulation 232
cross-validation 246–247, 249
crossover distance 134, 140, 148, 154
CRS 264–265
CSS 43, 46–47
CSV 45–48, 50, 176–177
CTD 327
cumulative distribution function 6
cumulative mass function 226
current 3, 30, 42, 53–62, 70, 73–76, 79, 96–97, 101–106,
108, 111, 113, 118, 121, 152–153, 157, 159–166,
174–175, 183, 241–242, 283, 286, 289–291, 308,
312–316, 318, 327–328, 336
current divider 59–60
current loop 79, 103, 105–106, 108
cycle 1–2, 4, 29, 41, 129, 135, 145, 147, 150, 166–167,
174–175, 187, 189, 279, 283, 305, 309–310, 324,
327, 344, 363–364

damping ratio 90–91, 93
DAS 109–110, 118–119, 155
data acquisition systems 108–109, 360
data circuit-termination equipment 33
data rate 29, 32, 37–38, 114, 138–143, 369
data storage tags 371; *see also* animal tracking
data structures 44
databases 3, 49, 251, 253, 255, 257, 259, 261–263, 265, 267,
269, 271, 273, 275, 277, 294, 328
attributes 209, 211, 224, 253–254, 256–259, 261–262,
264, 270–272, 274
data definition language 257
data manipulation language 257
data model 255–257, 285, 287–288
database management system 251
DDL 257
DML 257, 259
document type definition 262
DTD 262
entity relation diagrams 251, 255
foreign key 255
functional dependencies 253, 255
instance of a relation 253
keys 253, 255, 257–258, 276
normal forms 253
open database connectivity 251
primary key 255, 257–258
relational databases 251, 253

relations 1–2, 209, 221, 253–256, 259, 276
  schema 251, 254, 257, 262
datalogger 29, 33, 42, 109–111, 113–115, 117, 119–123,
    125, 127, 141, 143, 149–150, 162, 173, 175,
    251–254, 275, 301, 312, 314, 320–322, 324, 344,
    351, 363, 368–369, 371
  DIFF 110–111, 120–121, 123
dB 100–101, 105, 132–134, 136–137, 140, 148, 222,
    251–255, 257, 262–264, 274–276, 364
  dBi 137, 140, 153
  dBm 133–134, 140–141, 148–149, 152–153, 251
  dBZ 304–305
decision nodes 224, 243
decision theory 224
decision tree 224–225, 242
declination 167–169, 183
degrees of freedom 16, 231, 233, 390
DEM 140, 153, 268, 358; *see also* raster
dendrometer 89, 351
density 5–6, 10, 12–13, 15–16, 22, 93, 138, 149–150, 158,
    176, 178, 180–182, 195, 198, 210, 231, 248, 281,
    289, 298, 302, 309, 323, 347, 349–351, 358,
    361–362, 375–376, 378, 389–390
density of points 376
dependent 17, 37, 62, 218, 220, 233, 236, 240, 242, 323
depth of discharge 174
dielectric constant 93, 339, 341
differential 3, 20, 90, 95–97, 100, 104, 110, 114, 274,
    283–284, 296, 307–308, 325–326, 369, 389
differential optical absorption spectroscopy 296, 307–308
diffraction 130–132, 135, 290
diffraction grating 290
digital elevation model 23, 140, 185, 232, 268
digital repository 275, 277
diode 164
  bypass diode 164
  photodiode 289
  Schottky diode 165
dipole antenna 136–137
direct beam 170, 281
direct current 55
  DC 35, 39, 42, 55, 57, 60–62, 102, 120–121, 123, 134,
    140, 148, 165, 174, 341
direct-sequence 139
directional antenna 137, 140
directivity 137
discharge 149, 153, 174–175, 312–314, 326
discrete RV 224–226, 228, 242, 248
dispersive spectrometers 290
distance 8, 12, 22, 129–130, 132–135, 139–142, 146,
    148–150, 152–154, 185, 209–213, 240,
    271–272, 276–277, 298, 303–304, 310–311,
    316, 350, 355, 358, 364, 368–369, 378–384,
    386, 390–391
distribution 5–6, 8, 10, 12–17, 150, 177, 180–182, 193, 225,
    228, 230–232, 248, 272, 304, 307, 312, 327,
    349, 354, 356, 363, 371, 374–376, 378–379, 382,
    389–390
diversity 230
divisive 198, 209, 212
DMM 60–61
DNS 37
DOAS 251, 276, 284, 296–297, 299–300
Dobson units 280

domain name system 37
Doppler radar 304–305, 308, 312
doubly exponential 4, 285, 287–288, 306
DRAM 29
drone 22–23, 27, 185–186, 214, 373, 390
DTE 33–34, 120
dual polarization 304
duty cycle 145, 147, 150, 166–167, 174, 364
dynamic specifications 90, 93; *see also* static
    specifications
dynamics 3, 20, 23, 175, 279, 282, 284, 286, 288, 309–310,
    318, 324–326, 342, 347, 353–354, 360, 362,
    389, 392

earth 1–2, 22–24, 26, 141, 153, 158–159, 167, 169–171,
    185–189, 192, 194, 262–263, 279, 281–282,
    287, 301, 305–307, 309–311, 327, 358,
    361–362, 369
EC 115, 117, 122–123, 128, 152, 154, 236, 277, 309,
    313–317, 320–322, 327, 335, 341, 343–345,
    357, 361
Eckart-Young theorem 203, 206
ecoinformatics 275, 277
ecosystem 1–2, 23, 224, 279, 281, 309, 322, 336, 339,
    341–345, 347, 349, 351, 353, 355, 357–362;
    see also abiotic
eddy covariance 349–351, 360
edge 12, 30, 374–375, 379–380
edge density 375
effect size 15
electric circuits 53
  charge 53–54, 91, 93, 154, 164–165, 175, 290
  Kirchhoff's 57–58, 160
  loops 29, 43, 57, 103–104, 125
electric power 54
  hydropower 157, 175
  maximum power point 161
  power supply 34, 54, 62, 96, 106, 140, 174, 294–296
  received power 132–134, 137, 140, 153
  specific power 176, 178, 180
electrical conductivity 115, 152, 309, 315, 336, 341, 361
electrochemical 25, 79, 90–93, 314, 317
electrochemical sensors 79, 90–93, 317
electrode polarization 316
electromagnetic 23, 53, 101, 129, 157, 185, 324, 339, 391
EM 23, 129–131, 185–186, 306, 339, 357, 360, 363
empirical 12, 20, 22, 178–181, 224, 236, 373, 378–385,
    389, 391
empirical covariance 382–383, 385
enclosure 102–103, 118–119, 149, 151, 323
entity 1, 251, 253–257, 277–278, 350
entropy 229, 243
environmental specifications 79, 90–91; *see also* static
    specifications
EPROM 29, 41
equatorial 167, 187–188, 305
ERD 256–257, 276–278
ESP8266 143–144, 148, 152, 154
estimated 7, 16–17, 19–20, 64–65, 174, 176, 182, 231, 236,
    240–241, 284, 295, 305, 313, 322, 333, 344,
    350, 358–359, 386
Ethernet 31, 35–39, 42, 113, 140, 146, 150
Euclidian distance 209, 211, 213, 271, 378, 382
evaluation dataset 246

evapotranspiration 2, 339, 344, 360
    ET 344, 346–347, 354–355
    PET 344
    potential ET 344
    reference ET 344
event 5, 14–15, 25, 106, 111, 118, 150, 180, 210, 215–226, 230, 248, 275, 311, 319, 326, 344, 353, 368, 372, 376
    intersection 164–165, 183, 216, 218–219, 248, 274
    null event 215–216
    union 130, 216, 248
    universal event 215–216
expected cost 224
expected value 6–8, 10, 14–15, 180, 226–227, 231, 349, 378, 382, 386–387
explanatory 17, 215, 236, 243–244, 284
extensible markup language 251
extent 3, 21, 191, 264–265, 270–271
extraterrestrial radiation 169, 297

fade margin 137, 140
false color composite 191–193
false negative rate 222
fetch/execute 29
field-effect resistance 316
filter 4, 147–148, 166, 267, 279, 293
filtering 4, 95
first moment 6–7, 226
flash memory 39, 41–42
flow 1–3, 20, 29, 43, 53–54, 56, 73–75, 77, 102, 104, 107, 110, 157, 175, 178, 182–183, 301, 305, 309, 311–313, 316, 318, 326, 328, 336, 342, 347–348, 359, 361
fluorescence 291, 295–296, 319, 324, 336, 348–349
Fourier series 101, 310
Fourier transform 101, 291, 296
fractal dimension 375
fragmentation 363, 374, 389, 392
frame check sequence 38
frequency 3, 9, 12–13, 32, 91, 101–105, 107, 111, 129–133, 137–141, 143, 146, 148–149, 153–154, 157–158, 174–175, 177–178, 187, 232–233, 275, 289, 302, 304, 312, 316, 341, 349–350, 353, 356, 362–363, 365, 367–368, 371–372
    frequency-domain 101
    frequency hopping 139
Fresnel zones 135–136
friction coefficient 178
FSL 132, 140
FTP 150, 252–253
full scale 60
    FS 63
full width at half maximum 300
FWHM 300, 354

Gaussian 8, 12, 210, 329, 333
general purpose I/O 33
    GPIO 33
geodetic datum 188
geographic information system 24, 198, 214, 251, 253, 255, 257, 259, 261, 263, 265, 267, 269, 271, 273, 275, 277
geosphere 2
geostationary 187, 305, 308, 358

geostationary operational environmental satellite 305
Geostatistics 277, 381–382; see also kriging
GIS 198, 214, 251, 263–265, 268–271, 276–277, 310, 326, 369, 373, 376, 378
glass electrodes 92, 314, 318
global 2–4, 24, 26–27, 38, 141, 145, 188, 271, 279, 282–283, 286–289, 306, 308, 327–328, 341, 358, 360, 369
    global temperature 279, 282, 286–288, 306
GOES 2, 19, 54, 59, 61, 63, 118, 161, 168–169, 178, 281, 290, 295, 305–308, 322, 342, 349–350, 384
GPS 3, 42, 327, 352, 363, 369–370
greenhouse effect 2, 186, 281
ground 21–24, 27, 34–35, 42, 57, 82, 96, 102–104, 110, 113, 117, 120–121, 134, 140, 148, 150, 153, 178, 185–187, 207, 233–234, 279–281, 288, 294–295, 297, 299, 301, 319, 339, 351–352, 354–355, 357–360, 363, 373
ground penetrating radar 339, 357, 360
ground-truth 23, 186, 207
groundwater 110, 270, 309–310, 328, 334–335, 337, 339, 344

habitat 110, 154, 358, 361, 363, 365, 368, 371, 373–376, 389, 391–392
habitat suitability index 373
hall effect 107, 121, 178
harmonic 101–102, 108, 166, 310–311, 336
heat dissipation factor 315
hierarchical agglomerative clustering 209, 212
HMI 127
homogeneity 243
HSI 373–374, 376
HTML 26, 29, 38, 43, 45–51, 78, 154, 214, 260–261, 277, 307–308, 336, 360–361; see also web browsers; World Wide Web
human machine interface 127
hydraulic conductivity 342, 358–359, 362
hydrodynamics 309, 326
hydrograph 326
hydrographic stations 110
hydrological 2, 305, 309, 326, 361
hydrophone 368–369
hydrosphere 2, 309
hyper text markup language 29
hyperspectral 186, 188, 327, 358–359, 376; see also remote sensing
hypertext preprocessor 29
hypothesis 14–17, 20, 211, 223–224, 231–232, 236, 390
    hypothesis testing 14
    null hypothesis 14, 16–17, 231–232, 390
    power of test 15
    type I error 14–15
    type II error 15

I2C 43, 109, 113, 116–117, 127–128, 143, 152
I/O 29, 33, 39, 41–42, 44, 115
ICSP 41–42
IEC 26, 119
IEEE 802.11 38, 140; see also Wi-Fi
if-else 43
images 22–23, 45–46, 185–187, 189, 200, 207, 212, 214, 264, 266–267, 301, 305–307, 358–359, 363, 371–373, 376; see also remote sensing
impedance 57, 92–93, 96–97, 99–100, 105, 117, 136–137, 325

in-circuit serial programming 41, 51
incoming solar radiation 186, 279, 282; *see also* 1-sun
independence test 232
independent 10, 16–17, 20, 37, 41, 56, 64, 146, 216–217, 219, 224, 228, 232, 236, 239–242, 248, 284, 375
index plot 12–13
indices 185, 195–196, 230, 234, 257, 264, 268, 374, 391
infiltration rate 322, 342
infiltrometer 342–344, 359–362
information theory 214, 229–230, 243
infrared 101, 130–131, 186, 281, 295, 318, 348, 360, 362, 372
input 29, 31, 41–42, 58–60, 62–63, 66–67, 71, 92, 95–97, 100, 104–106, 110–111, 115, 120–123, 136, 166, 182, 256, 301, 305, 326, 344, 347, 359, 373
input impedance 92, 96, 100, 136
input/output 29
insolation 169, 175, 177, 302, 344
integrity constraints 255, 257–258
intelligent sensors 109, 113, 117
interferogram 291–292, 297
interferometer 291–292, 296–297
internet 31, 38–39, 42, 113, 130, 141, 143, 146–148, 155, 260
    internet of things 39, 130
    IoT 39, 130, 143–146, 148, 154, 251–252
interrupt 35, 118–119, 147
ion-selective electrodes 91
IPv4 37–38
IR 142–143, 186, 188, 236, 281–282, 290, 293–296, 302, 306, 310, 320, 324, 327, 348, 354–355, 357, 359, 372–373
isochronous 35
isolation 104
isotropic radiation 137

Java Script 29
    JS 29, 43, 49
Joule heating 56, 70, 74, 76

K-fold cross-validation 246
K-means 204, 209–214
kappa statistic 215
kriging 363, 376, 382, 385–389, 391
    cokriging 376
    kriging error 386–389
    nugget effect 384–385
    ordinary kriging 386
    semivariance 383–385, 389
    semivariogram 384, 391
    sill 384–385, 391

LAI 352–354, 358–360
Lambert-Beer 169, 281
Landsat 185, 187–192, 195, 214, 236, 247, 266–267, 358, 373
landscape 25, 192, 373–376, 389, 391–392
Langley extrapolation 299–300, 307
latitude 140, 153, 167–170, 183–184, 188–190, 256, 270, 359, 369
layers 36–38, 135, 141–142, 146–147, 186, 251, 264, 266–267, 270–274, 276, 279–280, 350, 353–354, 360, 376
leaf area 339, 351–352, 360, 362
leaf area index 352, 362
least significant bit 30, 105, 108, 112

length 3, 38, 56, 106, 115, 129, 136, 159, 170, 178, 268, 274, 291, 298, 316, 339, 352–353
LiDAR 23, 185, 358–359, 376
line of sight 132
linear 1, 7, 17, 20, 53, 55–56, 59–60, 63–66, 68–69, 72–75, 77, 79–81, 83, 85, 87–88, 97–99, 106–107, 137, 180, 200, 215, 224, 236, 239, 241, 249, 266, 283–284, 286, 289–290, 299, 307, 313–314, 317, 322–323, 329–330, 335, 385–386
linearized response 79
lines 270–274
lists 44, 46
lithosphere 2
load 30, 41, 54, 88, 103, 159, 164–165, 173–174, 183
loadings 200–201, 203, 205, 207, 213
long term 1, 3, 275
LoRa 145–146, 154
LoRaWAN 146, 154
LoS 132
lower limit of detection 21

MAC 36–38, 146–147, 153, 155
    MAC Address 36–38
    Media Access Control 36, 146
    Organizational Unique Identifier 36
machine learning 23, 185, 198, 215, 217, 219, 221, 223, 225, 227, 229, 231, 233, 235, 237, 239, 241, 243, 245, 247, 249
majority resampling 268
Mallows' Cp 241–242
marginal PMF 229
marginal probability 223
marked point pattern 381–382
master 34, 114–116, 127, 158, 173, 178, 180, 182, 184, 371
master/slave 34
matrix 200–207, 209, 211–215, 220, 222, 228, 233–234, 236–237, 241, 246–247, 249, 266, 269, 271, 330–331, 361, 388
    correlation matrix 200, 202, 204
    covariance matrix 200–203, 205, 213
    data matrix 200–201, 203–204, 209, 211–213
    diagonal matrix 200, 204
    dissimilarity 209, 383–384
    eigenvalues 200–202, 204–206, 210
    eigenvectors 200–202, 204–206, 210
    error matrix 220, 233
    Hadamard 266
    Schur 266
    singular 198, 204–205
    singular value decomposition 198
    symmetrical 8–9, 209, 271
    transpose 201
maximum a posteriori 223
maximum likelihood estimate 242
MCU 29–30, 33, 39–43, 95, 109, 113, 115–119, 127, 129, 141, 143–144, 149
mean 226
    annual mean 26, 289
    geometric mean 373
    mean square error 242
    monthly mean 4–5, 283–284
    population mean 14, 227
measurand 22, 30, 62–63, 66, 70, 79, 81, 84, 90, 94, 97, 99, 106–107, 292, 357

measurement and control 109, 118, 128
median 8–9, 12, 17, 182
memory 29, 39, 41–43, 109–110, 116, 118, 120, 122, 144
meq 322
mesh 141
metadata 251, 262, 275, 277
metapopulation 389
methane 279, 281–282, 308, 350–351, 360
microcontroller 29, 39–40, 51, 68, 312, 315, 339, 355, 372
microphone 367–369, 392
microprocessor 30, 39, 41, 107, 109, 118, 128, 295
microwave 23, 130–132, 185–186, 339, 358
Mie scattering 298
ML 198, 210, 215, 223–224, 233, 236, 246, 359
MLE 242
Modbus 114–115
mode 60, 79, 100, 114–115, 117, 119–120, 140, 150, 152,
    174, 251–252, 267, 291, 297, 304, 362, 372
model 8, 17, 19–20, 23–24, 26, 36–37, 39–40, 53, 63–66,
    70–72, 78, 80, 84, 86, 96, 98–100, 114, 132,
    134, 140, 146, 148, 157, 159–161, 163, 169, 178,
    180, 184–185, 189, 210, 212, 224, 228, 232, 236,
    240, 242, 245–247, 255–257, 268, 274–275,
    277, 279, 282–288, 303, 305–307, 309, 314,
    322–323, 325–326, 328–329, 331–334, 336,
    339, 342, 344, 349, 352, 361, 371, 373, 376,
    384–386, 388–389, 391–392
    B model 70, 72, 80, 84, 86, 98, 100 (see also
        Steinhart-Hart)
    empirical model 224
    exponential model 283–286, 306, 323, 391
    modeling 20, 23–24, 39, 140, 150, 182, 263–265,
        277–278, 284–285, 287–289, 310–311, 325,
        328–329, 373, 389, 391
    process-based model 20
    statistical model 210
    theoretical model 8
modulation 32, 138, 145
monochromatic 293, 295, 299, 318
MOSFET 113, 165
Moteino 143–145, 151–152, 154, 162, 174, 184
MPP 161–162, 164–167, 183
MPP tracker 167, 183
MPPT 166–167
MQTT 147–148, 152–155, 251–253
multi-hop 141, 143, 146–147
multimeter 60–61
multiparameter 320–321
multiple access 138
multiple linear regression 215, 236, 241, 249; see also
    simple linear regression
multiplexing 138–139
multispectral 186, 188, 200, 359
multivariate analysis 185, 198
mutually exclusive 215
MySQL 150, 251

NAAQS 27, 288–289, 308
Naïve Bayes 224
National Ambient Air Quality Standards 27, 288, 308
    primary standards 288
natural color composite 192, 194
NDIR 295–296
NDVI 195–199, 212–213, 251, 264–265, 267, 270–271

NDWI 196–198, 267
nearest neighbor analysis 272, 378
neighborhood 267, 373
NEMA 119
nephelometer 319
nephelometric 319–320
Nernst equation 91–92
net radiation 302, 346
net radiometers 302, 354
network interface controller 36
networks 23, 29, 31, 36–38, 94, 109, 111, 113–115, 117,
    119, 121, 123, 125, 127, 129, 131, 133, 135, 137,
    139–143, 145–149, 151, 153–155, 184, 260, 275,
    278, 289, 325–328, 347, 351, 358, 373
NEXRAD 304–305, 307–308, 326
Nibble 36, 112–113, 127
NOAA 4–5, 24, 26, 283–284, 305–308, 310–311, 313,
    328, 336
node 57–58, 114–115, 141–154, 157, 162, 174–175,
    217–218, 224, 242–247, 252, 272, 276, 295, 345
NodeMCU 143, 152, 162
noise 20–21, 79, 95, 101–102, 105, 108, 114, 138, 245, 329,
    333, 354
nominal 63–64, 79–82, 84–86, 97, 107, 120, 161, 163, 173
nominal operating cell temp 173
non-dispersive infrared 295
nonlinear regression 279, 285–286; see also simple linear
    regression
nonparametric 15–16
normal 8–10, 13–17, 168, 170, 178, 231, 240, 253, 255, 281,
    316, 365, 367
normalized difference 185, 195–196, 266
    normalized difference vegetation index 195
    normalized difference water index 196

OAS 284, 293, 297–300
ocean 2, 22, 24–26, 141, 185–186, 281, 286, 300, 309, 313,
    327–328, 336, 363
ODBC 251
OGC 251, 274–275, 277
Ohm's law 55–57, 60, 160, 315–316
ohmic 55–56
ohmmeter 60
omni-directional antenna 137
omnidirectional 365, 384
one-hour precipitation 305
one tail 15, 217
op-amp 95–100
open-collector 113
open-drain 113
open geospatial consortium 251
open systems interconnection 36
optical absorption spectroscopy 284, 296, 307–308
optical fiber 292, 300, 323, 354–355
option 25, 141, 154, 209, 224, 251, 319, 351, 371
ORP 93, 313, 317–318, 320
oscilloscope 62
OSI 36
outgoing radiation 186, 282
output 20, 29–31, 41, 53, 59–60, 62–63, 66–73, 77, 79–86,
    88–89, 92–93, 95–101, 105–107, 110–111,
    115–117, 121, 123–124, 127, 161–162, 164, 166,
    175, 183, 256, 302, 314, 326, 341, 347, 351, 355,
    357, 359

overall accuracy 234
overfitting 245
overlaid 264, 385
oxidation-reduction potential 93, 313
ozone 2, 21, 23, 25–27, 110, 186, 236–237, 240–241,
    244, 276, 279–281, 289, 293–295, 298–301,
    306–308, 351; *see also* stratosphere; total
    column; troposphere

p-value 14–16, 20, 25, 230–231, 233, 284, 287, 376,
    378, 390
PACF 332–333
packed 112
packet 32
    data packet 32, 35–36
    handshake packet 35
    start of frame packet 35
    token packet 35
padding 38
PAR 323, 359
parameter 8, 14, 16–17, 19–20, 23–24, 41, 53, 63–66,
    70–71, 73, 79, 83–86, 99–100, 138, 140, 178,
    180, 186, 198, 204, 210, 231, 236, 242, 246,
    270, 284–286, 299, 309, 313–314, 318, 320, 325,
    333, 351, 353, 369, 373, 385, 388
parametric 12, 15–16
partial autocorrelation function 332
partial or marginal coefficients 239
particulate matter 23, 185, 289
passive 53, 62, 73, 185, 314, 358, 361, 365
passive integrated transponder tags 365; *see also* animal
    tracking
passive sensors 62, 358
path 103–104, 130–132, 134–135, 137–138, 159–160,
    170–171, 173, 189–192, 217–218, 221, 291–292,
    296–300, 339, 351, 360, 371
path loss 132
payload 35, 38, 148, 155
PCA 198, 200, 203–204, 207–209, 211, 213
peak sun 169, 175
Pearson's 16, 230
Peltier 73–75
Penman-Monteith 344
per thousand unit 317
per unit rate of change 283–284
percentiles 8–9, 13
permittivity 93–94, 302, 324–325, 339–341, 357–358,
    361, 365
pH 92–93, 108, 110, 117, 293, 309, 313–314, 317–318, 320
phase shift 135, 304
phase-shift keying 138
photoelectric 157, 290–291
photosynthesis 2, 281, 322–324, 348, 351, 353–355
photosynthetically active radiation 323
photovoltaic 62, 150, 157, 323
PHP 29, 38, 43, 47–49, 147, 155, 308, 337, 360
physiological 363, 369, 371, 373
piezoelectric sensors 79, 94, 175
PIT 365
pixels 186–187, 192–193, 196–200, 207, 209, 212–213,
    215–216, 233–235, 248, 264–265, 290,
    353–354, 373
PL 132–134, 139–140, 153, 308
plant productivity 339

PMF 225–229, 242–244, 248
point clouds 358
point patterns 363, 376–377, 390
point-to-point 37, 141
points 10, 17, 19–21, 29, 37, 53, 57, 63, 68, 72–73, 77, 110,
    135, 138, 150, 164–165, 182–183, 200–201,
    209–212, 231, 236, 269–273, 284, 286,
    290, 299, 326, 371, 373, 376–380, 382–385,
    387–388, 390–391
Poisson 376, 378, 380–381
polar mount 169, 184
polarization 94, 137–138, 140, 304, 316
pollutant 27, 93, 279, 288–289, 293, 296–297, 309, 314,
    319, 351
polygons 264, 270, 272–273, 326
polynomial regression 65, 322–323
pop-up satellite archival tags 370; *see also* animal tracking
population 14, 227–228, 289, 335, 349, 363, 367, 376, 389
port numbers 37
positive prediction value 220
posterior probability 218, 223, 236
PostgreSQL 251, 275
PPV 220
practical salinity units 317
    PSU 317, 327
preamble 38
precision 1, 21–22, 25, 63, 78, 360; *see also* accuracy
principal components analysis 198
prior probability 218–219, 221–223
probability 5–6, 8–10, 14–16, 22, 25, 176, 180, 182–183,
    214–231, 233, 235–237, 239, 241, 243–245,
    247–249, 349, 371, 376, 378, 380, 390
    probability density function 5, 349
    probability mass function 225
    probability tree 217–218, 224
processes 2–3, 20, 23, 25, 29, 54, 93, 101, 108, 115,
    127, 130, 135, 150, 159, 170, 185, 241, 264,
    279–281, 283, 288, 298–299, 309, 318,
    322–323, 326, 336, 339, 348–349, 354, 361,
    363, 373
producer accuracy 235–236, 246
programming 20, 29, 31, 33, 35, 37–39, 41–45, 47, 49, 51,
    109, 120–121, 127, 152, 198, 242, 264
PROM 29, 41
propagation 129–130, 132, 134–135, 137, 140, 148–150,
    153, 155, 365, 367–368, 391
    free space loss 132
    free-space model 132
    propagation model 132, 134, 140, 148
    two-ray model 132, 134, 148
proportion of agreement 234
PSAT 370
PSK 138
publish/subscribe 147
pull-down 113, 115
pull-up 113, 115, 117, 121, 127
pulse sensor 107
pulse-width-modulated 165
purity 230, 243
push and pull 251–252
PV 62, 157–159, 161, 164, 169, 171, 174–176, 283,
    289, 302
PV cell 62, 157, 159, 161, 171, 174, 302
PWM 165–167

pyranometer 125, 175, 302, 354
Python 29, 43–44, 50–51, 117–118, 147, 264, 344
　　associative arrays 44
　　dictionaries 44

Q-Q Plot 13–14, 240
QGIS 198, 214, 251, 263–265, 277
quadrature 138
quantile 8–9, 13–15, 240
quantile-quantile 13
quantization noise 105
quantum sensor 302, 323, 354
quaternary 138
queries 251, 259, 264

radar 23, 185, 279, 304–305, 308, 312, 339, 357–358,
　　360, 376
radial velocity 305
radio 3, 23, 101, 129–132, 136–138, 140–151, 153, 155,
　　158, 174, 185, 328, 363–369, 390–391
radiometric resolution 23, 187, 192, 213
rain gage 120–121, 276, 301
RAM 29–30, 39–40, 43, 144
random error 21
random variables 1, 5, 215, 225, 228, 349
range 63
　　frequency range 130, 140, 368
　　measurand range 63
　　range semivariogram 384
　　signal range 30
　　spatial range 22, 185
　　temperature range 71–72, 75, 77, 314
　　tidal range 310
　　transducer range 66
　　transmission range 140
　　UV range 293, 320
　　voltage range 75, 105, 120
　　wavelength range 290, 348
rank 12–13, 16–17, 205
raspberry Pi 29–30, 33, 36, 39–40, 42, 44, 51, 115, 117,
　　148, 152–153, 300, 344
raster 186–187, 192, 198, 214, 251, 264–273, 276, 326, 358,
　　374, 376, 385
rate coefficient 283–287, 306
rate of change 3–4, 20, 53–54, 93, 210, 239,
　　283–284, 342
rating curve 182, 313
Rayleigh 180–182, 298–299
Rayleigh scattering 298–299
real time clock 109
　　RTC 39, 109, 111, 113, 118, 127, 143–144, 352, 364
real values 5, 225
reception 129, 150, 363, 369
reclassification 185, 196, 198
rectenna 174
recursively 242, 286
reference class 233
reference height 178
reflection 23, 130–132, 134–135, 148, 159, 185, 291–292,
　　327, 354–355
refraction 130–131, 135, 292, 365–366
refractive index 131, 292, 365
relative humidity 3, 91, 110, 125, 149, 302, 344
relative permittivity 94, 339, 361, 365

remote sensing 22–24, 26–27, 185–188, 195, 198, 200,
　　207, 209, 212, 214–215, 32–233, 246, 248–249,
　　263–264, 301, 308–309, 326–327, 337, 339, 358,
　　60–363, 369, 373, 376, 392
remote station 129, 141, 175, 275
repeatability 22
reproducibility 22
resampling 265, 267–268
reservoir-based hydroelectric 182
residual 18–20, 240, 242, 286, 329, 333
resistivity 56, 94
　　resistance 53, 55–66, 69–72, 79–87, 89, 92–93, 95–97,
　　　99–101, 105–108, 115, 159–160, 175, 314–316,
　　　342, 346
　　resistive temperature detector 63, 314
　　resistor 54, 56–58, 62, 66, 71–73, 81, 84, 86–87, 101,
　　　105–106, 113, 115, 117, 121, 127, 315
resolution 21, 23, 25, 30–31, 49, 63, 68–69, 105, 123, 129,
　　141, 187–189, 192, 213, 267, 290, 294, 300,
　　305–306, 326–328, 336, 354, 358, 360, 376,
　　385, 387
response 3, 17, 23, 44, 50, 60, 62–63, 65–69, 72, 77,
　　79–83, 85–88, 90–91, 93, 95, 97–99, 109,
　　114–115, 150, 182, 185, 187, 223, 236, 240,
　　243–244, 274, 290, 302, 314, 323–324, 335,
　　339, 354–355, 362, 365, 372, 392
response time 90, 93
retardation 291
return loss 136
RF 66–73, 77, 80–89, 97–98, 100, 107, 131, 136–138, 146,
　　149, 154, 157, 174–175, 184, 304, 357, 363, 365,
　　367, 370
RF identification 365
RFID 365
Ripley's 379
RISC 39, 41
river flow 182–183
ROM 29
roughness length 178
row 25, 149, 153, 189–192, 200, 204, 209, 213, 220, 228,
　　232–235, 237, 253–254, 264–268
row metric preserving 204
RS-232, 31, 33–35, 39, 109, 113–114, 120
RS-485, 109, 113–115, 152, 344
RSS 148–149, 152, 267, 276
RTD 106, 314–315
run-of-river hydroelectric 182
runoff 2, 182, 305, 309, 311, 319, 326, 336
RV 5–8, 10, 17, 25, 84, 215, 224–229, 242, 248, 376, 382

salinity 50, 317, 327, 343, 361, 368–369, 371
sample 5–9, 11–19, 21–22, 25, 91–92, 105, 122, 125,
　　180, 182, 207, 212, 215–217, 224, 226–228,
　　230–231, 238, 248, 271, 290–296, 317–318, 327,
　　349, 353–354, 374, 382, 384
　　sample correlation coefficient r 11, 16
　　sample covariance 11, 18, 349
　　sample mean 6–8, 13–16, 19, 21, 180, 227, 238,
　　　248, 382
　　sample space 5, 215–217, 224, 226, 248
　　sample standard deviation 7–8, 11, 13, 16, 19, 21, 228
　　sample variance 7–8, 16, 18–19, 228, 384
sap flow 77, 347, 361
SAR 322, 328, 335, 358

satellites 3, 22, 185, 188, 279, 305, 327–328, 358–359, 363, 369–370
SCADA 127
scale 2, 21, 24, 60–61, 107–108, 157, 180–181, 185, 193, 200, 202, 204, 206, 240, 264, 266, 270, 313, 317, 335, 339, 345, 347, 350–351, 353, 362, 373, 376, 384, 392
scatter plots 11, 193, 236, 329–330
scene 186, 189–192
scores 13, 200–203, 207
SDI-12, 109, 113, 117–118, 128, 343, 355
sea surface 25, 327
    SST 327–328
second central moment 7–8, 227
secondary emission 290–291
Seebeck 74–75
self-heating 3, 70–71, 301, 314–315
semidiurnal 310
sensitivity 63, 66–70, 72–73, 75, 77, 82–83, 85–86, 88, 91–92, 97–98, 107, 140, 149, 175, 187, 219–220, 222, 314–315, 355, 372, 374
sensor network 29, 38, 94, 109, 111, 113, 115, 117, 119, 121, 123, 125, 127, 129, 131, 133, 135, 137, 139, 141, 143, 145–147, 149, 151, 153–155, 157, 184, 260, 278, 344, 358, 373
    sensor nodes 141, 143, 146–150
sentinel 223, 324, 358–361
serial communication 29, 31–33, 109, 113, 115–116, 118, 295
    baud 32, 45, 117
    data terminal equipment 33
    error checking 115
    error rate 15, 138
    parity 32
    start 32, 35–36, 38, 44–45, 47, 53, 61, 79, 81, 83, 86, 93, 115, 117, 121, 150, 241, 260, 270, 305, 364, 369
    stop 32, 164, 241, 245–246, 301
server 38, 47, 115, 146–147, 149–150, 152, 251–253, 274–275, 300, 308
shape metric 375
shielding 79, 102, 108
shortwave radiation 186
Siemens 56, 316
signal conditioning 51, 76, 78–79, 81, 83, 85, 87, 89, 91, 93, 95, 97, 99, 101, 103–109, 118, 143, 292, 336; *see also* 4-20 mA
signal-to-noise ratio 105
    SNR 105
simple linear regression 17
    intercept 17–19, 68, 236–237, 242, 284, 299–300, 357, 384 (*see also* multiple linear regression)
simulation 20, 26, 140, 336, 381–382, 391
single board computer 29
    SBC 29–30, 33, 39, 42–43, 115, 117–118, 127, 129, 140–142, 146–147, 149–150, 173, 251–252, 275, 283, 300, 372
single-ended 104, 110
    SE 16, 19, 27, 110–111, 120–123, 270
slave 34, 114–117, 127, 371
slope 6, 17–20, 55, 67–68, 79, 81–83, 85, 97, 107, 164, 183, 240, 267–270, 276, 284, 299, 314, 345, 357; *see also* aspect; simple linear regression
small departure 79, 82, 84
SoC 39, 143–144
sodium adsorption ratio 322; *see also* water quality

soil 1–2, 21, 23–25, 73, 76–79, 89, 93–95, 110, 121, 123, 128, 147–152, 154, 185–186, 193, 248, 264, 266–267, 275, 277, 282, 309, 322, 339–344, 346–348, 355, 357–362
soil sodicity 322; *see also* sodium adsorption ratio
solar 3, 41, 51, 54, 62, 110, 120–121, 152, 157–159, 164–167, 169, 171, 173–178, 183–186, 237, 240, 244, 276, 279, 281–283, 300, 302, 306–307, 310, 321–324, 327, 344–345, 352, 354, 372; *see also* 1-sun
solar radiation 3, 41, 51, 62, 110, 120–121, 157–159, 167, 169, 171, 175–176, 178, 184–186, 240, 244, 276, 279, 281–283, 302, 306–307, 322, 324, 327, 344–345, 352, 354; *see also* 1-sun
sonic anemometer 302–304, 350, 359; *see also* anemometers
sound 101, 178, 303–304, 310, 312, 335, 367–368, 392
spatial 20–23, 25, 129, 141, 147, 185, 187–189, 251, 263–264, 275, 306, 326–328, 336, 343, 358, 363, 374, 376–377, 379, 382, 389–390
spatial resolution 25, 187–188, 306, 326, 336, 358, 376
Spearman's 17
specificity 219–220, 222
spectrum 23, 38, 101, 130–131, 139–140, 158–159, 185–186, 289, 291–292, 297, 319, 325, 354–356, 358
    spectra 138, 291, 299, 323–324, 355, 357, 360
    spectral 23, 138, 140, 159, 186–188, 207, 214, 279, 290, 300, 327, 354–355, 359
    spectral efficiency 138
    spectral resolution 187–188, 290, 300
    spectral width 138, 140
SPI 42–43, 109, 113, 115–116, 127–128, 143, 152
SQL 29, 49, 251, 257–260, 264, 276
    structured query language 251
SQNR 105, 108; *see also* A/D converter
square error 236, 242, 286; *see also* multiple linear regression; nonlinear regression; polynomial regression; regression tree
squared error 18, 286
SRAM 29, 41–42
stage 95, 104, 182, 291, 294, 310, 312–313
standard deviation 7–11, 13, 15–16, 19, 21, 25, 175, 202, 227–228, 240, 248, 390
standard error 8, 15–16, 19, 333; *see also* sample standard deviation; standard deviation
standard normal 9–10, 13–15, 17, 231
standing wave ratio 137
start frame delimiter 38
static specifications 62; *see also* dynamic specifications; environmental specifications
stationary 291, 382, 386
statistical inference 1, 12, 14
statistics 1, 5, 8, 12, 15, 24, 26, 63, 196, 214–215, 217, 219, 221, 223, 225, 227, 229, 231, 233, 235, 237, 239, 241, 243, 245, 247, 249, 265, 328, 360
Steinhart-Hart 65–66; *see also* B model
stepwise regression 215, 242; *see also* multiple linear regression
storm relative motion 305
storm total precipitation 305
strain gage 63, 79, 87–89, 309, 341, 351
stratopause 279
stratosphere 279–281; *see also* ozone

string multi-hop 141
subnet mask 37
sun elevation angle 168, 171, 184, 298
sun-synchronous 187–188, 305
supervised 198, 207, 215, 233, 246; *see also* machine
 learning
supervised classification 207, 215, 233, 246
SVD 198, 203–206
SWR 137
symbol rate 32
symbols 32, 45, 47, 138, 257, 277
sync 35–36
synchronous 29, 115–116, 147, 187–188, 305
syntax 43, 46, 123, 262
system on a chip 39
systematic error 21

tables 44–46, 49, 75–76, 78, 122, 125, 178, 215, 232,
 253–258, 276
tagname 45, 261
tags 45–47, 49, 260–262, 276, 363–365, 370–371, 373,
 391–392
TCP 31, 37–38, 113, 138, 144, 147, 173
TCP/IP 31, 37–38, 113, 144, 147
TDR 339–342, 358–359, 361; *see also* soil
telemetry 38, 109, 127, 129, 131, 133, 135, 137, 139, 141,
 143, 145, 147, 149, 151, 153, 155, 363, 365,
 367–368, 391–392
temporal 2, 20–21, 23, 129, 141, 187, 275, 277, 327–328,
 353, 368
temporal resolution 129, 141, 187
tensiometer 79, 94–95, 108, 121, 341–342, 360–361
terminal nodes 243–247
terrain 21, 130, 178, 192, 232, 265, 267–268, 326, 351; *see*
 also aspect; slope
terrestrial ecosystems 279, 339, 341–343, 345, 347, 349,
 351, 353, 355, 357–359, 361
thermistor 3, 53, 62–64, 66, 69–73, 77–86, 97–101, 107,
 301, 314
thermocouple 53, 62, 73–78, 104, 106, 302, 314, 347–348;
 see also cold-junction compensation
thermoelectric equation 75
thermometer 78, 355, 357, 359
thermopile 302
Thomson 73–75
throughput 41, 134, 138, 140–141
tides 110, 310–311, 336
tilt 168–169, 183, 300; *see also* PV
time constant 90
time-domain 101, 361
time domain reflectometry 339, 361
time series 4, 11–12, 182, 187, 309, 326, 328–329,
 333–334, 336, 349, 363, 371, 373, 389
topic 1, 22, 24, 79, 93, 109, 129, 141, 147–148, 152,
 154–155, 157, 185, 253, 309, 358
total column 23, 279–280, 297–298, 300–301, 307–308;
 see also ozone
trace gases 279, 281, 294, 299, 301, 360; *see also*
 greenhouse effect
training dataset 236
transceiver 115
transducer 30, 53, 55, 57–59, 61–63, 65–71, 73–75, 77–83,
 85, 90–91, 95, 104–110, 113, 118, 120, 142–143,
 301, 303, 309, 314–315, 324, 341, 343

balanced source divider 79, 85–86, 100
 full-bridge 79, 88, 98, 124
 half-bridge 79, 82, 86–88, 97–98, 121, 123
 linearity error 63, 66, 68–69, 71–73, 77, 85–86, 89,
 97–98
 linearized bridge 99
 linearized quarter-bridge 97, 99
 quarter-bridge 84–86, 97–101, 107
 Wheatstone bridge 79, 82, 84–88, 90, 95, 107
transmission 31–33, 38, 105–106, 113–115, 129, 132, 134,
 136–137, 139–141, 144, 149, 152, 291, 302, 352,
 354–355, 363–366
transmission control protocol/internet protocol 31, 113;
 see also internet
tropopause 279
troposphere 279–281; *see also* ozone
true color composite 194; *see also* remote sensing
true positive rate 219, 222
TTL 115
tuple 44, 253–254
two-argument arctangent 269; *see also* aspect
two-tail 14, 16

UART 32–33, 43, 109, 113, 115–117, 143–144
UAV 185, 373
ultrasonic sensors 309–310; *see also* anemometers
ultrasound 302, 367
ultraviolet 23, 101, 130–131, 186, 279, 308, 354
unbiased 387–388
uncompressed 112
underfitting 245–246
uniform 5–7, 46, 226, 228, 231, 275, 351, 376–377, 390
uniform resource locator 46
unipolar 104
universal asynchronous receiver transmitter 32, 109
Universal Transverse Mercator 189
 easting 191, 277
 northing 191, 277
 UTM 189–192, 213, 251, 265, 270, 277, 375
unmanned aerial vehicles 185, 327, 373, 392
unsupervised 198, 207, 210, 212, 359
unsupervised classification 207, 212, 359
URL 45–46, 48, 50; *see also* web browsers; World Wide
 Web
USB 31, 33–36, 39, 42, 113, 117, 144, 150, 152, 372
 Universal Serial Bus 31
 USB-C 34
 USB Type-C 34
user accuracy 235, 246–247
user datagram protocol 37
USGS 20, 24, 27, 187–190, 214, 310, 312–313, 328–329,
 334, 337, 358–359, 362

vadose 309, 339, 347, 358, 361
variance 7–9, 11, 14, 16, 18–19, 22, 25, 200–202, 205, 207,
 213, 227–228, 230–231, 239, 241, 248, 333, 349,
 378, 382–389
 variance inflation factor 241
 VIF 241
vector 49, 137, 204, 237–238, 251, 264, 270–274, 277, 286,
 305, 331, 376, 378, 386
video 35, 184, 324, 336, 371–373, 391–392
VOCs 292
volatile organic compounds 292

voltage 30
    analog voltage 30, 60
    input voltage 42, 58–60, 66, 71, 104, 120, 166
    output voltage 53, 59–60, 62, 66, 68–73, 77, 80–85, 93, 95, 97–99, 105–106, 127, 164, 166
    voltage divider 53, 58, 60, 62, 66–67, 69, 71–73, 77, 79–83, 85, 93, 107, 121, 123, 127, 165, 315
    voltage excitation 120–121
    voltage level 33, 114
    voltage regulator 164
    voltage resolution 30–31, 49, 68, 187
    voltage source 54, 57, 66, 70–71, 85, 93, 95, 97, 107
voltmeter 60; *see also* ammeter; DMM; ohmmeter
volumetric water content 123, 341
VWC 123, 341–342, 344–345, 359

water quality 23–25, 27, 93, 110, 117, 192, 219–220, 223, 270, 272, 277, 279, 309, 313–314, 316, 318, 320–322, 324, 326–328, 336–337, 369
    dissolved oxygen 110, 117, 260, 309, 318
    DO 309–310, 313–314, 318, 320, 322–324, 327, 369, 373
    nephelometric turbidity units 319
    NTU 319
    TDS 267, 270, 309, 313, 317, 328, 335
    total dissolved solids 270, 309
    total suspended solids 313
    TSS 313, 327
    turbidity 260, 309, 313, 318–320, 336, 373
water quantity 309, 328
water vapor 2, 4, 279, 283, 302, 306, 309, 345, 348, 350
water velocity 20, 107, 182, 311–313, 325, 373
waveguide 292, 339–341
wavelength 129–132, 135–136, 157–159, 186–187, 189, 281–282, 284, 290, 292–293, 295–296, 298–299, 301, 306, 318–320, 323, 348, 355, 357, 367–368, 390
waves 101, 129–132, 135–136, 138, 149, 157–158, 174, 186, 282, 304, 365, 367–369
weather 1, 3, 24, 26–27, 73, 110, 120, 124, 137, 140, 148–151, 175, 188, 279, 301, 304–308, 326, 341, 344–345, 347, 358
weather stations 110, 326
Weibull 180–181; *see also* wind speed
wet chemical method 293
WGS84 188–189, 265; *see also* GIS

Wi-Fi 38–39, 51, 139–141, 143–144, 146, 148, 153–154, 158; *see also* 2.4 GHz; 5 GHz
Wilcoxon 16–17; *see also* nonparametric
wildlife monitoring 3, 279, 363, 365, 367, 369, 371–373, 375, 377, 379, 381, 383, 385, 387, 389, 391
wind 3, 93–94, 107, 110, 119–121, 125, 157, 173, 175–176, 178–182, 237, 240, 244–245, 257–258, 302–305, 311–312, 344–346, 349–351, 359
wind speed 93, 107, 110, 120–121, 125, 157, 173, 175–176, 178–182, 302–303, 311, 344–346, 349–351, 359
window filter 267
wireless 24, 38, 129, 131, 133, 135, 137, 139–147, 149–151, 153–155, 157, 184, 278, 336; *see also* 2.4 GHz; 5 GHz
    wireless fidelity 38 *(see also* 2.4 GHz; 5 GHz)
    wireless sensor network 38, 129, 131, 133, 135, 137, 139, 141, 143, 145, 147, 149, 151, 153–155, 157, 184, 278 *(see also* 2.4 GHz; 5 GHz)
    WSN 129–130, 141–154, 157, 162, 173–175, 251–252, 272, 275–276, 344, 358, 372 *(see also* ZigBee)
work function 290–291
world geodetic system 188; *see also* GIS
World Wide Web 3
    sensor web 251, 274–275, 277
    web 3, 24, 27, 29, 38, 43–50, 152, 251, 260, 274–275, 277, 283, 286, 294, 300, 305, 307, 310, 312–313, 328, 335, 361
    web browsers 38
    web clients 38
    web server 38, 47, 152, 300
    web service 24, 251, 274
    WWW 26–27, 38, 51, 78, 108, 128, 154–155, 184, 214, 261, 274, 277–278, 307–308, 335–337, 360–361, 391–392
WRS 189; *see also* GIS

XML 251, 260–262, 276–277
    XML schema 262

Yule-Walker 329
YW 329, 332–333, 335

ZigBee 141, 143; *see also* WSN
zone 93, 131, 135–136, 153, 189–191, 213, 265, 267, 309, 339, 341, 347, 358, 361

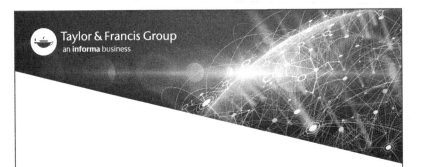

Printed in the United States
by Baker & Taylor Publisher Services